Applied Mathematics in Hydrogeology

Tien-Chang Lee

Department of Earth Sciences
University of California
Riverside, California

CRC Press
Taylor & Francis Group
Boca Raton London New York

CRC Press is an imprint of the
Taylor & Francis Group, an **informa** business

CRC Press
Taylor & Francis Group
6000 Broken Sound Parkway NW, Suite 300
Boca Raton, FL 33487-2742

First issued in paperback 2019

© 1999 by Taylor & Francis Group, LLC
CRC Press is an imprint of Taylor & Francis Group, an Informa business

No claim to original U.S. Government works

ISBN-13: 978-1-56670-375-8 (hbk)
ISBN-13: 978-0-367-40018-7 (pbk)

This book contains information obtained from authentic and highly regarded sources. Reasonable efforts have been made to publish reliable data and information, but the author and publisher cannot assume responsibility for the validity of all materials or the consequences of their use. The authors and publishers have attempted to trace the copyright holders of all material reproduced in this publication and apologize to copyright holders if permission to publish in this form has not been obtained. If any copyright material has not been acknowledged please write and let us know so we may rectify in any future reprint.

Except as permitted under U.S. Copyright Law, no part of this book may be reprinted, reproduced, transmitted, or utilized in any form by any electronic, mechanical, or other means, now known or hereafter invented, including photocopying, microfilming, and recording, or in any information storage or retrieval system, without written permission from the publishers.

For permission to photocopy or use material electronically from this work, please access www.copyright.com (http://www.copyright.com/) or contact the Copyright Clearance Center, Inc. (CCC), 222 Rosewood Drive, Danvers, MA 01923, 978-750-8400. CCC is a not-for-profit organization that provides licenses and registration for a variety of users. For organizations that have been granted a photocopy license by the CCC, a separate system of payment has been arranged.

Trademark Notice: Product or corporate names may be trademarks or registered trademarks, and are used only for identification and explanation without intent to infringe.

Visit the Taylor & Francis Web site at
http://www.taylorandfrancis.com

and the CRC Press Web site at
http://www.crcpress.com

PREFACE

This book has grown from my lecture notes for a ten-week graduate class. It is intended for students and professionals who want to know how to derive some of those familiar equations and formulas in hydrogeology. Demonstrated with numerous examples, the book compiles various means for solving differential equations. All problems in this book have been re-solved, and in many cases new solutions are provided. Step-by-step procedures are given when a method is introduced. Readers of other disciplines may find that the methods presented here are also applicable to their fields of interest.

Most textbooks on applied mathematics are structured according to a systematic methodology, and many students learn the subject matter by taking classes that are based on such an approach. The methods covered are usually fairly diversified and little time can be devoted to actual problem-solving applications. As a result, the mathematics and its applications still appear remotely related. On the other hand, a few textbooks are structured according to the subject areas with emphasis on the subjects rather than the development of mathematical skills. In such cases, important equations are often given without relevant derivations, thus necessitating many readers to take a detour to investigate how one equation leads to another. Such difficulty can lead to frustration that becomes more acute when reading journal articles, many of which provide only scanty derivations. I have therefore attempted in this book to bridge these two approaches at the risk of creating a haphazard presentation.

I feel that a modern-day, quantitatively oriented student of hydrogeology should have some training in analytical mathematics, numerical methods, and inversion methodology, plus perhaps probability-based statistics. However, under the constraint of limited time and resources, one needs to be selective. Representing my choice of subjects, this book is divided into ten chapters. Each chapter focuses on a few relevant topics and includes several examples for developing mathematical skills. These examples are followed by a set of problems, whose solution keys are often provided. It is not my intention to cover all important topics in hydrogeology, nor to provide an extensive literature review of the topics covered. The emphasis here is on how to set up application-oriented problems and to solve the resulting equations. I believe that once interested persons learn and master the mathematical skills, they can pursue topics of interest in journals and develop their own mathematical models. The contents of each chapter are summarized as follows.

Chapter 1 relates heat transfer, groundwater flow, advection, and solute transport through a set of fundamental equations that evolve from the principles of conservation of energy and mass. Different techniques are introduced to establish these governing differential equations.

Chapter 2 develops the principle of linear superposition. It introduces the concepts of convolution, Fourier transform, and z-transform. Their applications lead to the formulation of well functions (fluid flow toward pumping well) and solutions to a few heat-transfer and solute-transport problems.

Chapter 3 deals with the Laplace and Hankel transforms. It emphasizes the application of contour integration for developing the analytical solutions that are addressed in subsequent chapters. New and useful formulas are derived for the inverse Hankel transform. Various methods for analytical integration are demonstrated with examples. Also included are techniques of numerical transforms because frequently analytical or closed-form solutions may not be available or are too cumbersome to integrate.

Chapter 4 begins by rederiving the classical well functions that describe hydraulic response to pumping of a confined aquifer. It then offers new solutions that are based on leakage at the aquifer's boundary instead of invoking the conventional assumptions that treat leakage as a volumetric source of water. These new solutions are achieved by means of the analytical Hankel transform and the separation-of-variables technique.

Chapter 5 extends the methodologies presented in Chapter 4 to obtain drawdown responses in water-table aquifers. The resulting algorithms can be easily implemented to solve a variety of problems. The chapter ends by providing a generalized response function that accounts for wells of finite diameter, well-bore storage, head loss, slug height, and partial penetration. As such, it unifies many existing well functions that describe flows toward pumping wells in either confined or unconfined aquifers in the sense that they are special cases of this general solution.

Chapter 6 introduces advective heat transfer, which is a subject that deserves more attention than it currently receives in hydrogeology. Through modeling this chapter shows how groundwater flow can affect the subsurface temperature distribution in an otherwise conductive thermal regime. Two field examples are presented to show qualitatively how temperature distribution can be used to infer groundwater-flow patterns.

Chapter 7 incorporates existing analytical solutions for one-, two-, and three-dimensional problems in solute transport. It outlines how a solution can be built from others. Some solutions are improved upon for more efficient integration. Most importantly, it demonstrates how the z-transform technique can be used to numerically obtain transfer functions and system responses.

Chapter 8 describes methods for matrix inversion and serves as a prelude for material presented in Chapters 9 and 10. It deals with eigenproblems and differentiation with respect to vectors. Most of this chapter is devoted to singular value decomposition for evaluating the conditions of matrix inversion.

Chapter 9 introduces concepts of finite element analysis by formulating a one-dimensional algorithm for transient, axi-symmetric problems. The algorithm is then extended to two-dimensional problems. This chapter will be particularly useful to readers who use software developed by others and want a convenient review of the method.

Chapter 10 reviews linear and nonlinear inverse problems with emphasis on how to obtain model parameters with or without preprocessing constraints. This chapter covers only the essentials, but is considered important because inverse problems are encountered in every discipline of science.

I have adopted the notation of square brackets, like $f[x]$, to enclose the argument of a function. This is the convention used by Mathematica$^{\circledR}$. A list of notations is presented at the end of each chapter except for the last three chapters in which the notations are clear from the context. The first appearance of every variable or parameter in each chapter is defined and attached with a SI unit; its corresponding dimension is given in the list of notations. The notations are consistent within each respective chapter only. In order to track dimensional consistency during derivation, I have avoided introducing dimensionless variables and equations, unless they are necessary for reducing the number of independent variables and only after the derivation of a problem is completed. Some equations are boxed for emphasis.

Different subjects often require different problem-solving techniques. Most methods in this book are presented in the sequence of chance encounter, such as one may discover when reading professional papers. Readers who wish to see how a particular method can be applied to different problems should browse through the Appendix. In addition, the Index is useful for tracing methodologies to applications.

Acknowledgment

I have benefited from many individuals during the course of writing this book. Comments by Kathleen Aikin, Mark Apoian, Bob Bielinski, Heather Gledhill, Tom King, Kirk Williams, Jeff Zawila and other students on its roughest draft form are appreciated. Cin-Ty Lee provided the perspective of a non-enrolled student. Many figures have been modified from the draft figures provided by Heather Gledhill (2.1, 4.1, 4.2, 4.4, 4.5, 4.6, and 4.7) and Cin-Young Lee (1.1, 1.2, 1.3, 1.4, 2.3, 3.1, 3.3, 3.4, 3.5, 3.7, 3.10, 3.11, 3.13, 3.14, and 3.16). Field data for Figure 6.8 was provided by Bill Teplow and data for Figure 6.5 was collected by GSI Water, Inc., and provided by Tom Blackman. Brian Damiata provided the software for Stehfest's numerical scheme of inverse Laplace transform and for evaluating Bessel functions. He also called my attention to the integrals used in Problems 4 and 6 of Chapter 3. Professional comments by Yoram Eckstein, Robert A. Johns and Tom Perina are appreciated. I am indebted to Lewis Cohen, my colleague and critic, for his fine tuning of the entire manuscript.

Trademarks

Maple[R] is a registered trademark of Waterloo Maple Software. Matlab[R] is a registered trademark of The Math Works, Inc. Mathematica[R] is a registered trademark of Wolfram Research, Inc. Mathcad[R] is the trademark of MathSoft, Inc. Scientific WorkPlace[R] is the trademark of TCI Software Research, a division of Wadsworth, Inc.

The Author

Tien-Chang Lee is a Professor of Geophysics in the Department of Earth Sciences, University of California, Riverside, California, USA. He is a Registered Geologist, Geophysicist, and Certified Hydrogeologist in California. He earned his BS in geology from the National Taiwan University, MS in geology from the University of Idaho, and Ph.D. in geological sciences from the University of Southern California. He has taught at UCR since 1974.

He has authored or coauthored numerous journal articles in tectonophysics, seismology, electromagnetism, potential field theory, and hydrogeology. His work covers field-data collection, numerical modeling, and development of analytical solutions.

This book is dedicated to my wife, Zora, and sons, Cin-Ty and Cin-Young.

Contents

Chapter 1

CONSERVATION EQUATIONS

Heat transfer, fluid flow, and solute transport are governed by the principles of conservation of energy, momentum, and mass. Mathematically, these conservation principles, together with empirical laws, can be expressed as a set of partial differential equations. Subject to initial and boundary conditions as well as appropriate source functions, the equations can be solved analytically or numerically to interpret observations or predict certain phenomena.

In this chapter we will show how seemingly different physical processes can be modeled by differential equations that have the same functional forms. We will attempt to derive each equation with a different method.

1.1 Heat Conduction

1.1.1 Heat Energy

Heat flux is the rate of heat energy transferred through a unit area across the transport path. It is a vector and has a dimension of energy per unit area per unit time, $[W\ m^{-2}]$. The heat flux can be conductive (diffusive), convective, or radiative. In this section, we consider heat conduction problems only.

Let the positive x-coordinate be in the direction of heat flux. Given heat flux q at location x in a one-dimensional continuum, the heat flux at

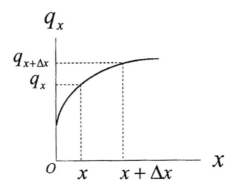

Figure 1.1: Heat flux $q_{x+\Delta x}$ at $x + \Delta x$ as extrapolated by the Taylor series from q_x at x, i.e., $q_{x+\Delta x} \approx q_x + (\partial q / \partial x)\Delta x$.

neighboring location $x + \Delta x$ is,

$$q_{x+\Delta x} = q_x + \left(\frac{\partial q}{\partial x}\right)_x \Delta x, \qquad (1.1)$$

which follows the first-order Taylor theorem (Figure 1.1). This relation means that if a functional value q_x and its slope $\partial q_x/\partial x$ at x are given, the functional value $q_{x+\Delta x}$ at $x + \Delta x$ can be extrapolated from the given value by following the slope to the desired location.

If $q_{x+\Delta x} > q_x$, the material in the space between $x + \Delta x$ and x loses energy by the amount of ΔE per unit volume over time interval Δt. This energy change is accompanied by temperature change ΔT. The two changes are empirically related by

$$\Delta E = \rho c_v \Delta T \qquad (1.2)$$

where the proportional constant ρc_v is the *volumetric heat capacity* $[Jm^{-3}K^{-1}]$, ρ is the mass density $[kg\ m^{-3}]$, and c_v is the *specific heat* at constant volume $[Jkg^{-1}K^{-1}]$ (the energy needed to raise the temperature of one kilogram of material by one degree Kelvin).

The principle of energy conservation says that the net flow through the enclosed surface of a control volume, or an elementary volume $(\Delta x \Delta y \Delta z)$, is equal to the rate of internal energy loss from the control volume (Figure 1.2)

$$\left(q_{x+\Delta x} - q_x\right)\Delta y \Delta z = -\frac{\Delta E}{\Delta t}\Delta x \Delta y \Delta z \qquad (1.3)$$

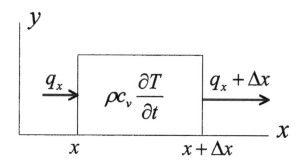

Figure 1.2: Energy conservation: net energy outflow equal to change in energy storage in the control volume.

where surface $\Delta y \Delta z$ is perpendicular to the flux in the x-direction. Simplifying the above relation yields

$$\frac{\partial q_x}{\partial x} = -\rho c_v \frac{\partial T}{\partial t}. \tag{1.4}$$

The negative sign indicates that the efflux is accompanied by decreasing temperature or internal energy.

In general the heat flux vector **q** has three independent, orthogonal components. Each contributes to the change in energy storage. Hence, the governing equation of heat conduction in three spatial dimensions is

$$\boxed{\frac{\partial q_x}{\partial x} + \frac{\partial q_y}{\partial y} + \frac{\partial q_z}{\partial z} = -\rho c_v \frac{\partial T}{\partial t}} \tag{1.5}$$

The left-hand side of this equation is the divergence of heat flux **q**, i.e.,

$$\nabla \cdot \mathbf{q} = \frac{\partial q_x}{\partial x} + \frac{\partial q_y}{\partial y} + \frac{\partial q_z}{\partial z}. \tag{1.6}$$

1.1.2 Fourier's Law

There are two unknowns, **q** and T, in equation (1.5). To solve for T we need another equation to relate the two variables. Empirically, the heat flux is shown to be proportional to the temperature gradient. For example,

$$\boxed{q_x = -k \frac{\partial T}{\partial x}} \tag{1.7}$$

in the x-direction, which is a constitutive relation known as *Fourier's law of heat conduction*. The proportional constant k is the thermal conductivity $[W\ m^{-1}K^{-1}]$. Here we adopt the sign convention that a vector component is positive if it points in the positive coordinate direction. The negative sign indicates that the heat energy flows in the direction of decreasing temperature, i.e. in the positive x-direction.

Combining the energy conservation equation and Fourier's law yields the *heat conduction equation*

$$\boxed{\frac{\partial}{\partial x}\left(k\frac{\partial T}{\partial x}\right) + \frac{\partial}{\partial y}\left(k\frac{\partial T}{\partial y}\right) + \frac{\partial}{\partial z}\left(k\frac{\partial T}{\partial z}\right) = \rho c_v \frac{\partial T}{\partial t}}. \qquad (1.8)$$

In the case of homogeneous material, of which k is independent of position, equation (1.8) becomes

$$\frac{\partial^2 T}{\partial x^2} + \frac{\partial^2 T}{\partial y^2} + \frac{\partial^2 T}{\partial z^2} = \frac{1}{\kappa}\frac{\partial T}{\partial t}, \qquad (1.9)$$

where $\kappa = k/\rho c_v$ is the *thermal diffusivity*, $[m^2\ s^{-1}]$.

In the presence of heat sources, a term for heat production rate per unit volume, $H\ [J\ m^{-3}\ s^{-1}]$, should be added to the left-hand side of equation (1.8) to complete the heat conduction equation

$$\frac{\partial}{\partial x}\left(k\frac{\partial T}{\partial x}\right) + \frac{\partial}{\partial y}\left(k\frac{\partial T}{\partial y}\right) + \frac{\partial}{\partial z}\left(k\frac{\partial T}{\partial z}\right) + H = \rho c_v \frac{\partial T}{\partial t}. \qquad (1.10)$$

This addition is permissible because the derivation of equation (1.8) is based on an elementary volume. The positive sign in H means that heat generation raises the temperature or internal energy in the absence of heat transfer.

The above derivation is for an isotropic medium (i.e., the property does not change with direction or orientation). In case of anisotropy, a temperature gradient in one direction affects heat flux in the other directions. In general,

$$q_x = -k_{xx}\frac{\partial T}{\partial x} - k_{yx}\frac{\partial T}{\partial y} - k_{zx}\frac{\partial T}{\partial z}. \qquad (1.11)$$

Here the conductivity is a third-rank tensor with nine components designated by the permutation of two subscripts: the first being orientation along which temperature gradient is measured and the second being the orientation of the flux. The tensor is symmetric (i.e., $k_{xy} = k_{yx}$). The coordinates can be rotated to diagonalize the tensor (see Chapter 8). When it is diagonalized, the axes of the coordinates are the principal directions, and the diagonal entries of the tensor are the principal conductivities. In terms of the principal directions and conductivities, the heat flux is

$$\mathbf{q} = -k_x \frac{\partial T}{\partial x} - k_y \frac{\partial T}{\partial y} - k_z \frac{\partial T}{\partial z}. \tag{1.12}$$

1.2 Groundwater Flow

1.2.1 Equation of Continuity

Consider fluid flow through an immobile porous medium. Let the volumetric flow rate through surface area A be Q. If the flow is perpendicular to the surface, then $v = Q/A$ represents the discharge rate averaged over area A (Figure 1.3) and it has a dimension of $[m^3 s^{-1} m^{-2}]$. Therefore v is a volume flux. The dimension of v can be simplified to have a dimension of velocity $[m\, s^{-1}]$; thus v is commonly named *specific discharge, discharge velocity,* or *Darcy's velocity.* Because the fluid occupies only the porous space and the flow paths are tortuous, the *average linear velocity* in the porous space is v/ϕ_{eff} where ϕ_{eff} ($\leq$ porosity) in the sense that only a fraction of the pore fluid is effectively mobile. This is also named *seepage velocity.*

For mass density ρ, the product ρv is the *mass flux* $[kg\, m^{-2}\, s^{-1}]$ of fluid because v is a volumetric flux through the porous medium. By the same token, if E is the energy density, Ev is the *energy flux.* Other fluid-driven fluxes can be defined likewise.

In a three-dimensional space, the mass flux $\rho\mathbf{v}$ is a vector. The mass flow rate through area element dS with an outward unit normal $\hat{n}$ is $\rho\mathbf{v}\cdot\hat{n}\,dS$. If ϕ is the porosity, then $\rho\phi$ is the fluid density averaged over the bulk volume (i.e., fluid mass per unit volume). Mass conservation requires that the rate of mass flowing out of an enclosed surface S equals the rate of mass lost from the control volume V

$$\iint\limits_{S} \rho\mathbf{v}\cdot\hat{n}\,dS = -\iiint\limits_{V} \frac{\partial(\rho\phi)}{\partial t}\,dV \tag{1.13}$$

with the minus sign indicating that the mass out-flux results in a decrease of mass in the control volume.

Applying the *divergence theorem,*

$$\iint_S \rho\mathbf{v} \cdot \hat{n}\, dS = \iint_V \nabla \cdot (\rho\mathbf{v})\, dV \, ,\tag{1.14}$$

to the mass balance equation yields the *continuity equation*

$$\nabla \cdot (\rho\mathbf{v}) = -\frac{\partial\,(\rho\phi)}{\partial t} \, .\tag{1.15}$$

The divergence of mass flux thus equals the declining rate of mass storage.

1.2.2 Darcy's Law

Again we have one equation (1.15) and two unknowns, $\rho\mathbf{v}$ and $\rho\phi$. We need an additional equation. By analogy to Fourier's law for heat conduction, we introduce a Darcy's law for fluid flow in porous medium.

First, let us deal with the mass flux term $(\rho\mathbf{v})$ in equation (1.15). Under isothermal conditions, fluid mass m possesses three types of mechanical energy: kinetic energy $mv^2/2$, potential energy mgz, and strain energy (pressure-volume work) $m\int_{p_o}^{p} V/m\, dp$ at pressure p, $[Pa]$, where v is the velocity, $[m\,s^{-1}]$, g is the gravitational acceleration, $[m\,s^{-2}]$, z is the elevation, $[m]$, V is the volume, and p_o is the reference pressure. The total mechanical energy per unit mass is

$$E = \frac{v^2}{2} + gz + \int_{p_o}^{p} \frac{dp}{\rho} \, ,\tag{1.16}$$

which is the well-known *Bernoulli equation* in fluid mechanics. This energy is conserved, meaning that E stays constant during fluid motion provided that there is no energy dissipation through viscous heating.

Groundwater typically flows at a velocity of a few tens of meters per year or less under natural conditions. The kinetic energy term is therefore negligibly small when compared to the other two terms. If the fluid is incompressible, an additional simplification can be made to obtain

$$E \approx gz + \frac{p - p_o}{\rho} \, .\tag{1.17}$$

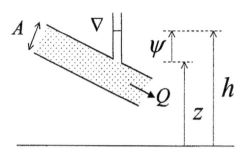

Figure 1.3: Hydraulic head h as the sum of pressure head ψ and elevation head z. Darcy's velocity $v = Q/A$ with Q being the total volumetric flow through cross-sectional area A.

Now, we introduce a new variable $h = E/g$ such that

$$\boxed{h = z + \psi}\,,\tag{1.18}$$

where

$$\psi = (p - p_0)/\rho g.\tag{1.19}$$

Parameter h is *hydraulic head,* and in this context, z and ψ are *elevation* and *pressure heads,* respectively (Figure 1.3). All heads have the dimension of length, $[m]$. As defined, the hydraulic head represents energy per unit weight of water. Usually the reference pressure p_0 is taken as zero, and in this case, p is the *gauge pressure.* A gauge pressure can be greater or less than zero, but an absolute pressure cannot be less than zero. Note that in areas with variable fluid density (e.g., brines or geothermal fluids), hydraulic head is ill-defined and the pressure itself should be used rather than the hydraulic head.

Empirically it has been found that the discharge velocity (specific discharge) in a porous medium is proportional to the gradient of hydraulic head

$$v_x = -K_x \frac{\partial h}{\partial x},\quad \text{etc.}\tag{1.20}$$

where K_x is the hydraulic conductivity, $[m\ s^{-1}]$, in the direction of coordinate x, assuming the coordinates align with the principal axes of the hydraulic conductivity tensor.

Darcy's law is generally accepted to represent the principle of conservation of momentum, but a satisfactory proof is still wanted. For an isotropic medium, *Darcy's law* is

$$\boxed{\mathbf{v} = -K\nabla h}.$$

(1.21)

and $\mathbf{v}$ is Darcy's velocity. Darcy's law is equivalent to Fourier's law in heat conduction and to Ohm's law in electric current flow.

1.2.3   Flow Equation

Here we combine Darcy's law with the mass conservation equation to formulate a flow equation. Expanding the right-hand side of the continuity equation (1.15) yields

$$\frac{\partial}{\partial t}(\rho\phi) = \phi\frac{\partial\rho}{\partial t} + \rho\frac{\partial\phi}{\partial t}.$$

(1.22)

The first term is the rate of change in fluid density, and the second term is the rate of porosity change.

● **Compressibility of Water**

The rate of mass density change can be expressed by

$$\phi\frac{\partial\rho}{\partial t} = \phi\frac{\partial\rho}{\partial p}\frac{\partial p}{\partial t} = \phi\beta\rho\frac{\partial p}{\partial t},$$

(1.23)

where β, $[1/Pa]$, is the compressibility of water,

$$\beta = \frac{\partial\rho}{\rho\partial p} = -\frac{\partial V}{V\partial p}.$$

(1.24)

● **Change in Porosity**

The rate change in porosity $\rho\partial\phi/\partial t$ depends on the stress that acts on the porous medium. The *total stress* p_{total} on a surface is shared by the stresses acting on the matrix and fluid (Figure 1.4). The former is the *effective stress* p_{eff} acting on the matrix, and the latter is the pore pressure p. The three terms are related by

$$\boxed{p_{\text{total}} = p_{\text{eff}} + p}.$$

(1.25)

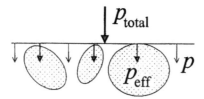

Figure 1.4: Total stress is shared by the pore pressure and efective stress on the matrix (dotted areas).

The porosity is defined as ratio of the void volume to the bulk volume,

$$\phi = (V_b - V_g)/V_b, \tag{1.26}$$

where V_g is the grain volume and V_b is the bulk volume. We assume that the grain is incompressible (i.e., $\partial V_g/\partial p_{\text{eff}} \approx 0$) and the total stress stays fairly constant at a given location such that the increase in effective stress is accompanied by a decrease in pore pressure only

$$\Delta p_{\text{eff}} = -\Delta p. \tag{1.27}$$

The rate of change in porosity is

$$
\begin{aligned}
\rho\frac{\partial \phi}{\partial t} &= \rho\frac{\partial \phi}{\partial p}\frac{\partial p}{\partial t} = -\rho\frac{\partial \phi}{\partial p_{\text{eff}}}\frac{\partial p}{\partial t} \\
&= -\rho\frac{\partial\left[(V_b - V_g)/V_b\right]}{\partial p_{\text{eff}}}\frac{\partial p}{\partial t} \\
&= \rho\left(1 - \phi\right)\beta_b\frac{\partial p}{\partial t},
\end{aligned}
\tag{1.28}
$$

where β_b is the compressibility of the bulk rock

$$\beta_b = -\frac{\partial V_b}{V_b\partial p_{\text{eff}}}. \tag{1.29}$$

• **Flow Equation**

After putting the rates of density and porosity changes into equation (1.22), the continuity equation (1.15) becomes

$$\nabla \cdot (\rho\mathbf{v}) = -\rho\left[\phi\beta + (1 - \phi)\beta_b\right]\frac{\partial p}{\partial t}, \tag{1.30}$$

where the rate of pore pressure change is expressible in term of change in hydraulic head,

$$\frac{\partial p}{\partial t} = \rho g \frac{\partial (h - z)}{\partial t} = \rho g \frac{\partial h}{\partial t}. \tag{1.31}$$

Recalling Darcy's velocity for equation (1.30) and assuming negligible spatial variation in ρ, we obtain the flow equation for a porous medium

$$\frac{\partial}{\partial x}\left(K_x \frac{\partial h}{\partial x}\right) + \frac{\partial}{\partial y}\left(K_y \frac{\partial h}{\partial y}\right) + \frac{\partial}{\partial z}\left(K_z \frac{\partial h}{\partial z}\right) = S_s \frac{\partial h}{\partial t}, \tag{1.32}$$

where

$$S_s = \rho g \left[\beta \phi + (1 - \phi) \beta_b\right] \tag{1.33}$$

is the specific storage, $[m^{-1}]$.

• Remarks

☐ Note that specific storage S_s has a dimension of inverse length. Its value depends on the compressibility of water and bulk rock, as weighted by the volume ratio of pore and grain. As defined, S_s is an elastic property, implying that a deformed aquifer will return elastically to its prestress state upon removal of the imposed stress. So strictly speaking, the common flow equation for a porous medium is only valid for an elastic aquifer. If an aquifer remains saturated during water extraction, the water is released through expansion of pore water and compression of pore space.

Land subsidence associated with excessive groundwater withdrawal cannot be fully restored by water injection, indicating an irreversible inelastic process occurs in aquifers or aquitards. Hence, the flow equation must be revised accordingly for subsidence studies.

☐ Like many empirical laws, Darcy's law is valid within certain limits of velocity. It is applicable in regions of slow laminar flow, which is typical for natural groundwater flow. It is not applicable when turbulent flow occurs. For example, flow near a pumping well, in a large fracture, or in a solution cavity, can be turbulent. However, we will assume that Darcy's law is valid for all cases considered in this book.

☐ Traditionally specific storage is defined as the volume of water released per unit volume of aquifer per unit decline in hydraulic head,

$$S_s = -\frac{1}{V_b} \frac{\partial V_w}{\partial h}. \tag{1.34}$$

Usage of this definition leads to a well-accepted relation (for example, Freeze and Cherry, 1979, p. 59)

$$S_\mathrm{s} = \rho g \left(\phi \beta + \beta_\mathrm{b} \right), \tag{1.35}$$

which differs from the S_s defined in equation (1.33) in the weighting of the bulk compressibility. The readers are referred to an interesting discussion of the two different relations by Domenico and Schwartz (1998). In practice it makes no difference which relation is used because the specific storage or storativity (specific storage times aquifer thickness) is usually determined by field-pumping tests rather than by computation.

□ The similarity in functional form between equations (1.8) and (1.32) indicates that solutions to heat conduction and groundwater flow equations are interchangeable under equivalent initial condition and boundary conditions and material properties.

1.3 Advective Heat Transfer

Given the volumetric heat capacity of water $(\rho c_\mathrm{v})_\mathrm{w}$, the *enthalpy density* (heat energy per unit volume, $[J\ m^{-3}]$) of water at temperature T above reference temperature T_o is $(\rho c_\mathrm{v})_\mathrm{w} \left(T - T_\mathrm{o} \right)$. The enthalpy density multiplied by the flow velocity $\mathbf{v}$ yields the *advective heat flux* (i.e., enthalpy flux), $(\rho c_\mathrm{v})_\mathrm{w} \left(T - T_\mathrm{o} \right) \mathbf{v}$, which is the amount of heat energy per unit area carried by the flowing water. Conductive and advective heat fluxes together constitute the total heat flux

$$\boxed{\mathbf{q} = -k\nabla T + (\rho c_\mathrm{v})_\mathrm{w} \left(T - T_\mathrm{o} \right) \mathbf{v}}. \tag{1.36}$$

For a fluid in a porous medium, this $\mathbf{v}$ is Darcy's velocity, not the linearized velocity $\mathbf{v}/\phi$.

The divergence of advective flux

$$\nabla \cdot \left[(\rho c_\mathrm{v})_\mathrm{w} \left(T - T_\mathrm{o} \right) \mathbf{v} \right] = (\rho c_\mathrm{v})_\mathrm{w} \nabla T \cdot \mathbf{v} + \left(T - T_\mathrm{o} \right) \nabla \cdot (\rho c_\mathrm{v})_\mathrm{w} \mathbf{v} \tag{1.37}$$

is the net outflow of advective heat flux through the enclosed surface of a control volume. Adding the net inflow (negative outflow) to the left-hand side of equation (1.8) yields another equation of heat transfer,

$$\frac{\partial}{\partial x} \left(k_x \frac{\partial T}{\partial x} \right) + \frac{\partial}{\partial y} \left(k_y \frac{\partial T}{\partial y} \right) + \frac{\partial}{\partial z} \left(k_z \frac{\partial T}{\partial z} \right)$$
$$-\nabla \cdot \left[(\rho c_\mathrm{v})_\mathrm{w} \left(T - T_\mathrm{o} \right) \mathbf{v} \right] = \rho c_\mathrm{v} \frac{\partial T}{\partial t}. \tag{1.38}$$

Note that ρc_v and k_x are the bulk properties of a water-solid mixture. They are generally but not always represented by

$$k = \phi k_w + (1 - \phi)k_g, \quad \rho c_v = \phi(\rho c_v)_w + (1 - \phi)(\rho c_v)_g \qquad (1.39)$$

with subscript g designating the solid grain. These relations are similar to the equivalent definition for the specific storage S_s.

If the mass flow is steady, i.e., if there is no time-dependency in equation (1.15), then

$$\nabla \cdot (\rho_w \mathbf{v}) = 0 \qquad (1.40)$$

and

$$\nabla \cdot \mathbf{v} = 0 \qquad (1.41)$$

for an incompressible fluid. Under these conditions, equation (1.38) is reduced to

$$\boxed{\frac{\partial}{\partial x}\left(k_x \frac{\partial T}{\partial x}\right) + \frac{\partial}{\partial y}\left(k_y \frac{\partial T}{\partial y}\right) + \frac{\partial}{\partial z}\left(k_z \frac{\partial T}{\partial z}\right) - (\rho c_v)_w \mathbf{v} \cdot \nabla T = \rho c_v \frac{\partial T}{\partial t}}.$$
$$(1.42)$$

1.4   Dispersion Equation

Similar to advective heat flux, solute flux $\mathbf{J}$, $[kg\ m^{-2}\ s^{-1}]$, consists of diffusive and advective parts

$$\mathbf{J} = -D\nabla C + C\mathbf{v}, \qquad (1.43)$$

where D is the mass diffusivity or diffusion coefficient $[m^2\ s^{-1}]$, and C is the solute concentration, $[kg\ m^{-3}]$. The relation without the advective term, $C\mathbf{v}$, is known as *Fick's first law of diffusion*.

In a porous medium, $\mathbf{v}$ is Darcy's velocity and the diffusion part is modified to account for the fact that the fluid occupies pore space only

$$\boxed{\mathbf{J} = -D\nabla(\phi C) + C\mathbf{v}}. \qquad (1.44)$$

As compared to diffusion in solution alone, the diffusion of solute in porous media is hindered by the obstacle of solid matrix and tortuosity along the flow path. Hence, D should be regarded as an effective diffusivity which is less than the diffusivity in solution alone. The two are empirically related by porosity and/or tortuosity. The actual relation could be more complicated; for example, dead-end pores also play a role in the diffusion process.

Mass balance of solute in a nonreactive, source-free porous system is

$$\nabla \cdot \mathbf{J} = -\frac{\partial}{\partial t}\left(\phi C\right), \tag{1.45}$$

which is the *continuity equation* (mass balance equation) for solute transport. Expanding the relation yields the solute transport equation

$$\boxed{\nabla \cdot \left(D\nabla C\right) - \frac{\mathbf{v}}{\phi} \cdot \nabla C = \frac{\partial C}{\partial t}} \tag{1.46}$$

for steady flow (i.e., $\nabla \cdot \mathbf{v} = 0$) with a time-invariant porosity. This is *Fick's second law of diffusion* if the advective part vanishes (i.e., $\mathbf{v} = 0$).

Empirically it has been found that the D in equation (1.46) depends on fluid velocity

$$\boxed{D_x = \alpha_x \frac{v_x}{\phi} + D^*}, \quad \text{etc.} \tag{1.47}$$

In this context, D_x is the *hydrodynamic dispersion coefficient* in the x-direction, D^* is the diffusion coefficient in a stationary solution, and α_x is the *dispersivity*, $[m]$. Dispersion arises from mechanical mixing of fluid along branched flow paths, change in path diameter or curvature, variation of fluid velocity across a flow tube, etc. Because D depends on flow direction, dispersion is always anisotropic even though the medium is otherwise isotropic. D is thus a tensor. One challenge in modeling or experimentation is how to assign or measure the tensor components. In addition, the dispersivity is space-scale dependent (see Chapter 7), i.e., it increases with the spatial dimension of the problem. Hence, care must be exercised in the application of a laboratory-determined dispersivity to field condition.

Note the similarity in functional form between the advective heat equation (1.42) and solute transport equation (1.46).

1.5 Boundary and Initial Conditions

All differential equations derived above are second order in space coordinates and first order in time domain. To solve these equations, two boundary conditions are required for each spatial coordinate, and one condition is needed for the temporal coordinate.

Solving a differential equation implies that we begin from the given initial and boundary conditions and use the differential equation to extrapolate our desired variable to the space and time of interest. The principle of mass conservation and Darcy's law together describe most groundwater flow situations, but there are voluminous solutions because the conditions vary with problems. How to match boundary conditions and cope with variations in material properties dictates how a differential equation or a boundary-value problem is solved.

Generally there are three types of boundary conditions. They are classified according to how the dependent variable or its gradient is specified at a boundary.

• Type 1 Condition – *Dirichlet* condition

Temperature, hydraulic head, pressure, or concentration as a dependent variable is specified on part of the boundary. A simple example is to let the dependent variable be zero at far distance from the source region. As another example for groundwater modeling, hydraulic heads are often specified at the boundary where observations of water level are available.

• Type 2 Condition – *Neumann* condition

Normal gradient of the dependent variable is specified. For example, along the groundwater divide (a hinge line) or an impermeable boundary, the normal gradient is set at zero. On the other hand, this condition can be made to represent input or output condition at the boundary.

• Type 3 Condition – *Cauchy* condition or mixed-type

Both the dependent variable and its gradient are coupled together as a condition along part of the boundary. It is usually imposed for a boundary with advective flow in solute transport problems.

• Remark

If the material is not homogeneous, additional conditions must be imposed at the interface of dissimilar media. Interface conditions usually require continuity of the dependent variable and its normal flux.

When the material properties are functions of the dependent variable (e.g., hydraulic conductivity is a function of hydraulic head in unsaturated flow), the differential equations become nonlinear and usually require iterative numerical solutions. This book deals with linear equations only.

1.6 Problems, Keys, and Suggested Readings

• Problems

1. Given heat production rate of uranium H [unit: $W\,kg^{-1}$], uranium content in rock U (weight fraction or percent %, [$kg\,kg^{-1}$]), mass density of rock ρ [$kg\,m^{-3}$], and specific heat of rock c_v [$Jkg^{-1}\,K^{-1}$], write a heat conduction equation that includes a heat production term.

2. Derive a heat conduction equation with anisotropic conductivity.

3. Explain why water in a confined aquifer is released during pumping through expansion of pore water and compression of aquifer.

4. The sorption of solute is usually measured by the ratio S_c of solute mass absorbed on the surface of the solid to the total mass of the solid matrix [unit: $kg\,kg^{-1}$]. Assuming that the sorbate concentration S_c is linearly proportional to the solute concentration C [unit $kg\,m^{-3}$] by an equilibrium relation

$$\boxed{S_c = k_d C}\,, \tag{1.48}$$

where k_d is the distribution coefficient, [$m^3 kg^{-1}$], describe how you may modify the dispersion equation to include the sorption term.

• Keys

Key 1

According to the principle of energy conservation, heat conduction without a heat source is governed by

$$\nabla \cdot \mathbf{q} = -\rho c_v \frac{\partial T}{\partial t}. \tag{1.49}$$

This equation is derived for a control volume. Each term has a unit of heating rate per unit volume, [Wm^{-3}]. We are to add a source term for the heat production of uranium.

The amount of uranium in the rock per unit volume is ρU, which produces heat energy at a rate of $H\rho U$ per unit volume. This should be subtracted from the left-hand side of the equation to yield

$$\nabla \cdot \mathbf{q} - H\rho U = -\rho c_v \frac{\partial T}{\partial t}. \tag{1.50}$$

The two negative signs imply that heat production $H\rho U$ is accompanied by an increase in energy storage $\rho c_v \partial T/\partial t$ if $\nabla \cdot \mathbf{q} = 0$.

Key 2

In case of anisotropy, the component of heat flux is

$$q_x = -k_x \frac{\partial T}{\partial x}, \quad \text{etc.}, \tag{1.51}$$

provided that the coordinate axes align with the principal axes of a conductivity ellipsoid. Substituting this relation into the energy conservation equation

$$\nabla \cdot \mathbf{q} = -\rho c_v \frac{\partial T}{\partial t} \tag{1.52}$$

gives

$$\frac{\partial}{\partial x}\left(k_x \frac{\partial T}{\partial x}\right) + \frac{\partial}{\partial y}\left(k_y \frac{\partial T}{\partial y}\right) + \frac{\partial}{\partial z}\left(k_z \frac{\partial T}{\partial z}\right) = \rho c_v \frac{\partial T}{\partial t}. \tag{1.53}$$

Key 3

The specific storage is traditionally defined as the water yield per unit aquifer volume per unit decline in hydraulic head

$$S_s = -\frac{\partial V_w}{V_b \partial h}. \tag{1.54}$$

According to the relation that $h = z + \psi$, a decline in hydraulic head $-\partial h$ induces a reduction in pore pressure head $-\partial \psi$, which in turn allows the water to expand.

According to the relation that $p_{\text{total}} = p_{\text{eff}} + p$, a reduction in pore pressure $-\partial p$ causes an increase in effective stress $+\partial p_{\text{eff}}$. Hence, the aquifer is compressed by pumping.

Key 4

In a nonreactive, source-free porous medium, the solute transport is governed by the principle of mass conservation,

$$\nabla \cdot \mathbf{J} = -\frac{\partial (\phi C)}{\partial t}. \qquad (1.55)$$

This represents the rate of mass change per unit volume in the control volume. We need to account for the solute removal from the solution and the adsorption of the removed solute onto the solid matrix.

Let $\rho_{\text{bulk}}^{\text{dry}}$ be the mass density of the dry solid matrix that adsorbs an amount $S_c \rho_{\text{bulk}}^{\text{dry}}$ of solute. Adding the rate of adsorption, $\partial(S_c \rho_{\text{bulk}}^{\text{dry}})/\partial t$, to the left-hand side of the mass balance equation yields

$$\nabla \cdot \mathbf{J} + \frac{\partial}{\partial t}\left(S_c \rho_{\text{bulk}}^{\text{dry}}\right) = -\frac{\partial (\phi C)}{\partial t}. \qquad (1.56)$$

The positive sign of $\partial(S_c \rho_{\text{bulk}}^{\text{dry}})/\partial t$ indicates that the rate of increasing sorption is accompanied by the declining rate of solute concentration in the solution.

At the equilibrium state of adsorption-desorption, let the absorption be proportional to the solute concentration

$$S_c = k_d C, \qquad (1.57)$$

where k_d is an equilibrium value (i.e., time invariant). Accordingly the divergence of the mass flux is

$$\nabla \cdot \mathbf{J} = -\left(k_d \rho_{\text{bulk}}^{\text{dry}} + \phi\right)\frac{\partial C}{\partial t}. \qquad (1.58)$$

Recall $\mathbf{J} = -D\nabla(\phi C) + C\mathbf{v}$ to yield

$$\nabla \cdot \nabla C - \frac{\mathbf{v}}{\phi} \cdot \nabla C = R_d \frac{\partial C}{\partial t} \qquad (1.59)$$

for a homogeneous, isotropic medium, where

$$\boxed{R_d = 1 + k_d \rho_{\text{bulk}}^{\text{dry}}/\phi} \qquad (1.60)$$

is the retardation factor (≥ 1). Rescaling the real time t to a new time scale t' ($= t/R_d$) yields

$$\nabla \cdot \nabla C - \frac{\mathbf{v}}{\phi} \cdot \nabla C = \frac{\partial C}{\partial t'}. \tag{1.61}$$

In the world of adsorption-time frame (t'), this equation is exactly the same in form as the adsorption equation in the real-time world (t) without adsorption. If an event happens at t', it occurs in real time at $t = R_d t'$. Because $R_d > 1$, the occurrence is delayed by a factor of R_d. Thus, adsorption retardation slows down contaminant migration and, ironically, it also retards clean-up processes.

Note that chromatography uses differences in the adsorption or retardation rates to differentiate different solutes.

• Suggested Readings

Many textbooks are available for further reading on how to formulate governing differential equations. New equations are constantly being published in research papers that modify existing ones with new physical insight or improved technique in numerical modeling. For general reading, we recommend Slattery's (1972) *Momentum, Energy, and Mass Transfer in Continuum*. For advanced pursuit of fluid flow in porous media, we recommend Bear's (1972) *Dynamics of Fluids in Porous Media* that provides rigorous treatment on various subjects covered in this chapter. We have not derived a flow equation for the unsaturated zone (such as Richardson's equation). Interested readers may consult Zaradny's (1993) *Groundwater Flow in Saturated and Unsaturated Soil*. For flow in fractured rocks, the readers are referred to the monograph on *Flow and Transport through Unsaturated Fractured Rock* edited by Evans and Nicholson (1987).

1.7 Notations

Symbol	Definition	SI Unit	Dimension
C	Solute concentration	$kg\,m^{-3}$	ML^{-3}
D, D^*, D_x	Mass diffusivity, diffusion coefficient, hydrodynamic dispersion coefficient	$m^2 s^{-1}$	$\mathrm{L}^2\mathrm{T}^{-1}$
E	Energy	J	$\mathrm{ML}^2\mathrm{T}^{-2}$
g	Gravitational acceleration	$m\,s^{-2}$	LT^{-2}
h	Hydraulic head	m	L
H	Heat production rate	Wm^{-3}	$\mathrm{ML}^{-1}\mathrm{T}^{-3}$
$J, \mathbf{J}$	Solute mass flux	$kg\,m^{-2}\,s^{-1}$	$\mathrm{ML}^{-2}\mathrm{T}^{-1}$
$k, k_x, ...$	Thermal conductivity	$Wm^{-1}K^{-1}$	$\mathrm{MLT}^{-3}\mathrm{K}^{-1}$
k_d	Distribution coefficient	$m^3 kg^{-1}$	$\mathrm{M}^{-1}\mathrm{L}^3$
$K, K_x, ...$	Hydraulic conductivity	$m\,s^{-1}$	LT^{-1}
$\hat{n}$	Unit outward normal	-	-
p, p_o	Pressure	Pa	$\mathrm{ML}^{-1}\mathrm{T}^{-2}$
$q, \mathbf{q}$	Heat flux	$W\,m^{-2}$	MT^{-3}
R_d	Retardation factor	-	-
S_c	Sorbate concentration	-	-
S_s	Specific storage	m^{-1}	L^{-1}
t, t'	Time	s	T
T, T_o	Temperature	K	K
$v, \mathbf{v}$	Velocity	$m\,s^{-1}$	LT^{-1}
V	Volume	m^3	L^3
z	Elevation head	m	L
$\alpha_x, ...$	Dispersivity	m	L
β, β_b	Compressibility	Pa^{-1}	$\mathrm{M}^{-1}\mathrm{LT}^2$
κ	Thermal diffusivity	$m^2 s^{-1}$	$\mathrm{L}^2\mathrm{T}^{-1}$
ρ	Mass density	$kg\,m^{-3}$	ML^{-3}
ρc_v	Volumetric heat capacity	$Jm^{-3}K^{-1}$	$\mathrm{ML}^{-1}\mathrm{T}^{-2}\mathrm{K}^{-1}$
ϕ, ϕ_{eff}	Porosity, effective porosity	-	-
ψ	Pressure head	m	L

Chapter 2

SOURCE FUNCTIONS AND CONVOLUTION

In this chapter we introduce the Theis well function, which is widely used for estimating aquifer transmissivity and storativity or predicting drawdown of water level due to pumping. It is equivalent to a line source solution in heat conduction. We will thus begin from an instantaneous point source solution in heat conduction, proceed to obtain a steady line source solution, then convert the line-source solution into the Theis well function. In the course of derivation, the concept of Green's function and convolution will be introduced to demonstrate how well functions can be superimposed to meet certain boundary conditions for a *linear system*. By linear, we mean that a differential equation does not contain products, powers, or any nonlinear functions of the dependent variable and its derivatives.

This chapter deals with Theis-related well functions only. Readers are referred to Chapters 4 and 5 for further discussion on aquifer responses to pumping.

2.1 Heat Sink and Source

2.1.1 Instantaneous Point Source

As shown in Chapter 1, heat conduction in a homogeneous, isotropic medium is governed by

$$\nabla^2 T = \frac{\partial^2 T}{\partial x^2} + \frac{\partial^2 T}{\partial y^2} + \frac{\partial^2 T}{\partial z^2} = \frac{1}{\kappa} \frac{\partial T}{\partial t}, \tag{2.1}$$

where T is the temperature rise, $[K]$, and κ is the thermal diffusivity, $[m^2 \, s^{-1}]$. If heat energy Q_T, $[J]$, is released at point (x', y', z') instantaneously, the temperature rise is

$$T[x, y, z, t] = \frac{Q_T}{\rho c_v \, (4\pi\kappa t)^{3/2}} \exp\left[-\frac{R^2}{4\kappa t}\right] \qquad (2.2)$$

with

$$R^2 = (x - x')^2 + (y - y')^2 + (z - z')^2$$

(Carslaw and Jaeger, 1959, p. 256) where ρc_v is the volumetric heat capacity at constant volume, $[J \, m^{-3} \, K^{-1}]$, and the distance R is measured from the source point. This solution for an instantaneous point source is our *Green's function* for heat conduction in a homogeneous, infinite medium. As demonstrated hereafter, many subsidiary solutions can be derived from this instantaneous point source solution.

2.1.2 Instantaneous Line Source

For a line source extending from z_a to z_b, we can integrate equation (2.2) with respect to z' over the line source to give

$$
\begin{aligned}
T[x, y, z, t] &= \frac{Q_T}{\rho c_v \, (4\pi\kappa t)^{3/2}} \exp\left[-\frac{r^2}{4\kappa t}\right] \int_{z_a}^{z_b} \exp\left[-\frac{(z - z')^2}{4\kappa t}\right] dz' \\
&= \frac{Q_T}{8\pi \rho c_v \kappa t} \exp\left[-\frac{r^2}{4\kappa t}\right] \\
&\quad \cdot \left\{ \mathrm{erf}\left[\frac{z - z_a}{\sqrt{4\kappa t}}\right] - \mathrm{erf}\left[\frac{z - z_b}{\sqrt{4\kappa t}}\right] \right\}, \qquad (2.3) \\
r^2 &= (x - x')^2 + (y - y')^2
\end{aligned}
$$

provided that Q_T is now regarded as heat release per unit length, $[J \, m^{-1}]$. Function

$$\mathrm{erf}[\xi] = \frac{2}{\sqrt{\pi}} \int_0^\xi \exp\left[-\eta^2\right] d\eta \qquad (2.4)$$

is the *error function*, which has the following properties,

$$0 \leq \mathrm{erf}[\xi] \leq 1 \quad \text{and} \quad \mathrm{erf}[-\xi] = -\mathrm{erf}[\xi]. \qquad (2.5)$$

Also, a closely related *complementary error function* is defined as

$$\text{erfc}[\xi] = 1 - \text{erf}[\xi].$$ (2.6)

For infinitely long line sources as $z_a \rightarrow -\infty$ and $z_b \rightarrow \infty$, the temperature distribution due to an instantaneous heat release is

$$T[r, t] = \frac{Q_T}{4\pi kt} \exp\left[-\frac{r^2}{4\kappa t}\right],$$ (2.7)

where $k = \rho c_v \kappa$ is the thermal conductivity, $[W\ m^{-1}K^{-1}]$.

2.1.3 Steady Line Source

If the heat is being generated continuously (i.e., Q_T is the heat generation rate per unit length, $[W\ m^{-1}]$), the temperature rise becomes

$$T[r, t] = \frac{1}{4\pi k} \int_0^t \frac{Q_T[t']}{t - t'} \exp\left[-\frac{r^2}{4\kappa(t - t')}\right] dt'.$$ (2.8)

If the heat generation rate is steady, the substitution of $\xi = r^2/4\kappa(t - t')$ yields

$$T[r, t] = \frac{Q_T}{4\pi k} E_1\left[\frac{r^2}{4\kappa t}\right],$$ (2.9)

where

$$E_1[x] = -\text{Ei}[-x] = \int_x^\infty \frac{\exp[-\xi]}{\xi} d\xi$$ (2.10)

is the *exponential integral*.

For small values of x, the exponential integral can be approximated by the series

$$\text{Ei}[-x] \approx \gamma + \ln x - x + \frac{1}{4}x^2 + ...,$$ (2.11)

where $\gamma = 0.5772$ is *Euler's constant*. Thus, for large time t,

$$T[r, t] \approx \frac{Q_T}{4\pi k} \left\{ \ln\left[\frac{4\kappa}{r^2}\right] - \gamma + \ln[t] \right\}.$$ (2.12)

The linear relation between temperature, T, and logarithm of time, $\ln t$, forms the theoretical basis for a transient *line-source* or *needle probe method* for measuring thermal conductivity in unconsolidated sediments or materials. The slope, $Q_T/4\pi k$, yields a measurement of conductivity,

$$k = \frac{Q_T}{4\pi \cdot \text{slope}} \qquad (2.13)$$

from the given Q_T.

In principle, thermal diffusivity can be determined from measured t_0 when the extrapolated temperature rise is zero,

$$\kappa = r^2 \exp[\gamma]/4t_0. \qquad (2.14)$$

However, the uncertainty in the determination of thermal diffusivity is relatively large because of the contact resistance around the probe and the inability to estimate an effective radius for a physical probe that is assumed to be a line source.

• **Remark**

The needle probe method for determining thermal conductivity is good for materials into which a needle can be easily inserted. Hole drilling is an inconvenient prerequisite for solids to be measured with a needle probe. An improvement has been made by sandwiching a line source between the flat surfaces of a sample and a reference material. The measured thermal conductivity k_T is the average conductivity for the two media,

$$k = \frac{1}{2}(k_1 + k_2), \qquad (2.15)$$

which is Lee's rule for the *half-needle probe method* (Lee, 1989). So the conductivity of the sample k_2 can be determined from the measured k and given k_1,

$$k_2 = 2k - k_1. \qquad (2.16)$$

See Problem 3 for the derivation.

2.2 Convolution

In the preceding section we have used the theorem of convolution to obtain solutions for different situations from a Green function without mentioning the term convolution. Now we introduce the concept.

2.2.1 Concept

For a physical system with an infinitely long line source, equation (2.7) represents the system response to an instant input of heat energy – *impulse response*. Equation (2.8) is the response to a general input function Q and represents the convolution of an input function with the impulse response as described below.

Let the impulse response be denoted by $I[t]$. If the input at time $t = 0$ is Q_0, the system output (response) is $Q_0 I[t]$; if another input with intensity $Q_{t'}$ appears at t', the corresponding system output is delayed by t' to have $Q_{t'} I[t - t']$. The total system response to these two input events is $Q_0 I[t] + Q_{t'} I[t - t']$ for $t > t'$.

Generalizing for a continuous input function per unit time, $Q[t]$, yields the system output

$$F[t] = \int_0^t Q[t'] I[t - t'] dt',$$
(2.17)

which is the *convolution integral* used in equation (2.8). It can also be written as

$$F[t] = \int_0^t Q[t - t'] I[t'] dt'.$$
(2.18)

Provided that the system or the differential equation is linear, the addition of two outputs for two inputs is called *linear superposition*. Convolution is a form of linear superposition. Equation (2.9) is a special case of system response: $T[r, t]$ is essentially the *step response* obtained by convolving impulse response $I[t]$ with a step-function input Q. The step response is the system output due to a step-function input.

If the input function lasts only from time 0 to t^*, the convolution integral is

$$F[t] = \begin{cases} \int_0^t Q[t'] I[t - t'] dt', & 0 \leq t \leq t^*; \\[2mm] \int_0^{t^*} Q[t'] I[t - t'] dt', & t > t^*. \end{cases}$$
(2.19)

A special case is worth noting. If $Q[t]$ is a step-pulse input, i.e.,

$$Q[t] = \begin{cases} Q, & 0 \leq t \leq t^*; \\ 0, & t \geq t^*, \end{cases}$$
(2.20)

then we have the step-pulse response,

$$F[t] = \begin{cases} F[t], & t \leq t^*; \\ F[t] - F[t - t^*], & t \geq t^*. \end{cases} \qquad (2.21)$$

A recovery test, as discussed later, is an example of practicing the step-pulse response.

- **Remark**

Note that any two functions can be convolved together. However, it has the meaning of a system output only when one function is an impulse response and the other is an input function.

2.2.2   Experimental Impulse Response

Usually an impulse response can be found for a linear differential equation and the convolution theorem can be invoked to deduce the system response for any input function. For a physical system that cannot be easily described with a differential equation, the impulse response needs to be obtained experimentally. However, it may be difficult to experimentally simulate an instantaneous input. Instead, experiments based on a step input, a square-wave input, or a sinusoidal input are generally conducted to obtain the impulse response.

Given step response $S[t]$, the impulse response is

$$\boxed{I[t] = \frac{\partial}{\partial t} S[t]}. \qquad (2.22)$$

For experimental data, the impulse response can be obtained from numerical differentiation of the step response. Because differentiation can amplify the error or noise in $S[t]$, care must be exercised to reduce the noise arising from differentiation. For example, smooth $S[t]$ with piecewise interpolation functions, then differentiate the interpolation functions to obtain $I[t]$.

See the Section 2.2.5 on z-transform for a better way to obtain impulse response from other types of input functions.

2.2.3 Numerical Convolution

The convolution integral in equation (2.17) can be represented by

$$F_t = \Delta t \sum_{t'=0}^{t} Q_{t'} I_{t-t'} \qquad (2.23)$$

for digital data processing, where Δt is the sampling interval. As t' progresses in steps from 0 to t, the $Q_{t'}$ proceeds likewise; whereas $I_{t-t'}$ retrocedes from t to 0 to form the product of $Q_{t'} I_{t-t'}$. The convolution thus can be viewed as a shifting product of two time series.

Graphically, the convolution of input

$$Q = [Q_0, Q_1, Q_2, Q_3, ...] \qquad (2.24)$$

with the impulse response

$$I = [I_0, I_1, I_2, I_{3,....}] \qquad (2.25)$$

can be viewed as follows. First, reverse the order of sequence I

$$I^{\mathrm{R}} = [...., I_3, I_2, I_1, I_0], \qquad (2.26)$$

then slide the entire I^{R} sequence step-by-step to the right to overlap with the Q sequence. The sum of the products of overlapped portions of Q and I^{R} gives the convolution at the time equal to the number of overlaps minus one. For example, the convolution at time step $t = 3$ has 4 overlapped positions (data points)

$$\begin{array}{cccccc} ...., & I_3, & I_2, & I_1, & I_0 \rightarrow \\ & \leftarrow Q_0, & Q_1, & Q_2, & Q_3, ... \end{array} \qquad (2.27)$$

Explicitly the convolutions for the first 4 time steps are

$$\begin{aligned} F_0 &= (Q_0 I_0)\Delta t, \\ F_1 &= (Q_0 I_1 + Q_1 I_0)\Delta t, \\ F_2 &= (Q_0 I_2 + Q_1 I_1 + Q_2 I_0)\Delta t, \\ F_3 &= (Q_0 I_3 + Q_1 I_2 + Q_2 I_1 + Q_3 I_0)\Delta t. \end{aligned} \qquad (2.28)$$

2.2.4   Transfer Function

Currently very few people would perform numerical convolution by the means described above. It is computationally more efficient to perform convolution via the Fourier transform.

The *Fourier transform* of the impulse response $I[t]$ is the *transfer function*

$$\overline{I}[\omega] = \int_{-\infty}^{\infty} I[t]e^{-i\omega t} dt \, . \tag{2.29}$$

$\overline{I}[\omega]$ is also known as the frequency spectrum of the impulse response. Its inverse Fourier transform is

$$I[t] = \frac{1}{2\pi} \int_{-\infty}^{\infty} \overline{I}[\omega]e^{i\omega t} d\omega \, . \tag{2.30}$$

Applying the Fourier transform to the convolution integral, equation (2.17), results in the product

$$\overline{F}[\omega] = \overline{Q}[\omega]\overline{I}[\omega] \, . \tag{2.31}$$

The above two relations demonstrate that convolution in the time domain is equivalent to multiplication in the frequency domain. (See Problem 7 for proof of the two relations and the concept of a Fourier delta function.) To obtain the convolution, simply take the inverse Fourier transform

$$F[t] = \frac{1}{2\pi} \int_{-\infty}^{\infty} \overline{F}[\omega]e^{i\omega t} d\omega . \tag{2.32}$$

Experimentally the impulse response $I[t]$ can be obtained from the step response $S[t]$ through

$$I[t] = \frac{\partial S[t]}{\partial t} = \frac{i}{2\pi} \int_{-\infty}^{\infty} \omega \overline{S}[\omega]e^{i\omega t} d\omega . \tag{2.33}$$

Therefore the transfer function is

$$\overline{I}[\omega] = i\omega \overline{S}[\omega] \tag{2.34}$$

in terms of the spectrum of step response $\overline{S}[\omega]$.

2.2.5 Fast Fourier Transform

Numerical Fourier transform is usually based on the method of Fast Fourier Transform (FFT),

$$
\begin{aligned}
\overline{I}[f_n] &= \int_{-\infty}^{\infty} I[t] e^{-i\omega t} dt \\
&\approx \Delta t \sum_{j=0}^{N-1} I_j \exp[-i2\pi f_n t_j]
\end{aligned} \tag{2.35}
$$

where I_js are $I[t]$ sampled at intervals of $\Delta t = T/N$ to have $N = 2^m$ data points (m is a positive integer), and t_j is the discrete time ($t_j = j\Delta t$). The nth frequency in units of cycles per unit time is

$$
f_n = \frac{\omega_n}{2\pi} = \frac{n}{N\Delta t}, \tag{2.36}
$$

where ω_n is the angular frequency (in unit of radian per unit time). There are N sampling intervals and N data points. The ending data point at $t = N\Delta t$ is excluded because of the implicit assumption that a finite-length time sequence is repeated periodically for the purpose of performing the Fourier transform.

Usually the discrete Fourier transform is taken by assuming that $\Delta t = 1$. Therefore

$$
\overline{I}_n = \sum_{j=0}^{N-1} I_j \exp[-i2\pi jn/N] \tag{2.37}
$$

and its inverse is

$$
I_j = \frac{1}{N} \sum_{n=0}^{N-1} \overline{I}_n \exp[i2\pi jn/N]. \tag{2.38}
$$

FFT algorithms in various forms are readily available. Most algorithms require 2^m data points. If the number of data points does not meet the 2^m requirement, interpolate the data set to have 2^m data points, or alternatively pad the data set with leading or trailing zeros to reach 2^m points. The padding of zeros will not affect the relative amplitude spectra. Readers should consult a text book on signal processing before actively using the transform technique for data interpretation.

• **Remark**

Some signals change slower as time increases. For example, the rate of drawdown due to pumping at a constant rate decreases with time. Drawdown thus can be sampled more frequently when it varies rapidly but less frequently when it varies slowly to reduce data storage and computation time. There are data loggers that allow sampling at equal intervals of logarithmic time in lieu of linear time. To do so, we need an FFT algorithm (Haines and Jones, 1988) that can process logarithmically sampled data. Such FFT for logarithmic time has been applied, for example, to analysis of slug tests by Marschall and Barczewski (1989). The vitality of such usage depends on successfully finding a scale parameter by trial-and-error.

2.2.6   Z-Transform

A time sequence $Q_t = [Q_0, Q_1, Q_2, ...]$ made from sampling $Q[t]$ at unit time intervals can be represented by

$$\mathcal{Z}\{Q[t]\} = Q_0 + Q_1 z + Q_2 z^2 + ... = \sum_{n=0}^{\infty} Q_n z^n, \qquad (2.39)$$

where superscript n designates sequential order of time and $\mathcal{Z}\{Q[t]\}$ is the z-transform of $Q[t]$. Convolution of $Q[t]$ and $I[t]$ can be obtained from the product of their respective z-transforms

$$\begin{aligned}
\mathcal{Z}\{Q[t]\}\mathcal{Z}\{I[t]\} &= \sum_{n=0}^{\infty} Q_n z^n \sum_{m=0}^{\infty} I_m z^m = \sum_{n=0}^{\infty}\sum_{m=0}^{\infty} Q_n I_m z^{n+m} \\
&= \sum_{k=0}^{\infty} F_k z^k = \mathcal{Z}\{F[t]\}.
\end{aligned} \qquad (2.40)$$

The term F_k is the sum of all combinations of Q_n and I_m for a given $k = m + n$. The sequence $\mathcal{Z}\{F[t]\}$ represents the *discrete convolution* of the two functions. Essentially a z-transform can be operated like a polynomial. For example,

$$\begin{aligned}
F_0 &= Q_0 I_0, \\
F_1 &= Q_0 I_1 + Q_1 I_0, \\
F_2 &= Q_0 I_2 + Q_1 I_1 + Q_2 I_0,
\end{aligned} \qquad (2.41)$$

as derived in equation (2.23) for $F_t = \Delta t \sum_{t'=0}^{\infty} Q_{t'} I_{t-t'}$.

If Q is the input and F is the output, then the transfer function of the linear system is

$$\mathcal{Z}\{I[t]\} = \frac{\mathcal{Z}\{F[t]\}}{\mathcal{Z}\{Q[t]\}}. \qquad (2.42)$$

Numerically one can obtain the impulse response from the following recursive relation,

$$I_n = F_n' - \sum_{j=1}^{n} Q_j' I_{n-j}, \qquad (2.43)$$

where $I_0 = F_0'$, $F_n' = F_n/Q_0$, and $Q_n' = Q_n/Q_0$. This is an alternative and superior way to obtain numerical impulse response. See Section 7.6 for application of the z-transform.

If we let $z = e^{-i\omega}$, the digitally sampled $Q[t]$ (discrete Q_t) in equation (2.39) becomes a Fourier transform (series)

$$
\begin{aligned}
\mathcal{Z}\{Q[t]\} &= \sum_{n=0}^{\infty} Q_n z^n = \sum_{n=0}^{\infty} Q_n \left(e^{-i\omega}\right)^n \\
&= \sum_{n=0}^{N-1} Q_n e^{-i2n\pi/N} = \mathcal{F}\{Q[t]\},
\end{aligned}
\qquad (2.44)
$$

where N is the total number of data points in the observation period T, i.e., $T = N\Delta t$ for a sampling interval of Δt and $\omega = 2\pi/N$.

Similarly let $z = e^{-s}$, equation (2.39) becomes

$$\Delta t \sum_{n=0}^{\infty} Q_n e^{-sn\Delta t} = \mathcal{L}\{Q[t]\} \qquad (2.45)$$

for $\Delta t = 1$. This is the discrete Laplace transform of Q_t. See Chapter 3 for details on the Laplace transform.

2.3 Theis Well Function

2.3.1 Assumptions

Aquifer properties are frequently determined by pumping tests that use Theis' solution or equivalently Jacob's solution to estimate transmissivity and storativity, although in reality very few field conditions meet the assumptions for its formulation and usage. Three sets of assumptions are used for the formulation.

☐ The first set regards the aquifer. The aquifer is horizontal with a constant thickness and extends laterally to infinity. It is homogeneous, isotropic, confined, and nonleaky.

☐ The second set is related to the well construction. The well penetrates vertically through the entire aquifer and is perforated or screened throughout the well bore within the aquifer. The well diameter is infinitesimally small so that water storage in the well bore is negligible and a line-source approximation is acceptable.

☐ The third set deals with the water. It requires that the hydraulic head is steady before the pumping test begins and the well discharge rate is constant. Water properties also are assumed to remain constant.

☐ Additionally, the drawdown is not so great as to turn the confined condition into a locally unconfined condition.

• Remark

Despite these stringent assumptions, the Theis solution remains very popular among practicing hydrogeologists due to its simplicity and low cost in estimating parameters. Frequently the Theis solution is applied to situations that do not meet the requisites. A practitioner should watch out for potential pitfalls in application.

Analyses of more sophisticated models for well functions may be marred by inadequate knowledge of aquifer characteristics (the knowledge that an aquifer pumping test is intended to determine) and compounded by missing well construction information. Without adequate data, the parameters for a given model or the models themselves may not be resolvable.

As used in this book, a *well function* is the hydraulic response to pumping and it describes groundwater flow toward a pumping well. The response is expressed in terms of *drawdown*, the deviation in hydraulic head from the prepumping head. Many well functions are described in Chapters 4 and 5. They are built upon one's ability to derive additional solutions for different assumptions or for improvement to meet relatively more realistic field conditions.

2.3.2 Derivation

The hydraulic head at the upper boundary of a *confined aquifer* is higher than the elevation head. This happens if and only if the pressure head at the upper boundary is positive. In comparison, the pressure head is zero at the water table of an *unconfined aquifer*. Overdraft of an aquifer can cause the aquifer to become partially unsaturated and make the initially confined aquifer become locally unconfined.

Under the assumptions for the Theis solution, groundwater in a confined aquifer flows radially toward a pumping well. Water is withdrawn from the well through compression of pore space due to increasing effective stress, and expansion of pore water due to decreasing pore pressure in response to decreasing hydraulic head. During pumping, the hydraulic head stays above the elevation of the upper boundary of the aquifer.

The governing differential equation for a homogeneous, isotropic aquifer is

$$\frac{\partial^2 h}{\partial x^2} + \frac{\partial^2 h}{\partial y^2} = \frac{S}{T_h}\frac{\partial h}{\partial t}, \tag{2.46}$$

where S is the storativity (specific storage S_s multiplied by aquifer thickness b, dimensionless) and T_h is the transmissivity (hydraulic conductivity k multiplied by aquifer thickness, $[m^2\ s^{-1}]$).

In cylindrical coordinates (Figure 2.1), the governing differential equation becomes

$$\boxed{\frac{\partial^2 h}{\partial r^2} + \frac{1}{r}\frac{\partial h}{\partial r} = \frac{S}{T_h}\frac{\partial h}{\partial t}}. \tag{2.47}$$

Its solution is subject to the initial condition of a constant head,

$$h(r,0) = h_0, \tag{2.48}$$

and the boundary conditions that

$$h\,[\infty, t] = h_0 \quad \text{and}$$
$$\lim_{r \to 0}\left(-2\pi rbk\frac{\partial h}{\partial r}\right) = -Q\,, \tag{2.49}$$

where Q is the steady rate of well discharge, $[m^3\ s^{-1}]$.

We will use the fact that the governing equations for heat conduction and groundwater flow are equivalent. The boundary and initial conditions qualify

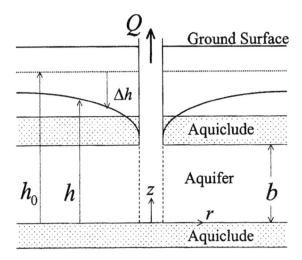

Figure 2.1: Sketch of a full-penetration pumping well in a confined aquifer. Well radius is zero for a Theis well function. Dotted horizontal line marks the pre-pumping piezometric surface.

the pumping well as a line source (sink) if the following pairs of equivalence are made

$$T \leftrightarrow h_0 - h, \quad \kappa \leftrightarrow \frac{T_h}{S}, \quad Q_T \leftrightarrow \frac{Q}{b},$$

$$\rho c_v \leftrightarrow S_s, \quad k = \kappa \rho c_v \leftrightarrow \frac{T_h}{b}. \tag{2.50}$$

Hence, the drawdown Δh can be directly obtained by analogy to the steady line-heat source solution, equation (2.9)

$$\boxed{\Delta h = h_0 - h = \frac{Q}{4\pi T_h} W[u]}, \tag{2.51}$$

where

$$W[u] = \int_u^\infty \frac{\exp[-\xi]}{\xi} d\xi = E_1[u], \quad u = \frac{r^2 S}{4t T_h}. \tag{2.52}$$

This drawdown formula is known as the Theis solution and $W[u]$ is the Theis well function, which is essentially an exponential integral. (See a Theis type curve in Chapters 4 and 5 along with other type curves for comparison with other well functions.)

See Problem 6 for the equivalent Theis solution when the horizontal hydraulic conductivity is anisotropic, and Section 4.3.2 for an alternative derivation of the Theis well function by the Laplace transform method.

2.3.3 Determination of Transmissivity and Storativity

Traditionally the transmissivity and storativity are determined from curve matching: First, a master curve of $\log[W]$ versus $\log[1/u]$ is prepared, preferably on a transparent film. Second, the observed data $(\Delta h, t)$ are plotted on log-log paper which has the same scale as the master curve. Third, match the two curves while keeping the t axis parallel to the $1/u$ axis. Fourth, from a convenient but arbitrary match point (this point does not have to be on the curves), find the equivalent values of $\Delta h_{\mathrm{p}} \leftrightarrow W[u_{\mathrm{p}}]$ and $t_{\mathrm{p}} \leftrightarrow u_{\mathrm{p}}$ pairs. Fifth, determine transmissivity T_{h} from the matched Δh_{p} and $W[u_{\mathrm{p}}]$ and the given Q via

$$\Delta h_{\mathrm{p}} = \frac{Q}{4\pi T_{\mathrm{h}}} W[u_{\mathrm{p}}], \tag{2.53}$$

then estimate the storativity S via

$$\frac{r^2 S}{4 T_{\mathrm{h}} t_{\mathrm{p}}} = u_{\mathrm{p}}. \tag{2.54}$$

The uncertainty in the determination of storativity is usually greater than that in determining transmissivity, especially when the pumping well is used as the only observation well.

• **Remark**

For multiwell observations, it is convenient to have drawdown data plotted as $\log[\Delta h]$ versus $\log[t/r^2]$ instead of $\log[t]$. Lateral inhomogeneity can often be spotted in such plots. Data reduction follows similar graphical procedures.

Graphical solutions can be easily obtained by manual curve matching but the goodness of fit cannot be quantified. A combination of curve matching on a computer monitor and inverse modeling is now the common practice for parameter determination. (See Chapter 10 on inverse modeling.)

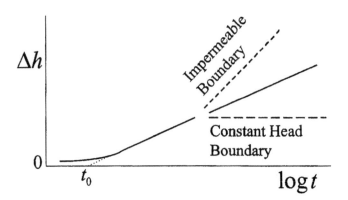

Figure 2.2: Semilog plot of drawdown versus time curve (solid curve). Dashed portions of the curves represent conditions of constant head boundary or impermeable boundary. Dotted line represents linear extrapolation to t_0 when the extrapolated Δh is zero.

2.3.4   Semilog Method

At large time t or small u (< 0.01), the Theis well function can be approximated by its series expansion,

$$\Delta h \approx \frac{Q}{4\pi T_\mathrm{h}} \left(-0.5772 - \ln \frac{r^2 S}{4 T_\mathrm{h} t} \right) \tag{2.55}$$

or

$$\boxed{\Delta h = \frac{Q}{4\pi T_\mathrm{h}} \left(2.303 \log t - 0.5772 - \ln \frac{r^2 S}{4 T_\mathrm{h}} \right)} \tag{2.56}$$

A plot of Δh versus $\log t$ for an observation well yields (Figure 2.2)

$$\text{slope} = \frac{2.303 Q}{4\pi T_\mathrm{h}} \quad \text{and} \quad \ln \frac{4 T_\mathrm{h} t_\mathrm{o}}{r^2 S} = 0.5772, \tag{2.57}$$

where t_o is the time when Δh is linearly extrapolated to zero drawdown, that is, when $\Delta h = 0$. A measurement of the slope and t_o from the semilog plot gives the transmissivity and storativity. This method is commonly known as the *semi-log* or *Jacob's method*. It does not require curve matching. Through regression analysis, it can provide a means of estimating the uncertainty of determination. As mentioned earlier, this relation is also used in line-source methods for measurement of thermal conductivity.

2.3.5 Radius of Pumping Influence

The radius of pumping influence at time t is the radial distance R where drawdown vanishes. According to equation (2.56),

$$R = \sqrt{\frac{4T_h t}{S \exp(0.5772)}}, \tag{2.58}$$

which is independent of the well discharge rate – a conclusion defying our common sense.

The radius of influence will depend on the discharge rate, however, if it is defined at the radial distance where drawdown has reached a predefined threshold value. See Problem 4 as an example for setting the threshold value and the determination of the radius of influence.

2.3.6 Recovery Test

According to equations (2.8), (2.19), and (2.51), if the pump is turned off at time t_{off}, the residual drawdown Δh at time beyond t_{off} is

$$\Delta h = \frac{Q}{4\pi T_h} \int_0^{t_{\text{off}}} \frac{1}{t - t'} \exp\left[-\frac{r^2 S}{4T_h(t - t')}\right] dt'. \tag{2.59}$$

Replacing $r^2 S / 4T_h(t - t')$ by ξ yields

$$
\begin{aligned}
\Delta h &= \frac{Q}{4\pi T_h} \int_{r^2 S/4T_h t}^{r^2 S/4T_h (t - t_{\text{off}})} \frac{\exp[-\xi]}{\xi} d\xi \\
&= \frac{Q}{4\pi T_h} \left\{ \int_{r^2 S/4T_h t}^{\infty} \frac{\exp[-\xi]}{\xi} d\xi - \int_{r^2 S/4T_h (t - t_{\text{off}})}^{\infty} \frac{\exp[-\xi]}{\xi} d\xi \right\} \\
&= \frac{Q}{4\pi T_h} \left\{ W[u] - W[u_{\text{off}}] \right\}, \tag{2.60} \\
u_{\text{off}} &= \frac{r^2 S}{4T_h(t - t_{\text{off}})}, \quad t > t_{\text{off}}.
\end{aligned}
$$

This is an example of step-pulse response.

The result shows that the drawdown after a pump is turned off can be simulated by superposition of two well functions at the same well site. $W[u]$ represents the real pumping well and it behaves as if the pumping continues, whereas $W[u_{\text{off}}]$ represents an image well with an injection rate of $-Q$ and a time delay of t_{off}. Thus the extraction and injection rates cancel each other for time greater than t_{off} and the two wells together indicate that the pump is indeed off.

There is no convenient graphical method to obtain T_{h} and S from a recovery test using the superposition of two Theis well functions. However, using the expression for the semi-log method (2.56) yields

$$\boxed{\Delta h \approx \frac{Q}{4\pi T_{\text{h}}} \ln \frac{t}{t - t_{\text{off}}}}, \quad t > t_{\text{off}} \tag{2.61}$$

which can be used to obtain transmissivity T_{h} from a plot of the residual drawdown Δh versus $\log[t/(t - t_{\text{off}})]$, but storativity S cannot be determined from the recovery test. This equation is used to construct the *Hornet plot* for petroleum reservoir engineering and the *Lachenbruch plot* for estimating formation temperature from thermal recovery data shortly after stopping drilling and mud circulation.

In the early part of pumping, data quality may suffer from setting and resetting operational parameters for finding an optimal pumping rate (thus creating an undesirable fluctuation in Q), especially when a single-well pumping test is conducted. Recovery tests provide checks on the T_{h} determined during the pumping period.

2.4   Linear Superposition of Well Functions

Any solutions to a linear differential equation can be combined or superimposed linearly to form a new solution provided that the new solution satisfies a set of prescribed initial and boundary conditions. The *method of images* is an example of linear superposition or convolution, as demonstrated for the recovery test. Following are more examples.

2.4.1   Pumping Near a Constant Head Boundary

Where a pumping well is located near a river, lake, or reservoir, the drawdown may become steady as if the supply reservoir, represented by a constant head, has an infinite capacity (Figure 2.2). This constant-head condition can be simulated by adding an image injection well (with negative Q) on the other

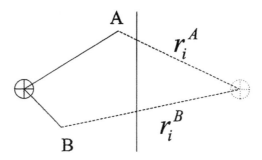

Figure 2.3: Plan view of an image well (dotted circle) across a constant head boundary. It is located at the intersection of two circles with different radii, r_i^A and r_i^B, centered at observation wells A and B (circles are not drawn). The boundary lies midway between the real and image wells.

side of the boundary (which is assumed to be vertical)

$$\Delta h = \frac{Q}{4\pi T_{\mathrm{h}}} \left(W[u] - W[u_{\mathrm{i}}] \right), \tag{2.62}$$

$$u = r^2 S/4T_{\mathrm{h}}t, \quad u_{\mathrm{i}} = r_{\mathrm{i}}^2 S/4T_{\mathrm{h}}t,$$

where r_{i} is the distance from the image well to the observation well.

Using the series expansion for the Theis well function (Jacob's semilog method, equation 2.56), the drawdown at large time is

$$\Delta h = \frac{Q}{2\pi T_{\mathrm{h}}} \ln \frac{r_{\mathrm{i}}}{r}. \tag{2.63}$$

This result says that at a given location, the drawdown is steady at large pumping time.

Thus, from the given Q, T_{h}, and r and the steady drawdown Δh, the radial distance r_{i} can be determined (for example, r_i^A in Figure 2.3 for observation well A). The image well is located on a circle with radius r_{i} around an observation well. Additional circles can be drawn around other observation wells. Ideally those circles intercept at one point—the image well site. The constant head boundary is located midway between the real well and the image well.

2.4.2 Pumping Near an Impermeable Boundary

The condition of no-flow occurs at an impermeable boundary. It can be simulated by zero hydraulic gradient across a boundary in a homogeneous medium. Assuming that this boundary is vertical and runs through the entire thickness of the aquifer, the drawdown can be simulated by adding an image discharge well

$$\Delta h = \frac{Q}{4\pi T_{\rm h}} \left(W[u] + W[u_{\rm i}] \right), \quad u_{\rm i} = \frac{r_{\rm i}^2 S}{4 T_{\rm h} t}. \tag{2.64}$$

Using the semi-log method yields

$$\Delta h = \frac{Q}{2\pi T_{\rm h}} \left(-\gamma - \ln \frac{S}{4 T_{\rm h} t} - \ln r r_{\rm i} \right). \tag{2.65}$$

This result gives a slope (of the Δh-versus-log$[t]$ curve) that is twice the slope for the case of a homogeneous medium (Figure 2.2).

Similar to the case of a constant head boundary, the impermeable boundary can be located from the $r_{\rm i}$ for two or more observation wells (Figure 2.3). It is cautioned that in this usage, $S r r_{\rm i} / 4 T_{\rm h} t$ must be made dimensionless for unit consistency.

2.4.3 Pumping Near Two Dissimilar Media

Pumping near an impermeable boundary represents an extreme example of the general case for two dissimilar media. Problem 3 addresses the line source solution for media with identical diffusivity but different conductivity. Lee (1989) provided a solution in integral form for media with different conductivity and diffusivity and established a theoretical basis for the half-needle probe method for measuring thermal conductivity, equation (2.15).

2.5 Evaluation of Exponential Function

We will first present two methods for series expansion of the exponential functions: one is useful for small argument u, while the other is for large u. An efficient method to evaluate the exponential function is then introduced.

• **Small u, Large r^2/t**

Rewrite the exponential integral in two parts

$$E_1[u] = \int_u^\infty \frac{e^{-x}}{x}\, dx = \int_0^\infty \frac{e^{-x}}{x}\, dx - \int_0^u \frac{e^{-x}}{x}\, dx. \qquad (2.66)$$

Then use the series expansion of e^{-x} for the second integral to yield

$$
\begin{aligned}
\int_0^u \frac{e^{-x}}{x}\, dx &= \int_0^u \frac{1}{x}\sum_{n=0}^\infty (-1)^n \frac{x^n}{n!}\, dx \\
&= \left. \ln x + \sum_{n=1}^\infty (-1)^n \frac{x^n}{n\, n!}\right|_0^u \\
&= -\ln 0^+ + \ln u + \sum_{n=1}^\infty (-1)^n \frac{u^n}{n\, n!}. \qquad (2.67)
\end{aligned}
$$

The poorly behaved $\ln 0^+$ requires extra treatment. According to the series expansion of $\ln x$

$$\ln x = -\sum_{n=1}^\infty \frac{(-1)^n (x-1)^n}{n}, \qquad (2.68)$$

we obtain the diverging series

$$\ln 0^+ = -\sum_{n=1}^\infty \frac{1}{n}. \qquad (2.69)$$

Now we need to evaluate the first integral of equation (2.66), which is a special case of the Gamma function

$$\Gamma[z] = \int_0^\infty x^{z-1} e^{-x}\, dx. \qquad (2.70)$$

That is,

$$\int_0^\infty \frac{e^{-x}}{x}\, dx = \Gamma[0^+] = -\gamma + \int_0^\infty \frac{e^{-x}}{1 - e^{-x}}\, dx \qquad (2.71)$$

(Feshbach and Mores, 1953, p. 422) where Euler's constant

$$\gamma = \sum_{n=1}^\infty \frac{1}{n} - \ln_{M\to\infty} M = 0.5772. \qquad (2.72)$$

Using the binomial expansion of $1/\left(1 - e^{-x}\right)$ yields

$$
\begin{aligned}
\Gamma[0^+] &= -\gamma + \int_0^\infty \sum_{n=0}^\infty e^{-(n+1)x} dx = -\gamma - \sum_{n=0}^\infty \frac{e^{-(n+1)x}}{n+1}\bigg|_0^\infty \\
&= -\gamma + \sum_{n=1}^\infty \frac{1}{n}.
\end{aligned}
\tag{2.73}
$$

Substituting these relations into equation (2.66) yields the series expansion of the exponential integral

$$
\int_u^\infty \frac{e^{-x}}{x} \, dx = -\gamma - \ln u - \sum_{n=1}^\infty (-1)^n \frac{u^n}{n\,n!}.
\tag{2.74}
$$

This forms the basis for Jacob's semilog method for small u, but the integral diverges logarithmically as $u \to 0$.

- **Large u, Small r^2/t**

For large u, we repeat integration by parts to obtain

$$
\begin{aligned}
\text{Ei}[-u] &= -E_1[u] = \int_\infty^u \frac{e^{-x}}{x} \, dx \\
&= -\frac{e^{-u}}{u} - \int_\infty^u \frac{e^{-x}}{x^2} \, dx = -\frac{e^{-u}}{u} + \frac{e^{-u}}{u^2} - 2\int_\infty^u \frac{e^{-x}}{x^3} \, dx \\
&= -\frac{e^{-u}}{u}\left[1 - \frac{1}{u} + \frac{2!}{u^2} - \frac{3!}{u^3} + ... + \frac{(-1)^n n!}{u^n}\right] \\
&\quad -(-1)^n (n+1)! \int_\infty^u \frac{e^{-x}}{x^{n+2}} \, dx.
\end{aligned}
\tag{2.75}
$$

This is an exact representation of $\text{Ei}[-u]$. Using the terms in the square brackets yields the approximation

$$
\text{Ei}(-u) \approx -\frac{e^{-u}}{u}\left[1 - \frac{1}{u} + \frac{2!}{u^2} - \frac{3!}{u^3} + ... + \frac{(-1)^n n!}{u^n}\right],
\tag{2.76}
$$

which is the *asymptotic series* representation of the exponential function. An asymptotic series converges as $u \to \infty$ for a given n.

A series that converges for any u as $n \to \infty$ is a *convergent series*. Being useful for small u only, the series in equation (2.74) is neither a converging series nor an asymptotic series. However, if u is replaced by $r^2 S/4Tt$, then the series part in equation (2.74) is qualified as an asymptotic series of t. One notes that the approximation

$$E_1[u] = \int_0^\infty \frac{e^{-x}}{x} \, dx \approx -\gamma - \ln[u] = -\gamma - \ln\left[\frac{r^2 S}{4T}\right] + \ln[t] \qquad (2.77)$$

is an asymptote to the exponential integral in terms of t/r^2 and the asymptote diverges as $t/r^2 \to \infty$.

- **Numerical Evaluation**

Exponential integrals can be evaluated by various means (Abramowitz and Stegun, 1968). None is accurate for the entire range of u. It is recommended that the series representation in equation (2.74) be used for small u ($u \leq 1$), in which the series terms can be neglected if $u < 0.01$.

For large u, use the Gauss-Laguerre quadrature formula (Tseng and Lee, 1998)

$$\begin{aligned}
E_1[u] &= \int_u^\infty \frac{e^{-x}}{x} dx = e^{-u} \int_0^\infty \frac{e^{-y}}{u+y} dy \\
&\approx e^{-u} \sum_{j=1}^N a_j f[y_j + u], \quad f[z] = \frac{1}{z}, \quad u \geq 1, \qquad (2.78)
\end{aligned}$$

where N refers to the desired Nth Laguerre polynomial, which is orthonormal over the interval $[0, \infty]$ with respect to the weighting function $\exp[-x]$, a_j and y_j are respectively the coefficients and roots of the Laguerre polynomial. The roots and the coefficients are available in Stroud and Secrest (1966).

2.6 Problems, Keys, and Suggested Readings

- **Problems**

1. The temperature rise $T[x, y, z, t]$ due to an instantaneous point source at (x', y', z') in a homogeneous, isotropic medium is often described by

$$T[x, y, z, t] = \frac{Q_T}{\rho c \, (4\pi\kappa t)^{3/2}} \exp\left[-\frac{(x-x')^2 + (y-y')^2 + (z-z')^2}{4\kappa t}\right], \qquad (2.79)$$

where κ is the thermal diffusivity and ρc is the volumetric heat capacity. Determine the physical meaning of Q_T by performing the integration

$$\int\int\int_{-\infty-\infty-\infty}^{\infty\infty\infty} \rho c T\,[x, y, z, t]\,dx\,dy\,dz \tag{2.80}$$

(assuming Q_T is a constant).

2. A north-south trending valley is bounded by two vertical groundwater barriers separated by a distance of w. A well located at distance b from the west barrier taps into a confined aquifer under the valley. It has full penetration and screening. If the well is pumped at a constant rate, determine the hydraulic head response in terms of the Theis well function.

3. Two semi-infinite media have different thermal conductivity but the same diffusivity. There is no thermal contact resistance at the interface between the two media. A point source is located at distance b from the interface. If heat is released instantaneously, determine the temperature distribution in the two media.

4. If the radius of influence of a pumping well is defined as distance R_c beyond which the drawdown is less than Δh_c, determine the radius of influence in terms of well discharge rate, transmissivity, storativity, and the threshold value Δh_c.

5. Two Theis wells tap the same aquifer. Each day Well A is to be pumped at rate Q_A for a duration of t_A and Well B is to be pumped at rate Q_B for a duration of t_B. Each pump has a rated capacity of Q_{max}. If the drawdown at the midpoint between the two wells is to be maintained at Δh_c or less for all time, describe how you may pump in terms of discharge rate, pumping duration, and recovery. (Note this problem does not have a unique answer.)

6. Groundwater flow in an anisotropic porous medium is governed by

$$\frac{\partial}{\partial x}\left(K_x\frac{\partial h}{\partial x}\right) + \frac{\partial}{\partial y}\left(K_y\frac{\partial h}{\partial y}\right) + \frac{\partial}{\partial z}\left(K_z\frac{\partial h}{\partial z}\right) = S_s\frac{\partial h}{\partial t} \tag{2.81}$$

(assuming that the coordinates are parallel to the principal axes of the hydraulic conductivity ellipsoid and the medium is homogeneous). Derive an equivalent Theis well function for a confined aquifer.

7. Use the Fourier transform pair to find the Fourier delta function.

8. Show that multiplication in the frequency domain is equivalent to convolution in the time domain.

• Keys

Key 1

The integral represents the total heat energy release of the instantaneous point source. We are to relate Q_T to energy release. One can use an integral table and perform the desired triple integration one by one. Here is a better way.

Let the coordinate origin be at the source point, i.e., set $(x', y', z') = (0, 0, 0)$. Then, use the spherical coordinates (r, ϕ, θ) to obtain the desired integration. Denote this integral as

$$
\begin{aligned}
A &= \int\!\!\!\int\!\!\!\int\limits_{-\infty}^{\infty} \rho c T\,[x, y, z, t]\,dx\,dy\,dz \\
&= \frac{Q_T}{(4\pi\kappa t)^{3/2}} \int_0^\pi \sin\theta\,d\theta \int_0^{2\pi} d\phi \int_0^\infty r^2 \exp\left[-\frac{r^2}{4\kappa t}\right] dr \\
&= \frac{4Q}{\sqrt{\pi}} \int_0^\infty \xi^2 e^{-\xi^2}\,d\xi
\end{aligned}
\tag{2.82}
$$

with

$$
\xi = \frac{r}{\sqrt{4\kappa t}}, \quad r^2 = x^2 + y^2 + z^2.
\tag{2.83}
$$

To get the last integral, let us play some tricks. First, set

$$
I = \int_0^\infty e^{-\xi^2}\,d\xi.
\tag{2.84}
$$

Then, square it and introduce u- and v-coordinates to have (Figure 2.4)

$$
I^2 = \int_0^\infty e^{-u^2}\,du \int_0^\infty e^{-v^2}\,dv,
\tag{2.85}
$$

which is integrated over one-quarter of the uv-space. In polar coordinates, I^2 becomes

$$
I^2 = \int_0^{\pi/2} d\phi \int_0^\infty e^{-R^2} R\,dR, \quad R^2 = u^2 + v^2.
\tag{2.86}
$$

Proceed further by letting $\zeta = R^2$ to obtain

$$
I^2 = \frac{\pi}{4} \int_0^\infty e^{-\zeta}\,d\zeta = \frac{\pi}{4}.
\tag{2.87}
$$

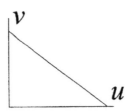

Figure 2.4: The integration of $\int_0^\infty \exp[-\xi^2]d\xi$ is squared to transform into the uv-plane to facilitate integration.

Therefore

$$I = \int_0^\infty e^{-\xi^2} d\xi = \frac{\sqrt{\pi}}{2}. \tag{2.88}$$

Now one can use this relation three times in rectangular coordinates to achieve the desired triple integrations. However, we will pursue the integration $\int \xi^2 \exp[-\xi^2]d\xi$ from $\int \exp[-\xi^2]d\xi$ to learn one more useful integration trick.

Let $\xi = \sqrt{a}\beta$, then

$$I = \int_0^\infty e^{-(\sqrt{a}\beta)^2} d\left(\sqrt{a}\beta\right) = \frac{\sqrt{\pi}}{2}. \tag{2.89}$$

Rearrange terms and redefine the reshuffled relation as

$$P[a] = \int_0^\infty e^{-a\beta^2} d\beta = \frac{I}{\sqrt{a}} = \frac{1}{2}\sqrt{\frac{\pi}{a}}. \tag{2.90}$$

Take the derivative

$$\frac{\partial P[a]}{\partial a} = -\int_0^\infty \beta^2 e^{-a\beta^2} d\beta = -\frac{1}{4}\sqrt{\frac{\pi}{a^3}}. \tag{2.91}$$

Finally let $a = 1$ to obtain

$$\int_0^\infty \beta^2 e^{-\beta^2} d\beta = \frac{\sqrt{\pi}}{4}. \tag{2.92}$$

Substitute this relation into equation (2.82) to yield $A = Q$. Thus Q represents the total heat energy release.

Key 2

This is a problem solvable by the method of images. Multiple reflections of the real well upon the two walls give an infinite number of image wells. In this problem all image wells are pumped at the same rate as the real well, Q.

Let us start by setting the west wall at $x = 0$ (Figure 2.5) so that the real well is located at ($x = b, \quad y = 0$), and the first image well across the west wall is located at $x = -b$. A no-flow condition is met at the west wall but the condition at the east wall is not satisfied. Hence, add the image of the first image across the east wall at $x = 2w + b$ (the second image at $w + b$ east of the east wall). The second image destroys the no-flow condition at the west wall. Hence, the third image across the west wall at $x = -(2w + b)$ is needed. Continuing the exercise, we obtain the following tabulated locations of image wells (left column).

x Coordinate Origin at West Wall	x' Coordinate Origin at East Wall
$-b$	a
$2w + b$	$-(2w + a)$
$-(2w + b)$	$2w + a$
$4w + b$	$-(4w + a)$
...	...

Ditto for the series of images that start with the east wall. Note that $x' = x - w$ and $a = w - b$.

Let the observation well be located at (x, y) within the valley. Based on the sequence of image well locations, the image wells can be grouped into 4 sets of distances between observation and image wells

$$r_n^{W1} = \sqrt{(x + 2nw + b)^2 + y^2}, \quad n = 0, 1, 2, \dots$$

$$r_m^{W2} = \sqrt{(x - 2mw - b)^2 + y^2}, \quad m = 1, 2, 3 \dots$$

$$r_n^{E1} = \sqrt{(x' - 2nw - a)^2 + y^2}, \quad n = 0, 1, 2, \dots$$

$$r_m^{E2} = \sqrt{(x' + 2mw + a)^2 + y^2}, \quad m = 1, 2, 3, \dots \qquad (2.93)$$

So the drawdown is the sum of the contributions from the real and image

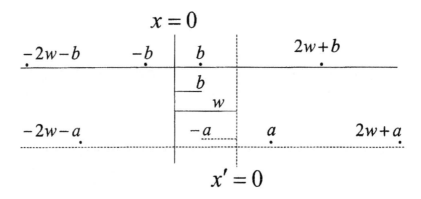

Figure 2.5: Locations of image wells. Valley width $= w$. The real well is located at $x = b$ with respect to the west wall and at $x' = -a$ with respect to the east wall. The first image well along the solid line reflects from the west wall, while the first image well along the dotted horizontal line reflects from the east wall.

wells

$$\Delta h[r,t] \;=\; \frac{Q}{4\pi T} \left\{ W[u] + \sum_{n=0}^{\infty} \left(W[u_n^{\mathrm{W1}}] + W[u_n^{\mathrm{E1}}] \right) \right.$$
$$\left. + \sum_{m=1}^{\infty} \left(W[u_m^{\mathrm{W2}}] + W[u_m^{\mathrm{E2}}] \right) \right\}, \tag{2.94}$$

where

$$u \;=\; \frac{r^2 S}{4Tt}, \quad r^2 = (x-b)^2 + y^2$$
$$u_n^{\mathrm{E1}} \;=\; \frac{S}{4Tt} \left(r_n^{\mathrm{E1}} \right)^2, \quad u_n^{\mathrm{W1}} = \frac{S}{4Tt} \left(r_n^{\mathrm{W1}} \right)^2$$
$$u_m^{\mathrm{E2}} \;=\; \frac{S}{4Tt} \left(r_m^{\mathrm{E2}} \right)^2, \quad u_m^{\mathrm{W2}} = \frac{S}{4Tt} \left(r_m^{\mathrm{W2}} \right)^2 \tag{2.95}$$

and the Theis well function is

$$W[u] = \int_u^{\infty} \frac{e^{-\xi}}{\xi} d\xi. \tag{2.96}$$

Now, we can unify the coordinates. Substitution of $x' = x - w$ and $a = w - b$ into r_n^{E1} and r_m^{E2} results in

$$r_n^{\mathrm{E1}} \;=\; \sqrt{(x - 2(n+1)w + b)^2 + y^2} = \sqrt{(x - 2mw + b)^2 + y^2} = r_m^{\mathrm{E1}},$$

$$r_m^{E2} = \sqrt{(x + 2mw - b)^2 + y^2}. \tag{2.97}$$

Note that the real well can be included as a term in the r_m^{W2} series such that the series becomes r_n^{W2}. The final solution is therefore

$$\Delta h[r, t] = \frac{Q}{4\pi T} \left\{ \sum_{n=0}^{\infty} \left(W[u_n^{W1}] + W[u_n^{W2}] \right) \right.$$

$$\left. + \sum_{m=1}^{\infty} \left(W[u_m^{E1}] + W[u_m^{E2}] \right) \right\} \tag{2.98}$$

The no-flow condition at $x = 0$ can be easily checked (i.e., $\partial \Delta h / \partial x = 0$ at $x = 0$) in the last formulation. It can also be shown to meet the no-flow condition at $x = w$.

Key 3

Let the interface be the xy-plane and the source point be at $(0, 0, b)$. We will use the method of images to solve the problem (Figure 2.6). According to equation (2.2), the temperature in the medium containing the source point is

$$T_1[r, z, t] = \frac{Q_T}{(\rho c)_1 (4\pi \kappa t)^{3/2}} \exp\left[-\frac{(z - b)^2 + r^2}{4\kappa t} \right]$$

$$+ \frac{A Q_T}{(\rho c)_1 (4\pi \kappa t)^{3/2}} \exp\left[-\frac{(z + b)^2 + r^2}{4\kappa t} \right], \tag{2.99}$$

where the first term represents the temperature due to the real source Q_T and the second term is due to the image point source $A Q_T$ located at $(0, 0, -b)$. Q_T is the total heat energy released, and $A Q_T$ is a to-be-determined image heat source, which represents partial reflection of heat energy flow.

The temperature distribution in the second medium due to the real source is

$$T_2[r, z, t] = \frac{B Q_T}{(\rho c)_2 (4\pi \kappa t)^{3/2}} \exp\left[-\frac{(z - b)^2 + r^2}{4\kappa t} \right], \tag{2.100}$$

where $B Q_T$ is another to-be-determined heat source (representing partial transmission of heat energy flow from the real source).

A and B can be determined from two boundary conditions at the interface:

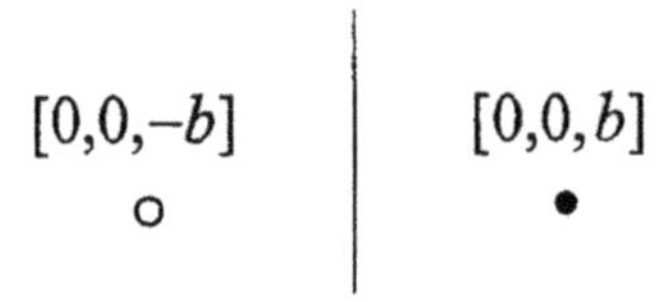

$[0,0,-b]$ $[0,0,b]$

o   ●

Figure 2.6: Real (solid dot) and image source points.

a) The continuity in temperature at $z = 0$,

$$T_1[r, 0, t] = T_2[r, 0, t] \tag{2.101}$$

leads to

$$\frac{A}{(\rho c)_1} - \frac{B}{(\rho c)_2} = -\frac{1}{(\rho c)_1}. \tag{2.102}$$

b) The continuity in normal heat flux at $z = 0$,

$$-k_1 \frac{\partial T_1}{\partial z} = -k_2 \frac{\partial T_2}{\partial z}, \tag{2.103}$$

leads to

$$\frac{k_1 A}{(\rho c)_1} + \frac{k_2 B}{(\rho c)_2} = \frac{k_1}{(\rho c)_1}. \tag{2.104}$$

Solving for A and B yields

$$A = \frac{k_1 - k_2}{k_1 + k_2}, \quad B = \frac{2k_1}{k_1 + k_2} \frac{(\rho c)_2}{(\rho c)_1}, \tag{2.105}$$

where A and B can be viewed as reflection and transmission coefficients respectively. Thus the temperature distributions are

$$\begin{aligned}
T_1[r, z, t] &= \frac{Q_T}{(\rho c)_1 (4\pi \kappa t)^{3/2}} \exp\left[-\frac{(z - b)^2 + r^2}{4\kappa t}\right] \\
&+ \frac{k_1 - k_2}{k_1 + k_2} \frac{Q_T}{(\rho c)_1 (4\pi \kappa t)^{3/2}} \exp\left[-\frac{(z + b)^2 + r^2}{4\kappa t}\right]
\end{aligned} \tag{2.106}$$

and

$$T_2[r, z, t] = \frac{Q_T}{(\rho c)_1 (4\pi \kappa t)^{3/2}} \frac{2k_1}{k_1 + k_2} \exp\left[-\frac{(z - b)^2 + r^2}{4\kappa t}\right]. \tag{2.107}$$

If the source point is at the interface (i.e., $b = 0$), the temperature distribution is

$$
\begin{aligned}
T_1[r, z, t] &= T_2[r, z, t] \\
&= \frac{Q_T}{(\rho c)_1 (4\pi\kappa t)^{3/2}} \frac{2k_1}{k_1 + k_2} \exp\left[-\frac{z^2 + r^2}{4\kappa t}\right].
\end{aligned}
\tag{2.108}
$$

This result indicates that both media have the same temperature distribution because the two media have the same thermal diffusivity. However, it can be shown that the energy partition is different because the heat capacities and thermal conductivities are different.

Key 6

For a full-penetration pumping well in a horizontal, confined aquifer, the flow is independent of z. Hence, the flow equation is

$$
\frac{\partial}{\partial x}\left(K_x \frac{\partial h}{\partial x}\right) + \frac{\partial}{\partial y}\left(K_y \frac{\partial h}{\partial y}\right) = S_s \frac{\partial h}{\partial t}.
\tag{2.109}
$$

We are to rescale the coordinates so that the equation in the new coordinate system behaves as if it were for an isotropic medium. There are several ways to accomplish the task. Let us divide the equation by the geometric average of the two principal conductivities

$$
K = \sqrt{K_x K_y}.
\tag{2.110}
$$

The equation becomes

$$
\frac{\partial}{\partial x}\left(\sqrt{\frac{K_x}{K_y}} \frac{\partial h}{\partial x}\right) + \frac{\partial}{\partial y}\left(\sqrt{\frac{K_y}{K_x}} \frac{\partial h}{\partial y}\right) = \frac{S_s}{K} \frac{\partial h}{\partial t}.
\tag{2.111}
$$

Let

$$
x' = x\left(K_y/K_x\right)^{1/4}, \quad y' = y\left(K_x/K_y\right)^{1/4},
\tag{2.112}
$$

then the flow equation in the primed coordinates is

$$
\frac{\partial^2 h}{\partial x'^2} + \frac{\partial^2 h}{\partial y'^2} = \frac{S_s}{K} \frac{\partial h}{\partial t}.
\tag{2.113}
$$

The drawdown for the modified Theis well function is thus

$$
\Delta h = \frac{Q}{4\pi T} W[u'],
\tag{2.114}
$$

where

$$
u' = r'^2 S/4Tt, \quad r'^2 = x'^2 + y'^2, \quad T = Kb.
\tag{2.115}
$$

Key 7

Substitute the Fourier transform

$$\overline{I}[\omega] = \int_{-\infty}^{\infty} I[t] e^{-i\omega t} dt \qquad (2.116)$$

into the inverse Fourier transform

$$
\begin{aligned}
I[t] &= \frac{1}{2\pi} \int_{-\infty}^{\infty} \overline{I}[\omega] e^{i\omega t} d\omega \\
&= \frac{1}{2\pi} \int_{-\infty}^{\infty} \left(\int_{-\infty}^{\infty} I[\tau] e^{-i\omega \tau} d\tau \right) e^{i\omega t} d\omega \\
&= \int_{-\infty}^{\infty} I[\tau] d\tau \left(\frac{1}{2\pi} \int_{-\infty}^{\infty} e^{-i\omega(\tau - t)} d\omega \right)
\end{aligned}
\qquad (2.117)
$$

For the above equality to be valid, the integral in the parentheses must be a delta function

$$\delta[\tau - t] = \frac{1}{2\pi} \int_{-\infty}^{\infty} e^{-i\omega(\tau - t)} d\omega. \qquad (2.118)$$

The *delta function* has the following properties

$$\delta[\tau - t] = \begin{cases} 0 & \text{if} \quad \tau \neq t \\ \infty & \text{if} \quad \tau = t \end{cases}, \qquad (2.119)$$

and

$$\int_{-\infty}^{\infty} \delta[\tau - t] d\tau = 1. \qquad (2.120)$$

By implication,

$$\int_{-\infty}^{\infty} I[\tau] \delta[\tau - t] d\tau = I[t]. \qquad (2.121)$$

Key 8

Consider the convolution integral as

$$F[t] = \int_{-\infty}^{\infty} Q[\tau] I[t - \tau] d\tau \qquad (2.122)$$

with the understanding that $F[t] = 0$ for t outside the range of $\int_0^t$. Take the Fourier transform of the convolution integral to obtain

$$
\begin{aligned}
\overline{F}[\omega] &= \int_{-\infty}^{\infty} \left(\int_{-\infty}^{\infty} Q[\tau] I[t-\tau] d\tau \right) e^{-i\omega t} dt \\
&= \int_{-\infty}^{\infty} Q[\tau] e^{-i\omega \tau} d\tau \int_{-\infty}^{\infty} I[t-\tau] e^{-i\omega(t-\tau)} dt \\
&= \overline{Q}[\omega] \overline{I}[\omega].
\end{aligned}
\tag{2.123}
$$

By similar procedures, one can take the inverse Fourier transform of the last expression to obtain the convolution integral.

• Suggested Readings

Carslaw and Jaeger's (1959) *Conduction of Heat in Solids* provides solutions for point, line, plane, spherical, and cylindrical sources.

The method of images is well developed in Smythe's (1968) *Static and Dynamic Electricity*. Van Nostrand and Cook's (1966) *Interpretation of Resistivity Data* gives many examples for wedge-shaped structural boundaries.

For the Fast Fourier transform we recommend Oppenheim and Schafer's (1975) *Digital Signal Processing* for further reading. For general reference about the z-transform we recommend Jury's (1982) *Theory and Application of the Z-transform Method*. The z-transform method has been widely used for processing of seismic reflection data. For example, see *Geophysical Signal Analysis* by Robinson and Treitel (1980).

2.7   Notations

Symbol	Definition	SI Unit	Dimension
b	Aquifer thickness	m	L
$\mathrm{E}_1[\xi]$, $-\mathrm{E}_i[-\xi]$	Exponential function	-	-
$\mathrm{erf}[\xi]$	Error function	-	-
$\mathrm{erfc}[\xi]$	Complementary error function	-	-
$F[t]$	System response		
$\overline{F}[\omega]$	Fourier transform		
h, Δh	Hydraulic head,drawdown	m	L
I	A dummy integral		
$I[t]$	Impulse response		
$\overline{I}[\omega]$	Transfer function		
k	Thermal conductivity	$W\,m^{-1}K^{-1}$	$\mathrm{MLT}^{-3}\mathrm{K}^{-1}$
K	Hydraulic conductivity	$m\,s^{-1}$	LT^{-1}
Q	Well discharge rate	$m^3 s^{-1}$	$\mathrm{L}^3\mathrm{T}^{-1}$
$Q[t]$	Input function		
Q_T	Heat energy	J	$\mathrm{ML}^2\mathrm{T}^{-2}$
R, r	Distance	m	L
S	Storativity	-	-
$S[t]$	Step response		
S_s	Specific storage	m^{-1}	L^{-1}
T	Temperature	K	K
T_h	Transmissivity	$m^2 s^{-1}$	$\mathrm{L}^2\mathrm{T}^{-1}$
t, t_0, t_off	Time	s	T
u	$= r^2 S/4T_\mathrm{h}t$	-	-
$W[u]$	Well function	-	-
x, y, z	Distance	m	L
$\mathcal{F}$	Fourier transform operator		
$\mathcal{L}$	Laplace transform operator		
$\mathcal{Z}$	z-transform operator		
$\Gamma[\xi]$	Gamma function	-	-
γ	Euler constant $= 0.5772$	-	-
Δh	Drawdown	m	L
κ	Thermal diffusivity	$m^2 s^{-1}$	$\mathrm{L}^2\mathrm{T}^{-1}$
ξ	Dummy argument		
ρc, ρc_v	Volumetric heat capacity	$J\,m^{-3}K^{-1}$	$\mathrm{ML}^{-1}\mathrm{T}^{-2}\mathrm{K}^{-1}$

Chapter 3

LAPLACE AND HANKEL TRANSFORMS

This chapter deals with inversion of the Laplace and Hankel transforms and their associated contour integration. The Laplace transform is a powerful method for solving a diffusion-like differential equation that contains a first order differential in time, for example, $\partial h / \partial t$. We will introduce fundamentals of the Laplace transform, perform contour integration for the inverse Laplace transform, and describe a numerical technique for the inverse Laplace transform. Applications of the Laplace transform in equation solving are demonstrated in Chapters 4 and 5 on well hydraulics and Chapter 7 on solute transport.

The Hankel transform is typically used in problems that can be described in terms of cylindrical coordinates. We will use contour integration to convert oscillating Bessel functions into exponential-decay-like Hankel functions for rapid convergence in numerical integration.

3.1 Fundamentals of the Laplace Transform

3.1.1 Transform Pairs

By definition, the *Laplace transform* of function $f[t]$ is

$$\boxed{\overline{f}[s] = \mathcal{L}\{f[t]\} = \int_0^\infty f[t]e^{-st}\,dt}, \tag{3.1}$$

where s is the transform parameter. The inverse of $\overline{f}[s]$ is

$$f[t] = \mathcal{L}^{-1}\{\overline{f}[s]\} = \frac{1}{i2\pi} \int_{a-i\infty}^{a+i\infty} \overline{f}[s]e^{st}\, ds, \qquad (3.2)$$

where a is a constant so large that all singularities lie to the left of the line $(a-i\infty,\ a+i\infty)$ on the complex s plane. A singularity occurs where function $\overline{f}[s]\exp[st]$ becomes infinity. This integral is also known as the *Bromwich integral*. We will refer to the pair of forward and inverse Laplace transforms as $f[t] \leftrightarrow \overline{f}[s]$.

3.1.2   Basic Transform Formulas

Following are some basic formulas for the Laplace transform.

☐ 1. A constant in time: If $f[t] = 1$, then

$$\overline{f}[s] = \int_0^\infty e^{-st}\, dt = \frac{1}{s}. \qquad (3.3)$$

☐ 2. A delta function: The delta function $\delta[t]$ has the property that

$$\delta[t - t_o] = \begin{cases} 0 & \text{if } t \neq t_o \\ \infty & \text{if } t = t_o \end{cases} \qquad (3.4)$$

and

$$\int_0^\infty \delta[t - t_o]\, dt = 1, \quad \int_0^\infty \delta[t - t_o]\, f[t]\, dt = f[t_o]. \qquad (3.5)$$

For example, if $f[t] = e^{-st}$, then

$$\overline{f}[s] = \int_0^\infty \delta[t - t_o]e^{-st}\, dt = e^{-st_o}. \qquad (3.6)$$

The inverse Laplace transform of e^{-st_o} is

$$\delta[t - t_o] = \frac{1}{i2\pi} \int_{a-i\infty}^{a+i\infty} e^{-s(t-t_o)}ds. \qquad (3.7)$$

This is one of many expressions for the delta function (see Problem 4).

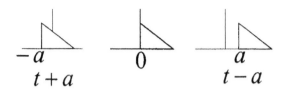

Figure 3.1: Negative (left) and positive (right) time shifting with respect to the central figure.

☐ 3. Time shifting: If $f[t] \leftrightarrow \overline{f}[s]$, then for a *positive* time shift (time advanced by $a > 0$, Figure 3.1)

$$f[t-a] \leftrightarrow \int_0^\infty f[t-a]e^{-st}dt = \int_{-a}^\infty f[\tau]e^{-s(\tau+a)}d\tau. \tag{3.8}$$

Because the Laplace transform is valid for positive time only (i.e., $f[t] = 0$ for $t < a$), the relation becomes

$$f[t-a] \leftrightarrow e^{-sa}\int_0^\infty f[\tau]e^{-s\tau}d\tau = e^{-sa}\overline{f}[s]. \tag{3.9}$$

Thus a forward time shift by step a is accompanied by its transform being reduced by a factor of e^{-sa}.

For a negative time shift,

$$\begin{aligned} f[t+a] &\leftrightarrow \int_0^\infty f[t+a]e^{-st}dt = e^{as}\int_a^\infty f[\tau]e^{-s\tau}d\tau \\ &= e^{as}\left\{\int_0^\infty f[\tau]e^{-s\tau}d\tau - \int_0^a f[\tau]e^{-s\tau}d\tau\right\}. \end{aligned} \tag{3.10}$$

In other words,

$$f[t+a] \leftrightarrow e^{as}\overline{f}[s] - e^{as}\int_0^a f[t]e^{-st}dt, \tag{3.11}$$

which contains one extra term in comparison with the result of positive time shift.

☐ 4. Parameter shifting:

$$\overline{f}[s-b] = \int_0^\infty f[t]e^{-(s-b)t}dt$$

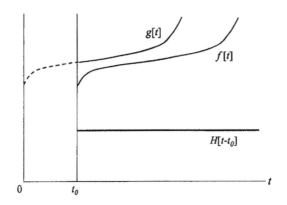

Figure 3.2: The role of Heaviside's unit step function $H[t - t_0]$. Function $f[t] = H[t - t_0]g[t - t_0]$ implies that the time origin of $g[t]$ is shifted to t_0, while $H[t - t_0]g[t]$ means $g[t]$ is zero for $t < t_0$ (dashed portion of $g[t]$ is zero).

$$= \int_0^\infty e^{bt} f[t] e^{-st} dt$$
$$\leftrightarrow e^{bt} f[t]. \qquad (3.12)$$

Similarly

$$\overline{f}[s + a] \leftrightarrow e^{-at} f[t]. \qquad (3.13)$$

□ 5. Step function: If $f[t] = H[t - t_o]g[t - t_o]$, where $H[t - t_o]$ is *Heaviside*'s unit step function (Figure 3.2)

$$H[t - t_o] = \begin{cases} 0 & \text{if } t < t_o, \\ 1 & \text{if } t > t_o, \end{cases} \qquad (3.14)$$

then

$$\overline{f}[s] = \int_0^\infty H[t - t_o]g[t - t_o]e^{-st} \, dt$$
$$= \int_{t_o}^\infty g[t - t_o]e^{-st} \, dt$$
$$= \exp(-st_o) \int_0^\infty g[t]e^{-st} \, dt = e^{-st_o}\overline{g}[s]. \qquad (3.15)$$

Thus, a time shifting of t_o is translated into a factor of $\exp[-st_o]$. This is equivalent to positive time shifting.

Note that $H[t - t_0]g[t - t_0]$ resets the origin time for $g[t]$ at $t = t_0$ while $H[t - t_0]g[t]$ means that $g[t]$ is zero for $t < t_0$, i.e., the dashed portion of $g[t]$ in Figure 3.2 is zero.

□ 6. Time derivative, $\partial f / \partial t$: Integration by parts yields

$$
\mathcal{L}\left\{ \frac{\partial f}{\partial t} \right\} = \int_0^\infty e^{-st} \frac{\partial f}{\partial t} dt = e^{-st} f\Big|_0^\infty + s \int_0^\infty e^{-st} f \, dt
$$
$$
= s\overline{f}[s] - f[0]. \tag{3.16}
$$

The initial condition, $f[0]$, is therefore imposed during the transform and requires no further implementation.

□ 7. Spatial derivative, $\partial^n f / \partial x^n$:

$$
\mathcal{L}\left\{ \frac{\partial^n}{\partial x^n} f[t] \right\} = \frac{\partial^n}{\partial x^n} \overline{f}[s]. \tag{3.17}
$$

□ 8. Integration:

$$
\mathcal{L}\left\{ \int_0^t f[\tau] d\tau \right\} = \int_0^\infty \left(\int_0^t f[\tau] d\tau \right) e^{-st} dt. \tag{3.18}
$$

Integration by parts yields

$$
\mathcal{L}\left\{ \int_0^t f[\tau] d\tau \right\} = \left(\int_0^t f[\tau] d\tau \right) \frac{e^{-st}}{-s} \Big|_0^\infty - \int_0^\infty f[t] \frac{e^{-st}}{-s} dt
$$
$$
= \frac{1}{s} \overline{f}[s]. \tag{3.19}
$$

A second way to obtain the transform is to view the integral $\int_0^t f[t] dt$ as the convolution of 1 and $f[t]$ in the time domain. Therefore, the transform is the product of $1/s$ and $\overline{f}[s]$.

A third way to obtain the transform is by means of switching the order of integration (see Problem 5).

□ 9. Convolution and multiplication: Convolution in the time domain is equivalent to multiplication in the Laplace transform domain. If $f[t] \leftrightarrow \overline{f}[s]$ and $g[t] \leftrightarrow \overline{g}[s]$, then the Laplace transform of the convolution is

$$
\mathcal{L}\left\{ \int_0^t f[\tau] g[t - \tau] \, d\tau \right\} = \overline{f}[s] \, \overline{g}[s]. \tag{3.20}
$$

☐ 10. Multiplication in time domain:

$$
\begin{aligned}
\mathcal{L}\{u[t]v[t]\} &= \int_0^\infty u[t]v[t]e^{-st}dt \\
&= \frac{1}{i2\pi}\int_0^\infty u[t]e^{-st}dt \int_{c-i\infty}^{c+i\infty} \overline{v}[p]e^{pt}dp \\
&= \frac{1}{i2\pi}\int_{c-i\infty}^{c+i\infty} \overline{v}[p]dp \int_0^\infty u[t]e^{-(s-p)t}dt \\
&= \frac{1}{i2\pi}\int_{c-i\infty}^{c+i\infty} \overline{v}[p]\,\overline{u}[s-p]dp.
\end{aligned}
\tag{3.21}
$$

☐ 11. Transfer function: If $g[t]$ is the impulse response of a linear system, its Laplace transform $\overline{g}[s]$ is the transfer function. To input function $f[t]$, the system response in the Laplace domain is

$$
\overline{h}[s] = \overline{f}[s]\,\overline{g}[s]
\tag{3.22}
$$

and in the time domain, the response is the convolution of $f[t]$ and $g[t]$.

3.2   Contour Integration

Since inversion of Laplace transforms often requires integration in the complex plane, we will review relevant information to prepare ourselves for examples of contour integration.

3.2.1   Singularity

Let $z = x + iy$ be a complex number. Function $f[z]$ is *analytic* at point z if it has a derivative. In other words, it is analytic if

$$
f'[z] = \frac{\partial f}{\partial z} = \lim_{\zeta \to 0} \frac{f[z+\zeta] - f[z]}{\zeta}
\tag{3.23}
$$

exists and is independent of the path by which the complex number $\zeta = x+iy$ approaches zero.

Consider two paths for an analytic $f[z]$ as $\zeta \to 0$: one along x and the other along iy. If

$$
f[z] = U[x,y] + iV[x,y]
\tag{3.24}
$$

with U and V being real, then

$$\frac{\partial f}{\partial z} = \frac{\partial U}{\partial x} + i\frac{\partial V}{\partial x} = \frac{1}{i}\frac{\partial U}{\partial y} + \frac{\partial V}{\partial y}. \tag{3.25}$$

Equating the real and imaginary parts gives the *Cauchy-Riemann* relations

$$\frac{\partial U}{\partial x} = \frac{\partial V}{\partial y}, \quad \frac{\partial V}{\partial x} = -\frac{\partial U}{\partial y}. \tag{3.26}$$

These two relations are sufficient and necessary conditions for $f[z]$ to be analytic, i.e., if $f[z]$ is analytic, then the Cauchy-Riemann relation holds; and vice versa.

A function is *regular* in region $\mathcal{R}$ if it is both analytic and single valued throughout $\mathcal{R}$. In a regular region, the line integral $\int_{z_1}^{z_2} f[z]\, dz$ is independent of the integration path. Hence, for a closed path, Cauchy's theorem says

$$\boxed{\oint f[z]\, dz = 0}. \tag{3.27}$$

In a regular region, function $f[z]$ has a *Taylor series* expansion around z_o:

$$f[z] = \sum_{n=0}^{\infty} a_n (z - z_o)^n, \tag{3.28}$$

$$a_0 = f[z_o], \quad a_n = \frac{1}{n!}\frac{\partial^n}{\partial z^n} f[z]\Big|_{z=z_o}$$

A Taylor series converges within a circle whose center is z_o and its radius extends to the nearest singular point (Figure 3.3).

If $f[z]$ is regular in the annulus between two concentric circles with center at z_o where $f[z_o]$ is singular, it has a *Laurent series* expansion around the singular point

$$f[z] = \sum_{n=-\infty}^{\infty} a_n (z - z_o)^n, \tag{3.29}$$

$$a_n = \frac{1}{i2\pi}\oint_C \frac{f[z]\, dz}{(z - z_o)^{n+1}},$$

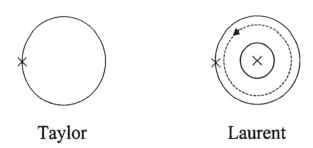

Taylor   Laurent

Figure 3.3: Left – Taylor series expansion around a regular point (center). It converges within the circle marked by the singular point ($\times$) nearest to the center. Right – Laurent series expansion around a singular point (center). It converges between two solid concentric rings. The outer ring is marked by the singular point nearest to the center. The inner ring can be as close as possible to the center.

where C is any closed path encircling the inner circle counterclockwise within the annular region. The outer radius is from z_o to the nearest singular point while the inner radius can be any size but zero. The point z_o is an isolated singularity if $f[z]$ is analytic in the vicinity of z_o. The highest m for which a_{-m} is not zero is the order of pole at $z = z_o$. For example, if $f[z] = 1/(z - z_o)^2$, then

$$
\begin{aligned}
a_{-m} &= \frac{1}{i2\pi} \oint \frac{dz}{(z - z_0)^{-m+1}(z - z_0)^2} \\
&= \frac{1}{i2\pi} \oint \frac{dz}{(z - z_0)^{-m+3}}.
\end{aligned}
\tag{3.30}
$$

The integrand is analytic for $m \geq 3$, i.e., $a_{-3} = a_{-4} = \ldots = 0$; therefore, function $f[z] = 1/(z - z_o)^2$ has a pole of order 2 at z_o (this is a trivial example).

In summary, the Taylor series centers around an analytic point with its circle of convergence extending to the nearest singular point; while the Laurent series centers around a singular point and is analytic within an annulus between two circles: the inner circle encloses the singular point and the outer circle extends to the next singular point.

3.2.2 Residue Theorem

The coefficient a_{-1} of the Laurent series is the *residue* of $f[z]$ at z_o,

$$a_{-1} = \frac{1}{i2\pi} \oint_C f[z] \, dz. \tag{3.31}$$

The integration of $f[z]$ along a path that encloses a regular region with a finite number of singularities at z_k is

$$\boxed{\oint_C f[z] \, dz = \pm i2\pi \sum_k \text{Residue}_k}, \tag{3.32}$$

where the sign is positive if the integration path is counterclockwise; otherwise it is negative.

The single-pole residue of $f[z]$ at z_o can be obtained through either Laurent series expansion or

$$\boxed{\text{Residue} = a_{-1} = \lim_{z \to z_o}(z - z_o)f[z]}. \tag{3.33}$$

If $f[z]$ is expressible as a fraction $f_n[z]/f_d[z]$ with $f_n[z]$ being free of root z_o, then we use l'Hopital's rule to obtain

$$\boxed{\text{Residue} = \lim_{z \to z_o}(z - z_o)\frac{f_n[z]}{f_d[z]} = \left.\frac{f_n[z]}{\partial f_d[z]/\partial z}\right|_{z=z_o}}. \tag{3.34}$$

For an nth order pole at z_0, the residue is

$$\boxed{a_{-1} = \frac{1}{(n-1)!} \lim_{z \to z_o} \left\{ \frac{d^{n-1}}{dz^{n-1}} \left[(z - z_0)^n f[z]\right] \right\}}. \tag{3.35}$$

• **Remark**

If the integration path runs through a singular point z_s, there are two ways to handle the situation: enclose it or exclude it (Figure 3.4). Either way yields the same answer.

□ To enclose the singular point inside the closed contour, take ($i\pi$·residue) at that point and as usual, add it to the right-hand side of equation (3.32).

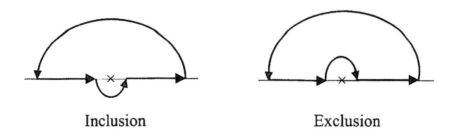

Figure 3.4: Detour of integration around a singular point on the integration path.

☐ To exclude the singular point, the contour integration must include a contribution from the integration along a semicircle around the singular point. This contribution is $i\pi \cdot$ (residue) but its sign is opposite to what is expected for the line integration C (if C is counterclockwise, then the semicircle path around the pole is clockwise and the sign is negative). This value is part of the left-hand side of equation (3.32) and, after it is shifted to the right-hand side, gives the same result as the integration for inclusion.

3.2.3   Multivalued Function

If a function can yield different values for a given argument in a complex plane, that function is *multivalued*. For example, given $f[z] = \sqrt{z}$, if $z = 1$, then $f[1] = \sqrt{1} = 1$. However, using the same value of $z = 1 = e^{i2\pi}$, the functional value is

$$f[e^{i2\pi}] = \sqrt{e^{i2\pi}} = e^{i\pi} = -1. \tag{3.36}$$

Thus, this $f[z]$ is a multivalued function of z.

Another example of a multivalued function is

$$\ln z = \ln(|z|\, e^{i2n\pi}) = i2n\pi + \ln|z|, \quad n = \text{integer}, \tag{3.37}$$

where $|z|$ is the absolute value of z.

In the presence of a multivalued function, care must be taken to ensure *single value* along the contour integration path. This is done by making a cut along a line or a curve. The cut can have finite length between two branch points or extend from one branch point to infinity. A *branch point* is the point on the complex plane where the multivalued function is zero or single valued. Function $f[z]$ is cut to have more than one sheet of plane and all sheets join at the branch point. Multivalue can arise from a term or a factor.

The integration path should not cross the cut. It goes around a branch point such that z makes a phase change of 2π from one sheet into another. Meanwhile the multivalued function $f[z]$ changes discontinuously from one value into another around the branch point.

A function bearing a factor of square root is not necessarily a multivalued function. For example, $f[z] = \sqrt{z}\sinh\sqrt{z}$ is not a multivalued function because

$$
\begin{aligned}
f[|z|\,e^{i2\pi}] &= \sqrt{|z|\,e^{i2\pi}}\sinh\sqrt{|z|\,e^{i2\pi}} \\
&= \sqrt{|z|}\sinh\sqrt{|z|} = f[|z|].
\end{aligned}
\tag{3.38}
$$

Thus, one must test whether an integrand is a multivalued function before carrying out the contour integration.

3.2.4 Example 1: Branch Cut, No Singularity

• **Problem**

Show that

$$
\boxed{\exp\left[-\sqrt{\frac{s}{\kappa}}\,x\right] \leftrightarrow \frac{x}{\sqrt{4\pi\kappa t^3}}\exp\left[-\frac{x^2}{4\kappa t}\right].}
\tag{3.39}
$$

• **Solution**

Let the inverse Laplace transform be

$$
I[t] = \frac{1}{i2\pi}\int_{a-i\infty}^{a+i\infty}\exp\left[-\sqrt{\frac{s}{\kappa}}x + st\right]ds, \quad a > 0.
\tag{3.40}
$$

The integrand has no singularity but it has a branch point at $s = 0$. Choose a branch cut along the negative real axis (Figure 3.5).

Since there is no singularity, the contour integration along the path shown in Figure 3.5 vanishes (Cauchy theorem), i.e.,

$$
\int_A^B + \int_{\mathrm{arc}BC} + \int_C^D + \int_{\mathrm{circle}DEF} + \int_F^G + \int_{\mathrm{arc}GA}\exp\left[-\sqrt{\frac{s}{\kappa}}x + st\right]ds = 0.
\tag{3.41}
$$

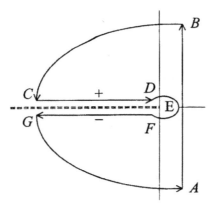

Figure 3.5: Contour integration path. The dashed line designates a branch cut.

Thus the desired integral $\int_A^B$ can be expressed in terms of other integrals, symbolically,

$$\int\limits_A^B = -\int\limits_{\text{arc}BC} - \int\limits_C^D - \int\limits_{\text{circle}DEF} - \int\limits_F^G - \int\limits_{\text{arc}GA} \tag{3.42}$$

provided that the integrals on the right-hand side can be easily evaluated.

☐ Step 1. Integration along great arcs BC and GA: We will examine the behavior of e^{+st} as the radius of BC goes to infinity. Let $s = \sigma e^{i\theta}$, we see

$$\text{as} \quad |s| = \sigma \to \infty, \quad \text{then} \quad \pi/2 < \theta < \pi \quad \text{and} \quad \cos\theta < 0. \tag{3.43}$$

Therefore

$$\begin{aligned} \exp\left[st\right] &= \exp\left[\sigma\left(\cos\theta + i\sin\theta\right)t\right] \\ &\to \exp\left[-\infty + i\infty\right] \to 0. \end{aligned} \tag{3.44}$$

In addition, the factor

$$\exp\left[-\sqrt{s/\kappa}x\right] \propto \exp\left[-|s|^{1/2}\right] \to 0. \tag{3.45}$$

Consequently

$$\exp\left[-\sqrt{s/\kappa}x + st\right] \to 0 \tag{3.46}$$

and the integration along arc BC vanishes.

Similarly the integration along arc GA vanishes.

☐ Step 2. Integration around small circle DEF: The integration around the small circle DEF vanishes as its radius shrinks to zero.

☐ Step 3. Integration along the branch cut: The desired integral along AB can be expressed in terms of integration along CD above the branch cut and FG below the cut,

$$\int_A^B = -\int_C^D - \int_F^G \tag{3.47}$$

☐ Step 3a. Let s along path CD be $\sigma e^{i\pi}$ $(\sigma > 0)$, then the phase angle of s changes from π along path CD to $-\pi$ along path FG for circling clockwise around the branch point to yield a total change of -2π. Thus s is $\sigma e^{-i\pi}$ along FG. That is, by letting s be $\sigma e^{i\pi}$ and $\sigma e^{-i\pi}$ along CD and FG, respectively, gives

$$
\begin{aligned}
I &= -\frac{1}{i2\pi} \int_\infty^0 \exp\left[-\sqrt{\frac{\sigma e^{i\pi}}{\kappa}}x + \sigma e^{i\pi}t\right] d\sigma e^{i\pi} \\
&\quad -\frac{1}{i2\pi} \int_0^\infty \exp\left[-\sqrt{\frac{\sigma e^{-i\pi}}{\kappa}}x + \sigma e^{-i\pi}t\right] d\sigma e^{-i\pi}
\end{aligned}
\tag{3.48}
$$

Note that in this polar form, σ is positive and the integration limit at infinity is ∞ (point C or D) rather than $-\infty$.

Taking the square root and noting that $e^{i\pi/2} = i$ and $e^{i\pi} = -1$ results in

$$I = -\frac{1}{i2\pi} \int_0^\infty \exp\left[-i\sqrt{\frac{\sigma}{\kappa}}x - \sigma t\right] d\sigma + \frac{1}{i2\pi} \int_0^\infty \exp\left[i\sqrt{\frac{\sigma}{\kappa}}x - \sigma t\right] d\sigma. \tag{3.49}$$

☐ Step 3b. For the convenience of integration, make the exponents into quadratic forms

$$
\begin{aligned}
I &= \frac{\exp\left[-x^2/4\kappa t\right]}{i2\pi}\left\{-\int_0^\infty \exp\left[-\left(\sqrt{\sigma t} + \frac{ix}{\sqrt{4\kappa t}}\right)^2\right] d\sigma \right. \\
&\quad \left. +\int_0^\infty \exp\left[-\left(\sqrt{\sigma t} - \frac{ix}{\sqrt{4\kappa t}}\right)^2\right] d\sigma\right\}
\end{aligned}
\tag{3.50}
$$

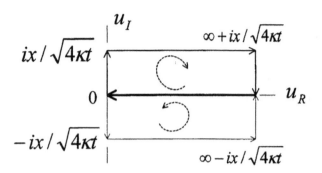

Figure 3.6: Contours for integration of [A] and [B].

Then, simplify it by letting $u = \sqrt{\sigma t} \pm ix/\sqrt{4\kappa t}$ and use $du = t/\sqrt{4\sigma t}\,d\sigma$ to obtain

$$
\begin{aligned}
I &= \frac{e^{-x^2/4\kappa t}}{i\pi t}\left\{
\begin{array}{l}
-\displaystyle\int_{ix/\sqrt{4\kappa t}}^{\infty+\,ix/\sqrt{4\kappa t}}\left(u-\frac{ix}{\sqrt{4\kappa t}}\right)e^{-u^2}du \\[4mm]
+\displaystyle\int_{-ix/\sqrt{4\kappa t}}^{\infty-\,ix/\sqrt{4\kappa t}}\left(u+\frac{ix}{\sqrt{4\kappa t}}\right)e^{-u^2}du
\end{array}
\right\} \\[6mm]
&= \frac{e^{-x^2/4\kappa t}}{i\pi t}\left\{
\underbrace{-\int_{ix/\sqrt{4\kappa t}}^{\infty+\,ix/\sqrt{4\kappa t}}ue^{-u^2}du+\int_{-ix/\sqrt{4\kappa t}}^{\infty-\,ix/\sqrt{4\kappa t}}ue^{-u^2}du}_{[A]}\right. \\[4mm]
&\qquad\left.+\frac{ix}{\sqrt{4\kappa t}}\underbrace{\left[\int_{ix/\sqrt{4\kappa t}}^{\infty+\,ix/\sqrt{4\kappa t}}e^{-u^2}du+\int_{-ix/\sqrt{4\kappa t}}^{\infty-\,ix/\sqrt{4\kappa t}}e^{-u^2}du\right]}_{[B]}\right\}. \quad (3.51)
\end{aligned}
$$

□ Step 3c. The integrands in the above equation are analytic. Applying the Cauchy theorem to the contour integration (Figure 3.6) yields

$$
[A] = \left[\int_{\infty+\,ix/\sqrt{4\kappa t}}^{\infty}+\int_{\infty}^{0}+\int_{0}^{ix/\sqrt{4\kappa t}}ue^{-u^2}du\right]
$$

$$- \left[\int\limits_{\infty - ix/\sqrt{4\kappa t}}^{\infty} + \int\limits_{\infty}^{0} + \int\limits_{0}^{-ix/\sqrt{4\kappa t}} ue^{-u^2} du \right]. \tag{3.52}$$

The first integrals in the two sets of square brackets vanish, and the second integrals cancel each other. Because $u \exp[-u^2]$ is an odd function, the third integrals vanish altogether

$$[A] = \int\limits_{0}^{ix/\sqrt{4\kappa t}} ue^{-u^2} du + \int\limits_{-ix/\sqrt{4\kappa t}}^{0} ue^{-u^2} du = 0. \tag{3.53}$$

Similarly, integration for $[B]$ yields

$$\begin{aligned}
[B] &= - \left[\int\limits_{\infty + ix/\sqrt{4\kappa t}}^{\infty} + \int\limits_{\infty}^{0} + \int\limits_{0}^{ix/\sqrt{4\kappa t}} e^{-u^2} du \right] \\
&\quad - \left[\int\limits_{\infty - ix/\sqrt{4\kappa t}}^{\infty} + \int\limits_{\infty}^{0} + \int\limits_{0}^{-ix/\sqrt{4\kappa t}} e^{-u^2} du \right] \\
&= -2 \int\limits_{\infty}^{0} e^{-u^2} du - \int\limits_{0}^{ix/\sqrt{4\kappa t}} e^{-u^2} du + \int\limits_{-ix/\sqrt{4\kappa t}}^{0} e^{-u^2} du = \sqrt{\pi} \quad (3.54)
\end{aligned}$$

because $\exp[-u^2]$ is an even function and the last two integrals cancel each other.

Therefore

$$I = \frac{x}{\sqrt{4\pi \kappa t^3}} \exp \left[\frac{-x^2}{4\kappa t} \right]. \tag{3.55}$$

3.2.5 Example 2: Branch Cut, Singularity

• **Problem**

Show that

$$\boxed{\frac{\exp \left[-\sqrt{s/\kappa} x \right]}{\sqrt{s/\kappa}} \longleftrightarrow \sqrt{\frac{\kappa}{\pi t}} \exp \left[-\frac{x^2}{4\kappa t} \right].} \tag{3.56}$$

There are two ways to obtain the inverse Laplace transform. The first one is straightforward while the second one uses a branch-cut integration.

• **Solution 1**

The answer can be obtained by integrating the result of Example 1 with respect to x from x to ∞.

• **Solution 2**

Follow the procedures in Example 1. Let the inverse Laplace transform be

$$I = \frac{1}{i2\pi} \int_{a-i\infty}^{a+i\infty} \frac{\exp\left[-\sqrt{s/\kappa}\,x + st\right]}{\sqrt{s/\kappa}} ds, \quad a > 0. \tag{3.57}$$

The integrand has one branch point and one pole of singularity (Figure 3.7). Both occur at $s = 0$.

□ Step 1. Select the cut along the negative real axis. Perform the contour integration that excludes the singular point.

□ Step 2. Let the line integration above the cut be positive and below it be negative. Because the integration path encloses no singular point, the contour integration vanishes in accordance with the Cauchy theorem. Thus,

$$I = -\frac{1}{i2\pi}\left\{\int_{-\infty}^{0} \frac{\exp\left[-\sqrt{s/\kappa}\,x + st\right]}{\sqrt{s/\kappa}} ds + \int_{0}^{-\infty} \frac{\exp\left[\sqrt{s/\kappa}\,x + st\right]}{-\sqrt{s/\kappa}} ds\right\}. \tag{3.58}$$

Note that the integration around the small circle vanishes because the residue at $s = 0$ is zero.

□ Step 2a. Change the sign of s to yield

$$I = -\frac{1}{i2\pi}\left\{\int_{0}^{\infty} \frac{\exp\left[-i\sqrt{s/\kappa}\,x - st\right]}{i\sqrt{s/\kappa}} ds + \int_{0}^{\infty} \frac{\exp\left[i\sqrt{s/\kappa}\,x - st\right]}{i\sqrt{s/\kappa}} ds\right\}. \tag{3.59}$$

Then, make the exponents into quadratic forms,

$$I = \frac{e^{-x^2/4\kappa t}}{2\pi}\left\{\int_{0}^{\infty} \frac{\exp\left[-\left(\sqrt{st} + ix/\sqrt{4\kappa t}\right)^2\right]}{\sqrt{s/\kappa}} ds\right.$$

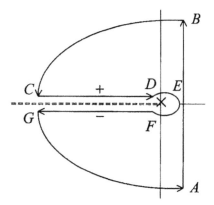

Figure 3.7: Contour integration path – one singular point (×) and one branch cut (dashed line).

$$+ \int\limits_{0}^{\infty} \frac{\exp\left[-\left(\sqrt{st} - ix/\sqrt{4\kappa t}\right)^2\right]}{\sqrt{s/\kappa}} ds \Bigg\} . \qquad (3.60)$$

□ Step 2b. Let $u = \sqrt{st} \pm ix/\sqrt{4\kappa t}$ and note that $du = \sqrt{t/4s}\, ds$. Then apply the Cauchy theorem to get (see Figure 3.6)

$$
\begin{aligned}
I &= \sqrt{\frac{\kappa}{t}} \frac{\exp\left[-x^2/4\kappa t\right]}{\pi} \left\{ \int\limits_{ix/\sqrt{4\kappa t}}^{\infty+ix/\sqrt{4\kappa t}} + \int\limits_{-ix/\sqrt{4\kappa t}}^{\infty-ix/\sqrt{4\kappa t}} \exp\left[-u^2\right] du \right\} \\
&= \sqrt{\frac{\kappa}{t}} \frac{\exp\left[-x^2/4\kappa t\right]}{\pi} \left\{ -\left[\int\limits_{\infty}^{0} + \int\limits_{0}^{ix/\sqrt{4\kappa t}} \exp\left[-u^2\right] du \right] \right. \\
&\qquad \left. - \left[\int\limits_{\infty}^{0} + \int\limits_{0}^{-ix/\sqrt{4\kappa t}} \exp\left[-u^2\right] du \right] \right\} \\
&= \sqrt{\frac{\kappa}{t}} \frac{\exp\left[-x^2/4\kappa t\right]}{\pi} 2 \int\limits_{0}^{\infty} \exp\left[-u^2\right] du \\
&= \sqrt{\frac{\kappa}{\pi t}} \exp\left[-\frac{x^2}{4\kappa t}\right] . \qquad (3.61)
\end{aligned}
$$

3.2.6   Example 3: Convolution 1

- **Problem**

Show that

$$\frac{\exp\left[-\sqrt{s/\kappa}x\right]}{s} \leftrightarrow \operatorname{erfc}\left[\frac{x}{\sqrt{4\kappa t}}\right]. \tag{3.62}$$

- **Solution 1**

Integrate the result of Example 2 with respect to x from x to ∞.

- **Solution 2**

Recall that multiplication in the Laplace domain is equivalent to convolution in the time domain. Recognizing that the transform is the product of $\exp\left[-\sqrt{s/\kappa}x\right]$ and $1/s$, we find

$$\exp\left[-\sqrt{\frac{s}{\kappa}}x\right] \leftrightarrow \frac{x}{\sqrt{4\pi\kappa t^3}}\exp\left[\frac{-x^2}{4\kappa t}\right] \tag{3.63}$$

from Example 1 and

$$\frac{1}{s} \leftrightarrow 1. \tag{3.64}$$

Perform the convolution

$$\frac{\exp\left[-\sqrt{s/\kappa}x\right]}{s} \leftrightarrow \int_0^t 1\cdot\frac{x}{\sqrt{4\pi\kappa\tau^3}}\exp\left[\frac{-x^2}{4\kappa\tau}\right]d\tau = I, \tag{3.65}$$

where I is the desired inverse.

Let $\xi = x/\sqrt{4\kappa\tau}$, then $d\xi = -(x/\sqrt{16\kappa\tau^3})\,d\tau$, and

$$I = \int_\infty^{x/\sqrt{4\kappa t}} \frac{-2}{\sqrt{\pi}}\exp\left[-\xi^2\right]d\xi = \operatorname{erfc}\left[\frac{x}{\sqrt{4\kappa t}}\right]. \tag{3.66}$$

• Solution 3

Alternatively if the Bromwich integral is applied, then

$$I = \frac{1}{i2\pi} \int_{a-i\infty}^{a+i\infty} \frac{\exp\left[-\sqrt{s/\kappa}x\right]}{s} \exp\left[st\right] ds, \qquad (3.67)$$

we have

$$I = 1 - \frac{2}{\pi} \int_0^\infty \frac{ds}{s} e^{-st} \sin\left[x\sqrt{\frac{s}{\kappa}}\right]. \qquad (3.68)$$

This closed-form solution is not as elegant as the complementary error function in solution 2 but incidently we have shown that

$$\frac{2}{\pi} \int_0^\infty \frac{ds}{s} e^{-st} \sin\left[x\sqrt{\frac{s}{\kappa}}\right] = \mathrm{erf}\left[\frac{x}{\sqrt{4\kappa t}}\right]. \qquad (3.69)$$

3.2.7 Example 4: Convolution 2

• Problem

Find the inverse Laplace transform of

$$\frac{\exp[-qx]}{s - \alpha}, \quad q = \sqrt{\frac{s}{\kappa}}. \qquad (3.70)$$

• Solution 1

This transform can be found from the following two transform pairs:

$$\frac{1}{s - \alpha} \leftrightarrow e^{\alpha t}, \quad \exp\left[-qx\right] \leftrightarrow \frac{x}{\sqrt{4\pi\kappa t^3}} \exp\left[-\frac{x^2}{4\kappa t}\right], \qquad (3.71)$$

where the latter is given in Example 1. Hence, multiplication of $1/(s - \alpha)$ and $\exp[-qx]$ in the Laplace domain leads to the convolution in the time domain,

$$\begin{aligned}
I &= \int_0^t e^{\alpha(t-\tau)} \frac{x}{\sqrt{4\pi\kappa\tau^3}} \exp\left[-\frac{x^2}{4\kappa\tau}\right] d\tau \\
&= \frac{x\exp(\alpha t)}{\sqrt{4\pi\kappa}} \int_0^t \frac{\exp\left[-x^2/4\kappa\tau - \alpha\tau\right]}{\tau^{3/2}} d\tau.
\end{aligned} \qquad (3.72)$$

Convert the exponent in the integrand into quadratic forms so that

$$I = \frac{x \exp(\alpha t)}{2\sqrt{4\pi\kappa}} \left\{ \exp\left[x\sqrt{\alpha/\kappa}\right] \int_0^t \frac{\exp[-(x/\sqrt{4\kappa\tau} + \sqrt{\alpha\tau})^2]}{\tau^{3/2}} d\tau \right.$$

$$\left. + \exp\left[-x\sqrt{\alpha/\kappa}\right] \int_0^t \frac{\exp[-(x/\sqrt{4\kappa\tau} - \sqrt{\alpha\tau})^2]}{\tau^{3/2}} d\tau \right\}. \qquad (3.73)$$

Let $u_{\pm} = x/\sqrt{4\kappa\tau} \pm \sqrt{\alpha\tau}$ to simplify the expression. Then, substitute

$$\frac{x}{\sqrt{4\kappa\tau^3}} d\tau = -2 \, du_{\pm} \pm \sqrt{\frac{\alpha}{\tau}} \, d\tau \qquad (3.74)$$

into I to yield

$$I = \frac{\exp[\alpha t]}{2\sqrt{\pi}} \left\{ \exp\left[x\sqrt{\alpha/\kappa}\right] \left[\begin{array}{c} -2 \int_{\infty}^{u_+} e^{-u_+^2} \, du_+ \\ + \int_0^t \sqrt{\frac{\alpha}{\tau}} \exp[-(\frac{x}{\sqrt{4\kappa\tau}} + \sqrt{\alpha\tau})^2] \, d\tau \end{array} \right] \right.$$

$$\left. + \exp\left[-x\sqrt{\alpha/\kappa}\right] \left[\begin{array}{c} -2 \int_{\infty}^{u_-} e^{-u_-^2} \, du_- \\ - \int_0^t \sqrt{\frac{\alpha}{\tau}} \exp[-(\frac{x}{\sqrt{4\kappa\tau}} - \sqrt{\alpha\tau})^2] \, d\tau \end{array} \right] \right\}. (3.75)$$

After simplifying the exponents, the two $\int_0^t$ cancel each other. The remaining semi-indefinite integrals are complementary error functions. Therefore

$$\boxed{\mathcal{L}^{-1}\left\{ \frac{\exp[-qx]}{s - \alpha} \right\} = I}, \qquad (3.76)$$

$$I = \frac{e^{\alpha t}}{2} \left\{ e^{x\sqrt{\alpha/\kappa}} \text{erfc}\left[\frac{x}{\sqrt{4\kappa t}} + \sqrt{\alpha t} \right] + e^{-x\sqrt{\alpha/\kappa}} \text{erfc}\left[\frac{x}{\sqrt{4\kappa t}} - \sqrt{\alpha t} \right] \right\}$$
$$(3.77)$$

(Carslaw and Jaeger, 1959, p. 495, formula 19).

• Solution 2

See Section 6.2.2 for an alternative solution.

3.2.8   Example 5: Rationalization

• **Problem**

Find

$$\mathcal{L}^{-1}\left\{\frac{\exp[-qx]}{q+h}\right\}, \quad q=\sqrt{\frac{s}{\kappa}}. \tag{3.78}$$

• **Solution**

Rationalize the denominator by multiplying the transform with $(q-h)/(q-h)$ and rearrange the result such that

$$\frac{\exp[-qx]}{q+h} = \frac{(q-h)\exp[-qx]}{s/\kappa-h^2} = \frac{-\kappa}{s-\kappa h^2}\left(\frac{\partial e^{-qx}}{\partial x}+he^{-qx}\right). \tag{3.79}$$

Now Example 4 is applicable for the second term in the parentheses.

The differential $\partial e^{-qx}/\partial x$ requires differentiation of the error function, which follows Leibnitz' rule

$$\boxed{\frac{\partial}{\partial a}\int_B^A g[x,a]dx = \int_B^A \frac{\partial}{\partial a}g[x,a]dx + g[A,a]\frac{\partial A}{\partial a} - g[B,a]\frac{\partial B}{\partial a}}. \tag{3.80}$$

Note the equivalence between κh^2 and α of Example 4 yields

$$\boxed{\frac{\exp(-qx)}{q+h} \leftrightarrow \sqrt{\frac{\kappa}{\pi t}}e^{-x^2/4\kappa t} - h\kappa e^{\kappa h^2 t}e^{hx}\mathrm{erfc}\left[\frac{x}{\sqrt{4\kappa t}}+h\sqrt{\kappa t}\right]} \tag{3.81}$$

(Carslaw and Jaeger, 1959, p. 494, formula 12).

3.2.9   Example 6: A Series Representation

• **Problem**

Find

$$\mathcal{L}^{-1}\left\{\frac{e^{-qx}}{s\left(1+\sigma e^{-ql}\right)}\right\}, \quad q=\sqrt{s/\kappa}, \quad |\sigma|<1, \quad x>0, \quad l>0.$$

• **Solution**

This function has a branch point at $s = 0$ and an infinite number of singularities. In this case we want to avoid the inversion through the Bromwich integral. Because $\left|\sigma e^{-qx}\right| < 1$, the expression $1/\left(1 + \sigma e^{-ql}\right)$ can be represented by a binomial series expansion

$$
\begin{aligned}
\frac{e^{-qx}}{s\left(1 + \sigma e^{-ql}\right)} &= \frac{e^{-qx}}{s}\sum_{n=0}^{\infty}\left(-\sigma e^{-ql}\right)^{n} \\
&= \sum_{n=0}^{\infty}(-\sigma)^{n}\frac{e^{-q(nl+x)}}{s}.
\end{aligned}
\tag{3.82}
$$

Using the result of Example 3 leads to (Lee, 1996)

$$
\boxed{\mathcal{L}^{-1}\left\{\frac{e^{-qx}}{s\left(1 + \sigma e^{-ql}\right)}\right\} = \sum_{n=0}^{\infty}(-\sigma)^{n}\mathrm{erfc}\left[\frac{x + nl}{\sqrt{4\kappa t}}\right]}.
\tag{3.83}
$$

3.2.10   Example 7: Roots, No Branch Cut

• **Problem**

Find

$$
I = \frac{1}{i2\pi}\int_{a-i\infty}^{a+i\infty}\frac{e^{st}\cosh[qx]}{s\left(q\sinh[ql] + h\cosh[ql]\right)}ds, \quad q = \sqrt{\frac{s}{\kappa}}.
\tag{3.84}
$$

• **Solution**

The integrand has a singularity at $s = 0$ with a residue of $1/h$. In addition, it is singular at every root of

$$
q\sinh[ql] + h\cosh[ql] = 0.
\tag{3.85}
$$

The roots are found as follows. Let $q = i\alpha$ (or $s = -\kappa\alpha^2$), then the root equation becomes

$$
\begin{aligned}
-\alpha\sin[\alpha l] + h\cos[\alpha l] &= 0 \\
\text{or}\quad \alpha_n\tan[\alpha_n l] &= h, \quad n = 1, 2, 3, ...
\end{aligned}
\tag{3.86}
$$

The residue at root α_n is

$$
\begin{aligned}
\text{Res}_n &= \left. \frac{\exp\left[st\right]\cosh\left[qx\right]}{s\frac{\partial}{\partial s}\left\{q\sinh\left[ql\right]+h\cosh\left[ql\right]\right\}} \right|_{s=-\kappa\alpha^2} \\
&= \frac{\alpha_n \exp\left[-\kappa\alpha_n^2 t\right]\cos[\alpha_n x]}{2\left\{(1+hl)\sin[\alpha_n l]+\alpha_n l\cos[\alpha_n l]\right\}} \\
&= \frac{\exp\left[-\kappa\alpha_n^2 t\right]\cos[\alpha_n x]}{2\left\{l\left(h^2+\alpha_n^2\right)+h\right\}\cos[\alpha_n l]}.
\end{aligned}
\tag{3.87}
$$

Note that the integrand is an even function of q and there is no branch cut. Therefore

$$
\boxed{I = \frac{1}{h} + \sum_{n=1}^{\infty} \text{Res}_n}.
\tag{3.88}
$$

3.2.11 Example 8: Branch Cut, Logarithm

• **Problem**

Show that

$$
\boxed{\frac{\ln s}{s} \leftrightarrow -\ln\left[e^\gamma t\right]},
\tag{3.89}
$$

where $\gamma = 0.5772$ is Euler's constant (Carslaw and Jaeger, 1959, p. 496, formula 32).

• **Solution**

The transform has one singular point and one branch point at $s = 0$. Select the branch cut along the negative real axis (Figure 3.7).

$$
\begin{aligned}
I &= \frac{1}{i2\pi} \int_{a-i\infty}^{a+i\infty} \frac{\ln s}{s} e^{st} ds \\
&= -\frac{1}{i2\pi} \left\{ \int_{\infty}^{\epsilon} \frac{\ln\left[re^{i\pi}\right]}{re^{i\pi}} \exp\left[tre^{i\pi}\right] dre^{i\pi} + \int_{\pi}^{-\pi} \frac{\ln\left[\epsilon e^{i\theta}\right]}{\epsilon e^{i\theta}} \exp\left[t\epsilon e^{i\theta}\right] d\epsilon e^{i\theta} \right. \\
&\quad \left. + \int_{\epsilon}^{\infty} \frac{\ln\left[re^{-i\pi}\right]}{re^{-i\pi}} \exp\left[tre^{-i\pi}\right] dre^{-i\pi} \right\}.
\end{aligned}
\tag{3.90}
$$

As $\epsilon \to 0$, then $\exp\left[t\epsilon e^{i\theta}\right] \to 1$, and the second integral along the small circle around $s = 0$ becomes

$$\int_\pi^{-\pi} \frac{\ln \epsilon + i\theta}{e^{i\theta}} e^{i\theta} i d\theta = -i2\pi \ln \epsilon - \int_\pi^{-\pi} \theta d\theta = -i2\pi \ln \epsilon. \qquad (3.91)$$

Simplifying the relation leads to

$$I = \int_\epsilon^\infty \frac{e^{-rt}}{r} dr + \ln \epsilon = \int_{\epsilon t}^\infty \frac{e^{-u}}{u} du + \ln \epsilon \quad \text{as } \epsilon \to 0. \qquad (3.92)$$

For small values of x, the exponential integral approaches

$$\mathrm{Ei}(-x) = -\int_x^\infty \frac{e^{-u}}{u} du \approx \gamma + \ln x - x + \frac{x^2}{4} + O(x^3). \qquad (3.93)$$

Therefore,

$$I \approx -(\gamma + \ln \epsilon t) + \ln \epsilon = -\ln\left[e^\gamma t\right]. \qquad (3.94)$$

3.2.12   Example 9: Bessel Functions – Delta Function

• **Problem**

Show that

$$\boxed{K_o[qx] \leftrightarrow \frac{\exp\left[-x^2/4\kappa t\right]}{2t}}, \quad q = \sqrt{\frac{s}{\kappa}}. \qquad (3.95)$$

• **Solution**

To invert the modified Bessel function of the second kind $K_o[qx]$, we need an integral representation (Carslaw and Jaeger, 1959, p. 490)

$$K_o[x] = \frac{1}{2} \int_0^\infty e^{-\xi - x^2/4\xi} \frac{d\xi}{\xi}. \qquad (3.96)$$

Substituting this relation into the inverse formula

$$I = \frac{1}{i2\pi} \int_{a-i\infty}^{a+i\infty} K_o[qx] e^{st} ds$$

$$= \frac{1}{i2\pi} \int\limits_{a-i\infty}^{a+i\infty} \left(\frac{1}{2} \int\limits_{0}^{\infty} e^{-\xi - sx^2/4\kappa\xi} \frac{d\xi}{\xi} \right) e^{st} ds$$

$$= \frac{1}{2} \int\limits_{0}^{\infty} \frac{e^{-\xi}}{\xi} d\xi \cdot \frac{1}{i2\pi} \int\limits_{a-i\infty}^{a+i\infty} \exp\left[\frac{st}{\xi} \left(\xi - \frac{x^2}{4\kappa t} \right) \right] ds. \qquad (3.97)$$

Let $u = st/\xi$ and $b = at/\xi$, then

$$I = \frac{1}{2} \int\limits_{0}^{\infty} \frac{e^{-\xi}}{\xi} d\xi \cdot \frac{1}{i2\pi} \int\limits_{b-i\infty}^{b+i\infty} \frac{\xi}{t} \exp\left[u \left(\xi - \frac{x^2}{4\kappa t} \right) \right] du. \qquad (3.98)$$

Noting the delta function

$$\delta[\xi - \frac{x^2}{4\kappa t}] = \frac{1}{i2\pi} \int_{b-i\infty}^{b+i\infty} \exp\left[u(\xi - \frac{x^2}{4\kappa t}) \right] du, \qquad (3.99)$$

we obtain

$$I = \frac{1}{2t} \int\limits_{0}^{\infty} e^{-\xi} \delta\left[\xi - \frac{x^2}{4\kappa t} \right] d\xi = \frac{\exp\left[-x^2/4\kappa t \right]}{2t}. \qquad (3.100)$$

3.3 Numerical Inversion

As demonstrated above, the inversion of the Laplace transform always results in a closed-form solution. However, a closed-form solution comprised of semi-definite integrals may be difficult to evaluate numerically, especially when root finding is a part of the numerical integration as exemplified by Example 7. For hydrogeological applications, the integrals frequently contain Bessel functions J_v and Y_v, which oscillate and converge slowly. Each transform problem often requires its own method of inversion and many inversions are much more complicated than the transforms themselves.

Numerical inversion is appealing in the sense that one coding scheme may fit all transforms. Many numerical inversion schemes are available (Davies and Martin, 1979). The accuracy of a scheme varies greatly with the functional behavior of the target transform; however, there is no sure way to know the limit. It is therefore advisable to test the desired inversion scheme with a closed-form solution that resembles your type of problem; a test for an arbitrarily selected transform pair may not be satisfactory. Following is one algorithm, the Gaver-Stehfest method (Stehfest, 1970), which is frequently used for well hydraulic problems.

3.3.1   Gaver-Stehfest Method

The *Stehfest algorithm* is based on asymptotic expansion and extrapolation

$$
f[t] \approx \frac{\ln 2}{t} \sum_{n=1}^{N} K_n \bar{f} \left[\frac{n \ln 2}{t} \right],
\tag{3.101}
$$

where

$$
K_n = (-1)^{n+N/2} \sum_{k=[]}^{\min(n,N/2)} \frac{k^{N/2}(2k)!}{(N/2 - k)!k!(k-1)!(n-k)!(2k-n)!}
\tag{3.102}
$$

with N being an even number and $[]$ being the integer part of $(n+1)/2$.

Note that N should be selected carefully such that

$$
\sum_{n=1}^{N} K_n \approx 0 \quad \text{and} \quad \sum_{n=1}^{N} \frac{K_n}{n} \approx 1.
\tag{3.103}
$$

The latter condition is the test result for a step function. An optimal N depends somewhat on hardware and software and is around 10 to 14 for a personal computer. An N less than the optimal is not desirable but, due to round-off error, a greater N may deteriorate the accuracy too! Use double precision for all computations.

The Stehfest algorithm is generally acceptable for the Theis-type well functions. However, it may fail at large u (the argument of a Theis well function), i.e., at small time or large distance (Tseng and Lee, 1998).

3.4 Hankel Transform

3.4.1 Bessel Functions

• **Transform and Functions**

A Hankel transform of $f[r]$ is obtained by multiplying $f[r]$ with $rJ_0[ar]$, then integrating the product from 0 to ∞. The transform pairs are

$$F[a] = \int_0^\infty r f[r] J_0[ar] dr, \tag{3.104}$$

$$f[r] = \int_0^\infty a F[a] J_0[ar] da, \tag{3.105}$$

where $J_0[z]$ is the *Bessel function* of the first kind and zeroth order. It is the first kind of solution to the Bessel equation

$$\frac{d^2W}{dz^2} + \frac{1}{z}\frac{dW}{dz} + \left(1 - \frac{\nu^2}{z^2}\right) W = 0 \tag{3.106}$$

for $\nu = 0$.

The second kind of solution is $Y_\nu[z]$. Their functional behaviors are illustrated in Figure 3.8. Also shown are the first-order functions, $J_1[z]$ and $Y_1[z]$. Higher order functions can be obtained through recurrence relations (Abramowitz and Stegun, 1965), for example,

$$J_{\nu+1}[z] = \frac{2\nu}{z} J_\nu[z] - J_{\nu-1}[z]. \tag{3.107}$$

When the Bessel equation is modified,

$$\frac{d^2W}{dz^2} + \frac{1}{z}\frac{dW}{dz} - \left(1 + \frac{\nu^2}{z^2}\right) W = 0, \tag{3.108}$$

the solutions are called the modified Bessel functions, $I_\nu[z]$ and $K_\nu[z]$. Their functional behaviors for orders 0 and 1 are illustrated in Figure 3.9.

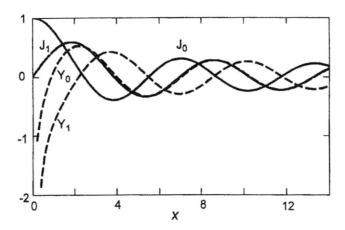

Figure 3.8: Bessel functions of the first kind, $J_0[x]$ and $J_1[x]$, and the second kind, $Y_0[x]$ and $Y_1[x]$.

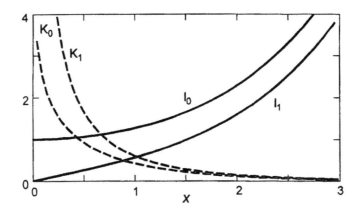

Figure 3.9: Modified Bessel functions of the first kind, $I_0[x]$ and $I_1[x]$, and the second kind, $K_0[x]$ and $K_1[x]$.

• Relations Among Bessel Functions

The two independent Bessel functions can be combined to form a pair of independent *Hankel functions,*

$$
\begin{aligned}
H_\nu^{(1)}[z] &= J_\nu[z] + iY_\nu[z], \\
H_\nu^{(2)}[z] &= J_\nu[z] - iY_\nu[z].
\end{aligned}
\tag{3.109}
$$

Conversely,

$$
\begin{aligned}
J_\nu[z] &= 0.5(H_\nu^{(1)}[z] + H_\nu^{(2)}[z]), \\
Y_\nu[z] &= -0.5i(H_\nu^{(1)}[z] - H_\nu^{(2)}[z]).
\end{aligned}
\tag{3.110}
$$

The two Hankel function are convertible to each other through the relations,

$$
\begin{aligned}
H_\nu^{(1)}[ze^{i\pi}] &= -e^{-i\nu\pi}H_\nu^{(2)}[z], \\
H_\nu^{(2)}[ze^{-i\pi}] &= -e^{i\nu\pi}H_\nu^{(1)}[z].
\end{aligned}
\tag{3.111}
$$

The Bessel (or Hankel) functions and the modified Bessel functions are related by

$$
\begin{aligned}
I_\nu[z] &= e^{-i\nu\pi/2}J_\nu[ze^{i\pi/2}], & -\pi < \arg z \le \pi/2, \\
I_\nu[z] &= e^{i3\nu\pi/2}J_\nu[ze^{-i3\pi/2}], & \pi/2 < \arg z \le \pi, \\
K_\nu[z] &= \frac{i\pi}{2}e^{i\nu\pi/2}H_\nu^{(1)}[ze^{i\pi/2}], & -\pi < \arg z \le \pi/2, \\
K_\nu[z] &= -\frac{i\pi}{2}e^{-i\nu\pi/2}H_\nu^{(2)}[ze^{-i\pi/2}], & -\pi/2 < \arg z \le \pi.
\end{aligned}
\tag{3.112}
$$

Bessel and Hankel functions of different orders are obtainable from the recurrence relations

$$
B_{\nu+1} = -B_{\nu-1} + \frac{2\nu}{z}B_\nu,
\tag{3.113}
$$

$$
\frac{\partial B_\nu[z]}{\partial z} = B_\nu'[z] = -B_{\nu+1}[z] + \frac{\nu}{z}B_\nu[z],
\tag{3.114}
$$

where B denotes J, Y, $H^{(1)}$, $H^{(2)}$ or any of their linear combinations. For example,

$$
J_0'[z] = -J_1[z].
\tag{3.115}
$$

Also, the recurrence relations for the modified Bessel functions are

$$C_{\nu+1}[z] = C_{\nu-1} - \frac{2\nu}{z}C_\nu[z],$$

$$C'_\nu[z] = C_{\nu+1}[z] + \frac{\nu}{z}C_\nu[z], \tag{3.116}$$

where C_ν denotes I_ν, $e^{i\nu\pi}K_\nu$ or any linear combinations of the two modified Bessel functions. For example,

$$I'_0[z] = I_1[z], \quad K'_0[z] = -K_1[z]. \tag{3.117}$$

If $y_1[x]$ and $y_2[x]$ are two independent solutions of a second-order differential equation, its *Wronskian* is

$$w[x] = y_1[x]y'_2[x] - y'_1[x]y_2[x] \neq 0. \tag{3.118}$$

The following Wronskian relations for the Bessel equations are useful,

$$J_{\nu+1}[z]J_{-\nu}[z] + J_{-\nu-1}[z]J_\nu[z] = -\frac{2\sin[\nu\pi]}{\pi z},$$

$$-I_{\nu+1}[z]I_{-\nu}[z] + I_{-\nu-1}[z]I_\nu[z] = -\frac{2\sin[\nu\pi]}{\pi z},$$

$$-J_\nu[z]Y_{\nu+1}[z] + J_{\nu+1}[z]Y_\nu[z] = \frac{2}{\pi z},$$

$$I_\nu[z]K_{\nu+1}[z] + I_{\nu+1}[z]K_\nu[z] = \frac{1}{z},$$

$$H^{(1)}_{\nu+1}[z]H^{(2)}_\nu[z] - H^{(1)}_\nu[z]H^{(2)}_{\nu+1}[z] = -\frac{4i}{\pi z}. \tag{3.119}$$

• A Series Representation

The following series representation

$$J_\nu[x] = \left(\frac{x}{2}\right)^\nu \sum_{k=0}^\infty \frac{(-x^2/4)^k}{k!\Gamma[\nu+k+1]} \tag{3.120}$$

is useful, where $\Gamma[y]$ is a Gamma function.

Function $J_\nu[z]$ oscillates and converges asymptotically like

$$J_\nu[z] \approx \sqrt{\frac{2}{\pi z}}\cos\left(z - \frac{\nu\pi}{2} - \frac{\pi}{4}\right) \tag{3.121}$$

as $|z| \to \infty$. If the factor $rf[r]$ or $aF[a]$ in the integrand of a Hankel transform does not converge much faster than J_0, the integration may converge so slowly (or worse, oscillate widely depending on the choice of a numerical ∞) that hundreds of terms may be needed to achieve a desired accuracy.

3.4.2 Example 1H: Hankel Transform – Using a Series

• **Problem**

Show that

$$A[r] = \int_0^\infty ae^{-ba^2} J_0[ar]da = \frac{1}{2b}e^{-r^2/4b}, \quad b > 0 \qquad (3.122)$$

(Carslaw and Jaeger, 1959, p. 490, formula 29).

• **Solution**

When various attempts to integrate a seemingly simple integral have failed, try a series or an integral representation. Here is one example that works. Use the series representation of the Bessel function to obtain

$$\begin{aligned}
A[r] &= \int_0^\infty ae^{-ba^2} \sum_{k=0}^\infty \frac{(-1)^k(ar)^{2k}}{4^k \, (k!)^2} da \\
&= \sum_{k=0}^\infty \frac{(-1)^k r^{2k}}{4^k \, (k!)^2} \int_0^\infty a^{2k+1} e^{-ba^2} da \qquad (3.123)
\end{aligned}$$

in which the Gamma function $\Gamma[k+1] = k!$ for integer k has been incorporated.

Let us do the integration by induction. Start by substituting $u = a^2$ to yield

$$\begin{aligned}
\int_0^\infty ae^{-ba^2} da &= \frac{1}{2} \int_0^\infty e^{-bu} du \\
&= \frac{1}{2b}, \quad k = 0. \qquad (3.124)
\end{aligned}$$

Differentiate the above relation with respect to b, resulting in

$$\int_0^\infty a^3 e^{-ba^2} da = \frac{1}{2b^2}, \quad k = 1. \qquad (3.125)$$

Differentiate it successively to obtain

$$\int_0^\infty a^5 e^{-ba^2} da = \frac{2}{2b^3}, \quad k = 2;$$

$$\int_0^\infty a^7 e^{-ba^2} da = \frac{1 \cdot 2 \cdot 3}{2b^4}, \quad k = 3. \tag{3.126}$$

Generalization leads to

$$\int_0^\infty a^{2k+1} e^{-ba^2} da = \frac{k!}{2b^{k+1}}. \tag{3.127}$$

Therefore, we obtain

$$\begin{aligned}
A[r] &= \sum_{k=0}^\infty \frac{(-1)^k r^{2k}}{4^k (k!)^2} \frac{k!}{2b^{k+1}} = \frac{1}{2b} \sum_{k=0}^\infty \frac{(-1)^k}{k!} \frac{r^{2k}}{4^k b^k} \\
&= \frac{1}{2b} \sum_{k=0}^\infty \frac{(-1)^k}{k!} \left(\frac{r^2}{4b}\right)^k \\
&= \frac{\exp[-r^2/4b]}{2b}. \tag{3.128}
\end{aligned}$$

3.4.3   Example 2H: Hankel Transform of $1/(a^2 + q^2)$

• **Problem**

Show that

$$\boxed{\mathcal{H}^{-1} \left\{ \frac{1}{a^2 + q^2} \right\} = K_0[qr]}. \tag{3.129}$$

• **Solution**

Let

$$A[r] = \mathcal{H}^{-1} \left\{ \frac{1}{a^2 + q^2} \right\} = \int_0^\infty \frac{a J_0[ar]}{a^2 + q^2} da. \tag{3.130}$$

For contour integration, it is preferable to integrate from $-\infty$ to $+\infty$. We will use to our advantage the fact that $J_0[ar]$ is an even function and the integrand is an odd function. Recall

$$J_0[z] = \frac{1}{2} \left(H_0^{(1)}[z] + H_0^{(2)}[z] \right) \tag{3.131}$$

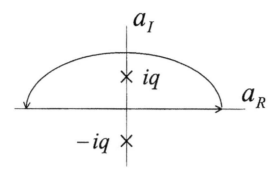

Figure 3.10: Contour integration around singular point iq.

to get

$$
\begin{aligned}
A[r] &= \frac{1}{2} \int_0^\infty \frac{a \left(H_0^{(1)}[ar] + H_0^{(2)}[ar] \right)}{a^2 + q^2} da \\
&= \frac{1}{2} \int_0^\infty \frac{a\, H_0^{(1)}[ar]}{a^2 + q^2} da + \frac{1}{2} \int_0^{-\infty} \frac{-a H_0^{(2)}[-ar]}{a^2 + q^2} d(-a). \quad (3.132)
\end{aligned}
$$

Since $H_0^{(2)}[-z] = -H_0^{(1)}[z]$, one gets

$$
A[r] = \frac{1}{2} \int_{-\infty}^\infty \frac{a H_0^{(1)}[ar]}{a^2 + q^2} da. \quad (3.133)
$$

Now, take the contour integration in the upper hemisphere (Figure 3.10) to yield

$$
\begin{aligned}
A[r] &= i2\pi \left. \text{Residue} \right|_{a=iq} = \frac{i\pi}{2} H_0^{(1)}[iqr] \\
&= K_0[qr]. \quad (3.134)
\end{aligned}
$$

The last equality is established because $H_0^{(1)}[iz] = (-i2/\pi)K_0[z]$.

3.4.4 Example 3H: Hankel Transform of $J_0[ar']/ \left(a^2 + q^2 \right)$

• **Problem**

Show that

$$
\boxed{\mathcal{H}^{-1} \left\{ \frac{J_0[ar']}{a^2 + q^2} \right\} = I_0[qr']K_0[qr]}, \quad r' \le r. \quad (3.135)
$$

• **Solution**

Let $A[r, r']$ be the inverse transform

$$A[r, r'] = \int_0^\infty \frac{a J_0[ar'] J_0[ar]}{a^2 + q^2} da. \tag{3.136}$$

Follow the procedure of the last example to convert $J_0[ar]$ into a Hankel function but keep $J_0[ar']$ intact. The result is

$$\begin{aligned} A[r, r'] &= \frac{i\pi J_0[iqr'] H_0^{(1)}[iqr]}{2} \\ &= I_0[qr'] K_0[qr], \quad r' \le r. \end{aligned} \tag{3.137}$$

It can also be shown that

$$A[r, r'] = I_0[qr] K_0[qr'], \quad r' \ge r. \tag{3.138}$$

3.4.5   Example 4H: Hankel Transform of $a J_1[ar']/(a^2+q^2)$

• **Problem**

Show that

$$\boxed{\int_0^\infty \frac{a^2 J_1[ar'] J_0[ar]}{a^2 + q^2} da = -q I_1[qr'] K_0[qr]}, \quad r' \le r. \tag{3.139}$$

• **Solution**

Take the derivative of $A[r, r']$ in Example 3H with respect to r',

$$\frac{\partial A[r, r']}{\partial r'} = -\int_0^\infty \frac{a^2 J_1[ar'] J_0[ar]}{a^2 + q^2} da. \tag{3.140}$$

Therefore,

$$\begin{aligned} \int_0^\infty \frac{a^2 J_1[ar'] J_0[ar]}{a^2 + q^2} da &= -\frac{\partial}{\partial r'} A[r, r'] \\ &= -q I_1[qr'] K_0[qr], \quad r' \le r. \end{aligned} \tag{3.141}$$

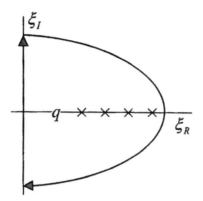

Figure 3.11: Integration along contour that encloses singular points q and ξ_n (marked by ×).

3.4.6 Example 5H: Inverse Hankel Transform 1

• **Problem**

Evaluate

$$
A[r] = \int_0^\infty \frac{a\cosh[\eta z]J_0[ar]}{(a^2+q^2)(c\eta\sinh[\eta b]+\cosh[\eta b])}da, \tag{3.142}
$$
$$
\eta^2 = (a^2+q^2)/k, \quad b>0,\ c>0,\ r>0,\ k>0.
$$

• **Solution**

The integrand is an odd function of a, so we follow Example 2H to integrate from $-\infty$ to $+\infty$,

$$
A[r] = \frac{1}{2}\int_{-\infty}^\infty \frac{a\cosh[\eta z]H_0^{(1)}[ar]}{(a^2+q^2)(c\eta\sinh[\eta b]+\cosh[\eta b])}da. \tag{3.143}
$$

Let $a = i\xi$ to give

$$
A[r] = -\frac{i}{\pi}\int_{-i\infty}^{i\infty} \frac{\xi\cos[\zeta z]K_0[\xi r]}{(\xi^2-q^2)(c\zeta\sin[\zeta b]-\cos[\zeta b])}d\xi, \tag{3.144}
$$
$$
\zeta^2 = (\xi^2-q^2)/k.
$$

Now we are ready to perform contour integration in the complex ξ plane (Figure 3.11). Note that

$$\zeta = \sqrt{\xi - q}\sqrt{\xi + q}/\sqrt{k} \tag{3.145}$$

is a multivalued function of ξ. Two branch cuts could have been drawn from q to $+\infty$ and from $-q$ to $-\infty$. (See Example 7H for a cut between $-q$ and $+q$.) Because the integrand is an even function of ζ, the integrations along the upper cut and lower cut cancel each other and effectively there is no branch cut.

Complete the contour integration clockwise in the right half of the ξ-plane. Apply the residue theorem

$$A[r] = -i2\pi \left(\text{Residue}|_{\xi=q} + \sum_{n=0}^{\infty} \text{Residue}_n \right). \tag{3.146}$$

The residue of a simple pole at $\xi = q$ (which makes the first factor ζ^2 in the denominator vanish) is

$$\text{Residue}|_{\xi=q} = \frac{i}{2\pi} K_0[qr]. \tag{3.147}$$

Other residues are obtained at $\xi = \xi_n$ that make the second factor

$$c\zeta \sin[\zeta b] - \cos[\zeta b] = 0. \tag{3.148}$$

In other words, we seek the root $\zeta_n b$ of

$$\tan[\zeta b] = \frac{b}{c\zeta b}. \tag{3.149}$$

For the chosen contour, we are interested in positive ζb. Hence,

$$\zeta_n b = n\pi + \theta_n, \quad 0 \leq \theta_n \leq \pi/2, \quad n = 0, 1, 2, ...; \tag{3.150}$$

$$\xi_n = \sqrt{\zeta_n^2 k + q^2}, \quad \theta_{n+1} < \theta_n.$$

Figure 3.12 depicts roots $\zeta_n b$ at the intersections of $y = \tan[\zeta b]$ and $y = (b/c)/\zeta b$. Also illustrated is the relation between $\zeta_n b$ and θ_n. It is obvious that $\theta_n \to 0$ as $n \to \infty$.

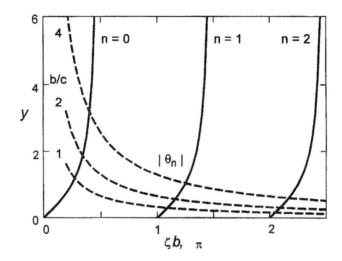

Figure 3.12: Roots $\zeta_n b$ at the intersections of $y = \tan[\zeta b]$ and $y = (b/c)/\zeta b$ for $b/c = 1$, 2, and 4. Note that the relation $\zeta_n b = n\pi + \theta_n$, $n = 0$, 1, 2, ... Also note that θ_n decreases with increasing n and decreasing b/c.

The residue at $\xi = \xi_n$ is

$$
\begin{aligned}
\text{Residue}_n &= -\frac{i}{\pi} \frac{\xi \cos[\zeta z] K_0[\xi r]}{(\xi^2 - q^2)\frac{\partial}{\partial \xi}\left(c\zeta \sin[\zeta b] - \cos[\zeta b]\right)}\bigg|_{\xi=\xi_n} \\
&= -\frac{i}{\pi} \frac{(-1)^n \sin[\theta_n] \cos[\zeta_n z]}{\zeta_n b + 0.5 \sin[2\theta_n]} K_0[\xi_n r].
\end{aligned}
\tag{3.151}
$$

Finally, putting these residues into $A[r]$ yields

$$
\boxed{A[r] = K_0[qr] - 2\sum_{n=0}^{\infty} \lambda_n K_0[\xi_n r]\cos[\zeta_n z]},
\tag{3.152}
$$

where

$$
\begin{aligned}
\lambda_n &= \frac{c}{\left(b + c + bc^2\zeta_n^2\right)\cos[\zeta_n b]} \\
&= \frac{(-1)^n \sin[\theta_n]}{n\pi + \theta_n + 0.5\sin[2\theta_n]}.
\end{aligned}
\tag{3.153}
$$

- **Root Finding**

Root finding is potentially the most time-consuming part in the computation of $A[r]$. However, root $\zeta_{n+1}b$ can be narrowly bracketed to reduce the computation time.

Since θ_n decreases with increasing n (Figure 3.12), a curve connecting the discrete points in the plot of θ_n versus n will be concave upward such that

$$\theta_{n-1} - \theta_n > \theta_n - \theta_{n+1} > 0, \quad n \geq 1. \tag{3.154}$$

Multiplying the inequality by -1 and adding $(n+1)\pi + \theta_n$ to each side yields

$$(n+1)\pi + \theta_n - (\theta_{n-1} - \theta_n) < (n+1)\pi + \theta_{n+1} < (n+1)\pi + \theta_n. \tag{3.155}$$

Now the root $\zeta_{n+1}b$ is bracketed as follows.

$$\zeta_n b + \pi - (\theta_{n-1} - \theta_n) < \zeta_{n+1}b < \zeta_n b + \pi. \tag{3.156}$$

The bracket width, $(\theta_{n-1} - \theta_n)$, becomes smaller as n increases.

3.4.7   Example 6H: Inverse Hankel Transform 2

- **Problem**

Find

$$\mathcal{H}^{-1}\left\{\frac{J_0[ar_w]}{a^2 + q^2}\frac{\sinh[\eta z']\cosh[\eta(z - b)]}{\sinh[\eta b]}\right\}, \quad \eta^2 = (a^2 + q^2)/k. \tag{3.157}$$

- **Solution**

Let the inverse Hankel transform be

$$A = \int_0^\infty aJ_0[ar]\frac{J_0[ar_w]}{a^2 + q^2}\frac{\sinh[\eta z']\cosh[\eta(z - b)]}{\sinh[\eta b]}da. \tag{3.158}$$

Since the integrand is an odd function of a, the same procedures for Example 5H are applicable here,

$$A = \frac{i}{\pi}\int_{-i\infty}^{i\infty}\omega K_0[\omega r]\frac{I_0[\omega r_w]}{\omega^2 - q^2}\frac{\sin[\zeta z']\cos[\zeta(z - b)]}{\sin[\zeta b]}d\omega, \tag{3.159}$$
$$\zeta^2 = (\omega^2 - q^2)/k.$$

There is no branch cut. So take a clockwise contour integration over the right half of the ξ-plane to enclose singular points at

$$\omega = q, \quad \text{and}$$
$$\omega = \omega_n, \quad \zeta_n b = n\pi, \quad n = 1, 2, 3, \dots \qquad (3.160)$$

for the root equation $\sin[\zeta b] = 0$. At $\omega = q$, the integrand is $0/0$ because $\zeta = 0$, too. Before the residue theorem can be applied, the zero in the numerator should be factored out by the relation that $\sin[\zeta z']/\sin[\zeta b] \approx z'/b$ as $\zeta \to 0$. In other words, $\omega = q$ is a first-order singular point despite the fact that either $\omega^2 - q^2 = 0$ or $\sin[\zeta b] = 0$ can contribute to the singularity. That is why we limit $\zeta_n b = n\pi$ for $n \geq 1$.

Use the residue theorem to obtain

$$A = \frac{z'}{b} I_0[qr_w] K_0[qr]$$
$$+ 2 \sum_{n=1}^{\infty} \frac{\sin[n\pi z'/b] \cos[n\pi z/b]}{n\pi} I_0[\omega_n r_w] K_0[\omega_n r] \qquad (3.161)$$

where

$$\omega_n^2 = \left(\frac{n\pi}{b}\right)^2 k + q^2. \qquad (3.162)$$

3.4.8 Example 7H: Integration Involving $1/\sqrt{1 - z^2}$

• Problem

Show that

$$I = \int_{-1}^{1} \frac{dx}{\sqrt{1 - x^2}\,(a + bx)} = \frac{\pi}{\sqrt{a^2 - b^2}}, \quad a > b > 0. \qquad (3.163)$$

• Solution 1

Let $x = \cos\theta$, then $\sqrt{1 - x^2} = \pm \sin\theta$ and the integral becomes

$$I = \int_0^{\pi} \frac{d\theta}{\pm(a + b\cos\theta)}. \qquad (3.164)$$

We shall choose the positive one of the dual-valued I because the original integral is positive. Since the integrand is even, we can evaluate I from $-\pi$ to π or equivalently from 0 to 2π for the convenience of integration.

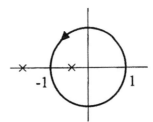

Figure 3.13: Counterclockwise contour integration around a unit circle. Singular points are located at $-a/b \pm \sqrt{a^2/b^2 - 1}$.

Now, let $z = e^{i\theta}$ and go into a complex plane. Recall $\cos\theta = (e^{i\theta} + e^{-i\theta})/2$ and set contour integration around a unit circle (Figure 3.13) to get

$$2I = \oint \frac{dz}{iz\left[a + b(z + z^{-1})/2\right]} = i2\pi \sum \text{Residues.} \qquad (3.165)$$

Simplify the integrand and factorize the denominator to give

$$2I = \oint \frac{-i2}{b\left[z + a/b - \sqrt{a^2/b^2 - 1}\right]\left[z + a/b + \sqrt{a^2/b^2 - 1}\right]} dz. \qquad (3.166)$$

Only one of the two singular points is located inside the unit circle. Therefore,

$$I = \frac{\pi}{\sqrt{a^2 - b^2}}. \qquad (3.167)$$

What we have learned is that the factor $\sqrt{1 - x^2}$ gives both positive and negative values and we have chosen the proper sign judiciously. By means of changing variables, we have also avoided the branch-cut integration.

• **Solution 2**

Extend the original integral into complex z plane

$$\oint \frac{dz}{\sqrt{1 - z^2}\,(a + bz)} \qquad (3.168)$$

around a closed path. It has three singular points at $z = +1, -1$, and $-a/b$. Also the first two are branch points. Since each of $\sqrt{1 - z}$ and $\sqrt{1 + z}$ is

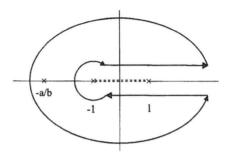

Figure 3.14: Integration along branch cut between -1 and +1 (dashed line).

a multivalued function, the product $\sqrt{1 - z^2}$ is also a multivalued function. Hence, our method for testing multivaluedness is inadequate here.

Now, draw a branch cut between -1 and $+1$. (The branch cut is equivalent to the combination of one cut from -1 to ∞ and another cut from $+1$ to ∞. The overlapped parts nullify each other.) Perform the contour integration along the path shown in Figure 3.14,

$$\oint \frac{dz}{i\sqrt{(z-1)(z+1)}(a+bz)} = i2\pi \sum \text{Residues.} \qquad (3.169)$$

The contribution along the great circular path vanishes; the integrations along the real axis from $+1$ to ∞ and from ∞ to $+1$ cancel out each other. The residues at $z = -1$ and $+1$ are zero.

The remainder of the contour integration is

$$\int_1^{-1} + \int_{-1}^1 = i2\pi \cdot \text{Residue}\big|_{z=-a/b}$$

$$= \frac{2\pi}{\pm b\sqrt{(a/b)^2 - 1}} = \frac{2\pi}{\pm\sqrt{a^2 - b^2}}. \qquad (3.170)$$

The two integrals along the branch cut do not cancel each other because they have different phases (signs). If $\int_1^{-1}$ along the lower cut is negative, then $\int_{-1}^1$ along the upper cut is positive and hence, the positive sign of $\pm$ is chosen; on the other hand, if $\int_1^{-1}$ is positive, then $\int_{-1}^1$ is negative and the chosen sign is negative. The choice of negative sign is contrary to the conventional sign assignment that produces a positive sign for the residue if the contour

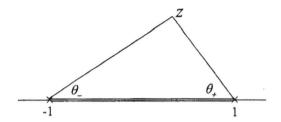

Figure 3.15: Definition of θ_+ and θ_-.

integration is counterclockwise. This viewpoint is clarified below (Figures 3.14 and 3.15).

Let

$$z + 1 = |x + 1|\, e^{i(2n\pi + \theta_-)}, \quad z - 1 = |x - 1|\, e^{i(2m\pi + \theta_+)} \tag{3.171}$$

where n and m are integers (Figure 3.15). Then, rewrite the integrand along the cut

$$\frac{1}{i\sqrt{|x^2 - 1|}\,(a + bx)} \exp\left[-i\left(n\pi + \theta_-/2\right) - i\left(2m\pi + \theta_+/2\right)\right].$$

We want the integrand along the upper part of the cut to be positive. Because $\theta_- = 0$ and $\theta_+ = \pi$, we can somewhat arbitrarily set $n = 1$ and $m = 0$ to achieve the goal.

Now, along the lower part of the cut, $\theta_- = 0$ and $\theta_+ = -\pi$ (keeping the same n and m) and the integrand becomes negative. Therefore

$$-\int_1^{-1} + \int_{-1}^{1} = 2\int_{-1}^{1} \frac{dx}{\sqrt{1 - x^2}(a + bx)} = \frac{2\pi}{\pm\sqrt{a^2 - b^2}}. \tag{3.172}$$

The choice of positive sign is obvious.

If we choose $n = m = 0$ (as is usually done), the integration is negative along the upper path while it is positive along the lower path. In this case,

$$\int_1^{-1} - \int_{-1}^{1} = -2\int_{-1}^{1} = -2I = \frac{2\pi}{\pm\sqrt{a^2 - b^2}}, \tag{3.173}$$

and the chosen sign is negative.

3.4.9 Numerical Hankel Transform

If a Hankel transform cannot be inverted analytically, one needs to use numerical integration. Owing to the oscillating nature of $J_0[x]$, numerical integration may converge slowly and show erratic results if an insufficient number of terms are used. Hence, the results should be carefully monitored.

Usually the numerical integration is performed by dividing the integral into numerous subintegrals, each of which is integrated from one root of the Bessel function to the next neighboring root,

$$\int_0^\infty r f[r] J_0[ar] dr = \sum_{n=0}^\infty \int_{r_n}^{r_{n+1}} r f[r] J_0[ar] dr, \qquad (3.174)$$

where ar_n is the nth root of $J_0[ar]$ except that $r_0 = 0$. To increase integration efficiency, the summation is often aided with an extrapolation scheme to increase the convergence rate.

Various techniques using the fast Fourier transform for numerical Hankel transforms can be found in the literature (e.g., Agnesi, Reali, and Patrini, 1993). What follows are two examples.

• 2D FFT

Consider a two-dimensional Fourier transform,

$$F[u, v] = \int_{-\infty}^\infty \int_{-\infty}^\infty f[x, y] e^{-i(ux+vy)} dx dy. \qquad (3.175)$$

If function $f[x, y]$ is axi-symmetric (circular symmetric) so that it is expressible as $f[r]$ with $r^2 = x^2 + y^2$, then $F[u, v]$ is also axi-symmetric and is expressible as $F[w]$ with $w^2 = u^2 + v^2$.

Let

$$x + iy = r e^{i\theta}, \quad u + iv = w e^{i\phi}. \qquad (3.176)$$

Then, in terms of cylindrical coordinates, we get

$$\begin{aligned}
F[w] &= \int_0^\infty r f[r] dr \int_0^{2\pi} e^{-irw(\cos\theta\cos\phi + \sin\theta\sin\phi)} d\theta \\
&= \int_0^\infty r f[r] dr \int_0^{2\pi} e^{-irw\cos(\theta-\phi)} d\theta. \qquad (3.177)
\end{aligned}$$

If $\phi = 0$ (i.e., $v = w \sin[\phi] = 0$), then

$$\int_0^\infty r f[r] J_0[rw] dr = \frac{1}{2\pi} F[w] \tag{3.178}$$

(Bracewell, 1986; Ferrari, 1995) because

$$J_0[z] = \frac{1}{2\pi} \int_0^{2\pi} e^{-iz \cos \psi} d\psi. \tag{3.179}$$

Now, we have demonstrated that the Hankel transform can be obtained from the Fourier transform

$$F[w] = F[u, 0]. \tag{3.180}$$

That is,

$$\begin{aligned}
\int_0^\infty r f[r] J_0[ar] dr &= \frac{1}{2\pi} \int_{-\infty}^\infty \int_{-\infty}^\infty f[r] e^{iux} dx dy \\
&= \frac{1}{2\pi} \int_{-\infty}^\infty e^{iux} dx \int_{-\infty}^\infty f[\sqrt{x^2 + y^2}] dy. \quad (3.181)
\end{aligned}$$

The y integration for each x can be achieved numerically or by taking the mean value of its Fourier transform (the first term of the numerical Fourier transform). The x integration is usually accomplished by the FFT method.

• **Fourier-Abel Transform**

The *Abel transform* of $f[r]$ is defined by

$$\overline{f}[\rho] = \int_\rho^\infty \frac{2f[r]r}{\sqrt{r^2 - \rho^2}} dr. \tag{3.182}$$

We are to show that the Fourier transform of the Abel transform is the Hankel transform (Bracewell, 1986; Hansen, 1985).

Let us carry the Fourier transform of $\overline{f}[\rho]$

$$\begin{aligned}
F[a] &= \int_{-\infty}^\infty \overline{f}[\rho] e^{-ia\rho} d\rho \\
&= \int_{-\infty}^\infty e^{-ia\rho} d\rho \int_\rho^\infty \frac{2f[r]r}{\sqrt{r^2 - \rho^2}} dr \\
&= \int_{-\infty}^\infty e^{-ia\rho} d\rho \int_0^\infty \frac{2f[r]rU[r^2 - \rho^2]}{\sqrt{r^2 - \rho^2}} dr, \tag{3.183}
\end{aligned}$$

where $U[r^2 - \rho^2]$ is Heaviside's unit step function

$$U[r^2 - \rho^2] = \begin{cases} 1 & \text{if} \quad r^2 > \rho^2, \\ 0 & \text{if} \quad r^2 < \rho^2. \end{cases} \tag{3.184}$$

Exchange the order of integration to yield

$$\begin{aligned} F[a] &= \int_0^\infty 2r f[r] dr \int_{-\infty}^\infty \frac{U[r^2 - \rho^2]}{\sqrt{r^2 - \rho^2}} e^{-ia\rho} d\rho \\ &= \int_0^\infty 2r f[r] dr \int_{-r}^r \frac{e^{-ia\rho}}{\sqrt{r^2 - \rho^2}} d\rho. \end{aligned} \tag{3.185}$$

Let $\zeta = \rho/r$, then

$$F[a] = \int_0^\infty 2r f[r] dr \int_{-1}^1 \frac{e^{-iar\zeta}}{\sqrt{1 - \zeta^2}} d\zeta. \tag{3.186}$$

Next, let $\zeta = \cos\theta$ to yield

$$\begin{aligned} F[a] &= \int_0^\infty 2r f[r] dr \int_0^\pi e^{-iar\cos\theta} d\theta \\ &= \int_0^\infty 2r f[r] dr \int_0^\pi \cos[ar\cos\theta] d\theta \\ &= 2\pi \int_0^\infty r f[r] J_0[ar] dr, \end{aligned} \tag{3.187}$$

where the relation

$$J_0[z] = \frac{1}{\pi} \int_0^\pi \cos[z\cos\theta] d\theta \tag{3.188}$$

has been used.

Now we have reached our goal

$$\int_0^\infty r f[r] J_0[ar] dr = \frac{F[a]}{2\pi}. \tag{3.189}$$

The difficulty in implementing the Fourier-Abel transform lies in the Abel transform.

3.5   Problems, Keys, and Suggested Readings

- **Problems**

   1. Find the Laplace transform of $\frac{1}{b}\sin[bt]$.
   2. Use the Bromwich integral (contour integration) to find the inverse Laplace transform for the result in Problem 1.
   3. Find the inverse Laplace transform of

$$\frac{1}{s}K_0\left[\sqrt{\frac{s}{\kappa}}x\right]. \tag{3.190}$$

Then, express your answer in terms of the exponential integral. (Hint: use Example 9.)

   4. Show that

$$\frac{1}{i2\pi}\int_{a-i\infty}^{a+i\infty}e^{-s(t_0-t)}ds \tag{3.191}$$

is a delta function, $\delta[t_0 - t]$.

   5. Show that

$$\mathcal{L}\left\{\int_0^t g[\tau]d\tau\right\} = \frac{\overline{g}[s]}{s} \tag{3.192}$$

by means of switching the order of integration.

   6. Integrate

$$F[a] = \int_0^a \frac{t\exp[t^2]}{\sqrt{a^2-t^2}}\,\mathrm{erf}[t]dt. \tag{3.193}$$

   7. Evaluate

$$\int_0^\infty \frac{a^{2n+1}}{a^2+q^2}J_0[ar]da. \tag{3.194}$$

   8. Find

$$\mathcal{H}^{-1}\left\{\frac{ar'J_1[ar']}{a^2+q^2}\frac{\sinh[\eta z']\cosh[\eta(z-b)]}{\sinh[\eta b]}\right\}, \quad \eta^2 = \left(a^2+q^2\right)/k. \tag{3.195}$$

   9. Assume that the initial temperature is zero. At time $t = 0$, a constant temperature T_0 is imposed on $x = 0$. Find the temperature distribution $T[x,t]$ and the heat flux into the medium at $x = 0$. (Hint: Use 1D heat conduction equation, apply the Laplace transform, solve an ordinary differential equation, then invert the solution in the Laplace domain.)

10. The conditions are the same as in Problem 9 except that the heat flux at $x = 0$ is fixed at a constant value q_0 instead of the temperature being fixed at T_0. Find the temperature distribution $T[x, t]$.

11. Show that

$$\int_0^b r J_0[ar] dr = \frac{b}{a} J_1[ba].$$
(3.196)

(Hint: Use a series representation of $J_\nu[ar]$.)

12. If $F[\rho]$ is the Hankel transform of $f[r]$, show that

$$\int_0^\infty e^{-\kappa \rho^2 t} F[\rho] \rho J_0[\rho r] d\rho = \frac{1}{2\kappa t} \int_0^\infty a e^{(r^2 + a^2)/4\kappa t} I_0 \left[\frac{ra}{2\kappa t} \right] f[a] da.$$
(3.197)

(Hint: Substitute the defining integral for $F[\rho]$ into the left-hand side, then use formula 6.633.2 in Gradshteyn and Ryzhik, 1994.)

• **Keys**

Key 1

Let the Laplace transform be

$$
\begin{aligned}
A &= \mathcal{L}\left\{ \frac{1}{b} \sin[bt] \right\} \\
&= \int_0^\infty \left(\frac{1}{b} \sin[bt] \right) e^{-st} dt.
\end{aligned}
$$
(3.198)

Recall the relation $\sin\theta = \left(e^{i\theta} - e^{-i\theta} \right)/i2$ to get

$$
\begin{aligned}
A &= \frac{1}{i2b} \int_0^\infty \left(e^{-(s-ib)t} - e^{-(s+ib)t} \right) dt \\
&= \frac{1}{i2b} \left[\frac{\exp\left[(-s + ib) t \right]}{-s + ib} - \frac{\exp\left[(-s - ib) t \right]}{-s - ib} \right] \Bigg|_0^\infty \\
&= \frac{1}{s^2 + b^2}.
\end{aligned}
$$
(3.199)

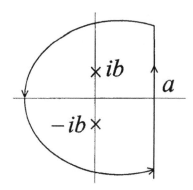

Figure 3.16: Contour integration that encloses poles at $s = ib$ and $s = -ib$.

Key 2

Let the inverse Laplace transform be (Figure 3.16)

$$
\begin{aligned}
B &= \mathcal{L}^{-1}\left\{\frac{1}{s^2 + b^2}\right\} \\
&= \frac{1}{i2\pi}\int_{a-i\infty}^{a+i\infty}\frac{\exp[st]}{s^2 + b^2}ds.
\end{aligned}
\tag{3.200}
$$

The integrand has two poles at ib and $-ib$. Use of the residue theorem yields

$$
\begin{aligned}
B &= \frac{i2\pi}{i2\pi}\sum\text{Residues} \\
&= \frac{\exp[st]}{2s}\bigg|_{s=ib} + \frac{\exp[st]}{2s}\bigg|_{s=-ib} \\
&= \frac{\sin[bt]}{b}.
\end{aligned}
\tag{3.201}
$$

Key 3

Let the inverse Laplace transform be

$$
\begin{aligned}
B &= \mathcal{L}^{-1}\left\{\frac{K_0\left[\sqrt{\frac{s}{\kappa}}x\right]}{s}\right\} \\
&= \frac{1}{i2\pi}\int_{a-i\infty}^{a+i\infty}\frac{K_0\left[\sqrt{\frac{s}{\kappa}}x\right]}{s}e^{st}ds.
\end{aligned}
\tag{3.202}
$$

Recognizing that the transform $K_0[\sqrt{s/\kappa}x]/s$ is the product of two transforms, $K_0[\sqrt{s/\kappa}x]$ and $1/s$, we can employ the convolution theorem. From the following transform pairs

$$1 \longleftrightarrow \frac{1}{s}, \quad \frac{\exp[-x^2/4\kappa t]}{2t} \longleftrightarrow K_0\left[\sqrt{\frac{s}{\kappa}}x\right], \tag{3.203}$$

one obtains

$$B = \int_0^t 1 \cdot \frac{\exp[-x^2/4\kappa\tau]}{2\tau} d\tau. \tag{3.204}$$

Now, introduce a new variable $\xi = x^2/4\kappa\tau$ to yield $d\xi = -\xi d\tau/\tau$ and

$$B = \frac{1}{2} \int_{x^2/4\kappa t}^{\infty} \frac{e^{-\xi}}{\xi} d\xi, \tag{3.205}$$

which is the exponential integral (excluding the factor $1/2$).

Key 4

Let

$$I = \frac{1}{i2\pi} \int_{a-i\infty}^{a+i\infty} e^{s(t-t_0)} ds. \tag{3.206}$$

Let $s = a + iw$, then

$$\begin{aligned} I &= \frac{e^{a(t-t_0)}}{2\pi} \int_{-\infty}^{\infty} e^{iw(t-t_0)} dw \\ &= e^{a(t-t_0)} \delta[t - t_0]. \end{aligned} \tag{3.207}$$

This delta function $\delta[t - t_0]$ is defined in terms of the Fourier transform. The two forms of the delta function (one in the inverse Laplace transform and the other in the Fourier transform) behave essentially the same when they are applied to an integral. That is, using $e^{a(t-t_0)}\delta[t - t_0]$ or $\delta[t - t_0]$ leads to the same result in practice.

Key 5

By definition, the desired Laplace transform is

$$\mathcal{L}\left\{ \int_0^t g[t]dt \right\} = \int_0^{\infty} \left(\int_0^t g[\tau]d\tau \right) e^{-st} dt. \tag{3.208}$$

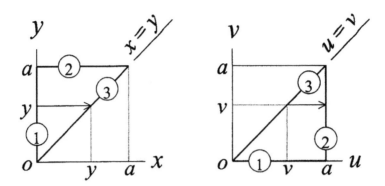

Figure 3.17: Left: integration over the upper triangle in the xy-plane. Right: integration over the lower triangle in the uv-plane.

We are to switch the order of integration. (If t were not an integration limit, the switch would be simple.) Let us rewrite the integral by letting $x = \tau$ and $y = t$,

$$\mathcal{L}\left\{\int_0^t g[t]dt\right\} = \lim_{a\to\infty} \int_0^a dy \int_0^y g[x]e^{-sy}dx. \qquad (3.209)$$

Referring to Figure 3.17, one can see that the integration is over the upper triangular area bounded by three heavy lines.

Solution 1:   For a given y (between 0 and a), the x integration is from 0 to line $y = x$. Switch the order of integration. Then, for a given x between 0 and a, the y integration is from line $y = x$ to $y = a$ in the upper triangle. As a result,

$$\begin{aligned}
\lim_{a\to\infty} \int_0^a e^{-sy}dy \int_0^y g[x]dx &= \lim_{a\to\infty} \int_0^a g[x]dx \int_x^a e^{-sy}dy \\
&= \int_0^\infty \frac{g[x]\exp[-sx]}{s}dx \\
&= \bar{g}[s]/s. \qquad (3.210)
\end{aligned}$$

Solution 2:   Alternatively, we can make a coordinate transform by letting

$$u = y, \quad v = x. \qquad (3.211)$$

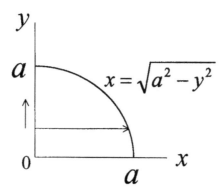

Figure 3.18: Integration over the area of one-quarter circle.

The upper triangle in the xy-plane is mapped into the lower triangle in the uv-plane

$$
\begin{array}{llll}
\bigcirc 1: & x = 0, & v = 0; \\
\bigcirc 2: & y = a, & u = a; \\
\bigcirc 3: & x = y, & u = v.
\end{array}
\tag{3.212}
$$

The integration is thus performed over the lower triangular area in the uv-plane.

The area element $dxdy$ needs to be transformed to the area element $dudv$. This is accomplished by multiplying the determinant of the Jacobian transformation matrix

$$
dxdy = \det \begin{vmatrix} \frac{\partial x}{\partial u} & \frac{\partial x}{\partial v} \\ \frac{\partial y}{\partial u} & \frac{\partial y}{\partial v} \end{vmatrix} dudv = dudv.
\tag{3.213}
$$

Thus the integral becomes

$$
\begin{aligned}
\mathcal{L}\left\{ \int_0^t g[t]dt \right\} &= \lim_{a \to \infty} \int_0^a dv \int_v^a g[v]e^{-su}du \\
&= \int_0^\infty g[v]dv \int_v^\infty e^{-su}du = \frac{\overline{g}[s]}{s}.
\end{aligned}
\tag{3.214}
$$

Key 6

Use the integral expression of the error function to have

$$
F[a] = \frac{2}{\sqrt{\pi}} \int_0^a \frac{t\exp[t^2]}{\sqrt{a^2 - t^2}}dt \int_0^t \exp[-x^2]dx.
\tag{3.215}
$$

Let

$$y = \sqrt{a^2 - t^2}, \tag{3.216}$$

then

$$dy = -\frac{t}{\sqrt{a^2 - t^2}} dt, \tag{3.217}$$

and $F[a]$ becomes

$$F[a] = \frac{2e^{a^2}}{\sqrt{\pi}} \int_0^a e^{-y^2} dy \int_0^{\sqrt{a^2-y^2}} e^{-x^2} dx. \tag{3.218}$$

For any y between 0 and a in the y integration, the x integration begins at 0 and terminates at circle $x^2 = a^2 - y^2$. That means the integration is over one-quarter of a circular area (Figure 3.18). As such, it is easier to perform the integration in polar coordinates. Since $dxdy = rd\theta dr$ and $r^2 = x^2 + y^2$, we have

$$\begin{aligned} F[a] &= \frac{2e^{a^2}}{\sqrt{\pi}} \int_0^{\pi/2} d\theta \int_0^a re^{-r^2} dr \\ &= \frac{\sqrt{\pi}}{2} \left(e^{a^2} - 1 \right). \end{aligned} \tag{3.219}$$

Note that r is the determinant of the Jacobian for transforming the rectangular coordinates into the polar coordinates.

Key 7

Let

$$A = \int_0^\infty \frac{a^{2n}}{a^2 + q^2} a J_0[ar] da. \tag{3.220}$$

Follow Example 2H to convert $J_0[ar]$ to $H_0^{(1)}[ar]$,

$$A = \frac{1}{2} \int_{-\infty}^\infty \frac{a^{2n+1} H_0^{(1)}[ar]}{a^2 + q^2} da. \tag{3.221}$$

Take the contour integration in the upper hemisphere that encloses a pole at $a = iq$ to obtain

$$\begin{aligned} A &= i2\pi \left. \text{Residue} \right|_{a=iq} = i2\pi \frac{1}{2} \left. \frac{a^{2n} a H_0^{(1)}[ar]}{2a} \right|_{a=iq} \\ &= (-1)^n q^{2n} K_0[qr]. \end{aligned} \tag{3.222}$$

Key 8

By analogy to Example 4H for the $J_1[ar]$ factor, differentiating the result of Example 6H with respect to r' yields the desired solution,

$$\int_0^\infty a J_0[ar] \frac{ar' J_1[ar']}{a^2 + q^2} \frac{\sinh[\eta z'] \cosh[\eta(z-b)]}{\sinh[\eta b]} da = -\frac{z'}{b} q r' I_1[qr'] K_0[qr]$$

$$-2 \sum_{n=0}^\infty \frac{\sin[n\pi z'] \cos[n\pi z/b]}{n\pi} \xi_n r' I_1[\xi_n r'] K_0[\xi_n r], \qquad (3.223)$$

$$\xi_n b = \sqrt{(n\pi)^2 k + q^2 b^2}.$$

Key 9

The one-dimensional heat conduction equation is

$$\frac{\partial^2 T}{\partial x^2} = \frac{1}{\kappa} \frac{\partial T}{\partial t}, \qquad (3.224)$$

where κ is the thermal diffusivity. The initial and boundary conditions are

$$\begin{aligned} T[x,0] &= 0, \\ T[0,t] &= T_0, \\ T[\infty,t] &= 0. \end{aligned} \qquad (3.225)$$

General Solution: Let the Laplace transform of $T[x,t]$ be $\overline{T}[x,s]$, i.e.,

$$\overline{T}[x,s] = \int_0^\infty T[x,t] e^{-st} dt. \qquad (3.226)$$

Then, take the Laplace transform of the heat conduction equation and use the basic transform formulas in Section 1 to obtain

$$\frac{d^2 \overline{T}}{dx^2} = \frac{s}{\kappa} \overline{T} - T[x,0]. \qquad (3.227)$$

The initial condition $T[x,0] = 0$ is implemented here during the Laplace transform. This second-order ordinary differential equation with constant coefficients has the general solution

$$\overline{T}[x,s] = A e^{\sqrt{s/\kappa}\, x} + B e^{-\sqrt{s/\kappa}\, x}, \qquad (3.228)$$

where the coefficients $A[s]$ and $B[s]$ are to be determined from the boundary conditions.

Boundary Conditions: As $x \to \infty$, the term $\exp[\sqrt{s/\kappa}] \to \infty$. To avoid blowing up at infinity, we set $A = 0$ to satisfy the condition that $\overline{T}[\infty, s] = 0$. The remaining coefficient B is determined from the condition of constant temperature at $x = 0$. In the Laplace transformed domain,

$$\overline{T}[0, s] = B = \frac{T_0}{s}. \tag{3.229}$$

So our solution in the Laplace domain is

$$\overline{T}[x, s] = T_0 \frac{e^{-\sqrt{s/\kappa}x}}{s}. \tag{3.230}$$

Inverse Laplace Transform: Take the inverse Laplace transform. According to Example 3, one can write immediately

$$\boxed{T[x, t] = T_0 \operatorname{erfc}\left[\frac{x}{\sqrt{4\kappa t}}\right].} \tag{3.231}$$

By means of Leibnitz' rule of differentiation, the heat flux is simply

$$-k\frac{\partial T}{\partial x} = \frac{kT_0}{\sqrt{\pi\kappa t}}e^{-x^2/4\kappa t}. \tag{3.232}$$

Key 10

The solution is similar to Problem 9 except the boundary condition at $x = 0$ is changed from specified temperature to specified flux. In the Laplace domain, the condition of constant flux is

$$-k\frac{\partial \overline{T}}{\partial x}\bigg|_{x=0} = \frac{q_0}{s}. \tag{3.233}$$

This constraint gives

$$k\sqrt{\frac{s}{\kappa}}B = \frac{q_0}{s}. \tag{3.234}$$

Substituting the B into the general solution yields our solution in the Laplace domain

$$\overline{T}[x, s] = \frac{\sqrt{\kappa}q_0}{k}\frac{\exp[-\sqrt{s/\kappa}x]}{s^{3/2}}. \tag{3.235}$$

You may now want to consult a Laplace transform table for the solution in the time domain. But, let's earn our credit by using three different methods to get the inverse.

Inverse Laplace Transform 1: Start from the known transform pair

$$\frac{e^{-\sqrt{s/\kappa}x}}{s} \leftrightarrow \text{erfc}\left[\frac{x}{\sqrt{4\kappa t}}\right]. \qquad (3.236)$$

Then, integrate both sides with respect to x from x to ∞,

$$\int_x^\infty \frac{e^{-\sqrt{s/\kappa}y}}{s}dy \longleftrightarrow \int_x^\infty \text{erfc}\left[\frac{y}{\sqrt{4\kappa t}}\right]dy, \qquad (3.237)$$

which yields

$$\frac{\sqrt{\kappa}e^{-\sqrt{s/\kappa}x}}{s^{3/2}} \longleftrightarrow \sqrt{4\kappa t}\int_{x/\sqrt{4\kappa t}}^\infty \text{erfc}\left[\xi\right]d\xi. \qquad (3.238)$$

Therefore,

$$T[x,t] = \frac{q_0}{k}\sqrt{4\kappa t}\,\text{ierfc}\left[\frac{x}{\sqrt{4\kappa t}}\right], \qquad (3.239)$$

where ierfc stands for the first repeated complementary error function

$$\text{ierfc}[\xi] = \int_\xi^\infty \text{erfc}[\zeta]d\zeta = \frac{1}{\sqrt{\pi}}e^{-\xi^2} - \xi\,\text{erfc}[\xi]. \qquad (3.240)$$

Other *repeated complementary error functions* can be obtained through the recurrence relation

$$\text{i}^n\text{erfc}[x] = \int_x^\infty \text{i}^{n-1}\text{erfc}[\xi]d\xi = \frac{1}{2n}\left(\text{i}^{n-2}\text{erfc}[x] - 2x\text{i}^{n-1}\text{erfc}[x]\right) \qquad (3.241)$$

with $\text{i}^0\text{erfc}[x] = \text{erfc}[x]$.

Inverse Laplace Transform 2: Instead of integrating the transform pair in equation (3.236), let us take the derivative of the transform pair with respect to x to obtain

$$\frac{1}{\sqrt{s\kappa}}\exp\left[-\sqrt{s/\kappa}x\right] \leftrightarrow \frac{1}{\sqrt{\pi\kappa t}}e^{-x^2/4\kappa t}. \qquad (3.242)$$

Here we see that our desired transform is a product of two transforms

$$\frac{e^{-\sqrt{s/\kappa}x}}{s^{3/2}} = \frac{1}{s} \cdot \frac{e^{-\sqrt{s/\kappa}x}}{s^{1/2}}, \qquad (3.243)$$

where the inverse for the second factor is given in Example 2. Recall the convolution theorem to get

$$T[x,t] = \frac{q_0}{k}\sqrt{\frac{\kappa}{\pi}}\int_0^t \frac{1}{\sqrt{\tau}}e^{-x^2/4\kappa\tau}d\tau. \qquad (3.244)$$

Now, let us pause for a moment to check for dimensional consistency: T with $[K]$, q_0 with $[Wm^{-2}]$, k with $[Wm^{-1}K^{-1}]$, κ with $[m^2s^{-1}]$, and t or τ with $[s^{-1}]$. Every term is dimensionally consistent. There are two choices for executing the integral: one by brute force – numerical integration, and the other by simplifying it further.

Let us use integration by parts to simplify the integral. Set

$$u = \exp[-x^2/4\kappa\tau] \quad \text{and} \quad dv = d\tau/\sqrt{\tau} \qquad (3.245)$$

to obtain

$$du = \frac{x^2}{4\kappa\tau^2}e^{-x^2/4\kappa\tau}d\tau \quad \text{and} \quad v = 2\sqrt{\tau}. \qquad (3.246)$$

Then

$$\begin{aligned}
T[x,t] &= \frac{q_0}{k}\sqrt{\frac{\kappa}{\pi}}\left\{2\sqrt{\tau}e^{-x^2/4\kappa\tau}\Big|_0^t - \int_0^t 2\sqrt{\tau}\frac{x^2}{4\kappa\tau^2}e^{-x^2/4\kappa\tau}d\tau\right\} \\
&= \frac{q_0}{k}\left\{\sqrt{\frac{4\kappa t}{\pi}}e^{-x^2/4\kappa t} - \frac{x^2}{2\sqrt{\pi\kappa}}\int_0^t \frac{1}{\tau^{3/2}}e^{-x^2/4\kappa\tau}d\tau\right\}. \qquad (3.247)
\end{aligned}$$

Now, let $\xi^2 = x^2/4\kappa\tau$ to get $d\tau/\tau = -2d\xi/\xi$ and

$$\begin{aligned}
T[x,t] &= \frac{q_0}{k}\left\{\sqrt{\frac{4\kappa t}{\pi}}e^{-x^2/4\kappa t} - \frac{x^2}{2\sqrt{\pi\kappa}}\int_\infty^{x/\sqrt{4\kappa t}}\left(-\frac{2}{\xi}\right)\frac{e^{-\xi^2}}{\sqrt{\tau}}d\xi\right\} \\
&= \frac{q_0}{k}\left\{\sqrt{\frac{4\kappa t}{\pi}}e^{-x^2/4\kappa t} - \frac{2x}{\sqrt{\pi}}\int_{x/\sqrt{4\kappa t}}^\infty \frac{x}{\sqrt{4\kappa\tau}}\frac{e^{-\xi^2}}{\xi}d\xi\right\} \\
&= \frac{q_0}{k}\left\{\sqrt{\frac{4\kappa t}{\pi}}e^{-x^2/4\kappa t} - x\,\mathrm{erfc}\left[\frac{x}{\sqrt{4\kappa t}}\right]\right\} \\
&= \frac{q_0}{k}\sqrt{4\kappa t}\left\{\frac{1}{\sqrt{\pi}}e^{-x^2/4\kappa t} - \frac{x}{\sqrt{4\kappa t}}\,\mathrm{erfc}\left[\frac{x}{\sqrt{4\kappa t}}\right]\right\} \\
&= \frac{q_0}{k}\sqrt{4\kappa t}\,\mathrm{ierfc}\left[\frac{x}{\sqrt{4\kappa t}}\right]. \qquad (3.248)
\end{aligned}$$

Thus we have obtained the same answer.

Inverse Laplace Transform 3:   Now let us pick the product

$$\frac{e^{-\sqrt{s/\kappa}\,x}}{s^{3/2}} = \frac{1}{s^{1/2}} \cdot \frac{e^{-\sqrt{s/\kappa}\,x}}{s} \,. \tag{3.249}$$

These two have the transform pairs

$$\frac{1}{\sqrt{s}} \leftrightarrow \frac{1}{\sqrt{\pi t}}, \quad \frac{e^{-\sqrt{s/\kappa}\,x}}{s} \leftrightarrow \text{erfc}\left[\frac{x}{\sqrt{4\kappa t}}\right]. \tag{3.250}$$

Hence, by the convolution theorem, we obtain

$$T[x,t] = \frac{q_0}{k\sqrt{\pi}} \int_0^t \frac{1}{\sqrt{t-\tau}} \text{erfc}\left[\frac{x}{\sqrt{4\kappa\tau}}\right] d\tau. \tag{3.251}$$

This is a closed-form solution but it is not in terms of a well-recognized function. It demonstrates that we can always obtain a closed-form solution (integral) for the inverse Laplace transform although some integrals may be very difficult to evaluate.

• Suggested Readings

Abramowitz and Stegun's (1968) *Handbook of Mathematical Functions* and Gradshteyn and Ryzhik's (1994) *Table of Integrals, Series, and Products* are two excellent sources for functional relations and integrals. SIAM's (Society for Industrial and Applied Mathematics) journals are good places to scout for exotic formulas. We have used several of Carslaw and Jaeger's (1959) tabulated formulas as examples for the inverse Laplace transform.

3.6   Notations

Symbol	Definition	SI Unit	Dimension
$\leftrightarrow$	Laplace transform pair, as in $f[t] \leftrightarrow \overline{f}[s]$		
overbar	As in $\overline{g}[s]$, transformed variable		
$\prime$ prime	Derivative, as in $f'[z]$		
a	Dumy variable, or Hankel transform parameter	m^{-1}	
a_{-1}	Residue		
$-\text{Ei}[-x]$, $E_1[x]$	Exponential integral	-	-
$\text{erf}[x]$	Error function	-	-
$\text{erfc}[x]$	Complementary error function	-	-
$H_\nu^{(1)}[z]$	Hankel function, first kind	-	-
$H_\nu^{(2)}[z]$	Hankel function, second kind	-	-
I	A symbol designating an integral		
$I_\nu[x]$	Modified Bessel function, 1st kind	-	-
$i^n\text{erfc}[x]$	nth repeated complementary error function	-	-
$J_\nu[x]$	Bessel function, first kind, ν	-	-
k	Thermal conductivity	$Wm^{-}K^{-1}$	$\text{MLT}^{-3}\text{K}^{-1}$
$K_\nu[x]$	Modified Bessel function, 2nd kind	-	-
q	$\sqrt{s/\kappa}$		
q_0	Heat flux	Wm^{-2}	MT^{-3}
s	Laplace transform parameter	s^{-1}	T^{-1}
t	Time	s	T
T	Temperature	K	K
$Y_\nu[x]$	Bessel function, second kind	-	-
z	Complex number		
$\mathcal{H}$	Hankel transform operator		
$\mathcal{L}$	Laplace transform operator		
κ	Thermal diffusivity	$m^2 s^{-1}$	L^2T^{-1}
γ	Euler's constant		
$\delta[t - t_0]$	Delta function		

Chapter 4

DRAWDOWN IN CONFINED AQUIFERS

Hydraulic responses to pumping depend on aquifer characteristics, fluid properties, well construction, and pumping rate. A *well function* is used here to represent the hydraulic responses (drawdown) to pumping or to describe the flow toward a pumping well under a set of idealized conditions. In application, we wish to determine the hydrogeological properties from observed responses such as hydraulic head variations. Once the properties are determined, one can predict the responses due to pumping or use the properties to formulate models of groundwater flow.

The nature of these properties and hence, the number of property parameters depend on how you set up the governing differential equation and boundary and initial conditions. The equation and conditions depend in turn on how you conceptualize the hydraulic system and whether you can define a solvable problem. Numerous well functions have been published in the literature (Dawson and Istok, 1991). This chapter deals only with a few problems that have analytic or closed-form solutions.

A *confined aquifer* is an aquifer in which the pressure head at its top boundary (the interface with its overlying confining layer) is positive such that the hydraulic head lies above the elevation head at the interface. The water level in a piezometer tapping the aquifer just beneath the interface will rise above the interface (excluding the capillary fringe above the interface). A potentiometric surface represents such a piezometric water level. Defined in terms of pressure head or hydraulic head, the portion of a confined aquifer near the pumping well can become unconfined when excessive

113

pumping causes the hydraulic head to drop below the elevation head at the top interface. If naturally recharged, a confined aquifer laterally becomes unconfined in the recharge area.

Note that the term *drawdown* used in this book is the difference between the prepumping head and hydraulic head, i.e., $\Delta h[r, z, t] = h[r, z, 0] - h[r, z, t]$.

The prepumping head is assumed to be uniform throughout the entire aquifer. Drawdown is a measurable value in a piezometer in confined or unconfined aquifers. Drawdown does not have to be a physical cone of depression that separates saturated from unsaturated zones. As such, the drawdown at a given distance from the pumping well can vary with depth; it is actually a hydraulic head with reference to the prepumping head.

We use the following basic procedures for equation solving. First, find a solution that satisfies initial and boundary conditions in the Laplace-transform domain, then convert it into a time-domain solution numerically or analytically. Whenever a condition is changed, a new problem is formulated and solved. These conditions will be named and numbered. The number for a changed condition will be affixed with an alphabet to track changes in condition or expression. The unchanged conditions will not be restated for a new problem.

4.1 Theis Well function

The Theis well function is the most widely used formula for estimating transmissivity and storativity. It also serves as an asymptote to various well functions for confined and unconfined aquifers. Many newly developed well functions have been compared with the Theis solution.

Let us review the assumptions for the Theis solution (as derived from the viewpoint of a line source in Section 2.3), then modify these assumptions to obtain other functions of hydraulic responses.

• Assumptions

The assumptions regarding the aquifer are confined, horizontal with uniform thickness, infinite lateral extent, homogeneous, isotropic, and nonleaky with respect to the overlying and the underlying confining aquicludes. Also the aquifer is elastically compressible and source free.

The requirements for the well are vertical, full penetration, full screen (perforation), infinitesimal well radius, and no storage.

The density and viscosity of groundwater are constant and the flow follows Darcy's law.

For mathematical derivation, we assume that the prepumping hydraulic head is constant throughout the aquifer, the well discharge rate is steady, and there is no head loss across the well filter pack and screen. In practice, a pumping test is conducted in the presence of regional groundwater flow. This effect can be significant at small pumping rate, for long-term pumping, or for far-distant observation; but it can be compensated by linearly superposing the regional hydraulic head variation onto the drawdown field. If prepumping head is influenced by pumping at distant wells, detrending of hydraulic head variation based on prepumping observations should be included in the data processing.

• **Solution**

Drawdown $\Delta h[r, t]$ in a fully penetrating well in a confined aquifer is

$$\Delta h = \frac{Q}{4\pi T} W[u], \qquad (4.1)$$

where the dimensionless Theis well function is an exponential integral,

$$W[u] = \int_u^\infty \frac{e^{-\xi}}{\xi} d\xi, \quad u = \frac{r^2 S}{4Tt}, \qquad (4.2)$$

Q is the well discharge rate $[m^3 s^{-1}]$ and is positive for discharge, T is the transmissivity $[m^2 s^{-1}]$, S is the storativity $[-]$, r is the distance $[m]$ from the pumping well, and t is the time $[s]$.

At large time t or small $u\,(\leq 0.01)$, the Theis solution can be approximated by

$$\Delta h = \frac{Q}{4\pi T}\left(-\gamma - \ln\frac{r^2 S}{4Tt}\right), \quad \gamma = 0.5772, \qquad (4.3)$$

which is known as Jacob's solution or semilog method.

4.2   Steady-State Solution

A steady state, as compared to a transient state, implies that the physical state does not change with time. Whether a state is steady or transient depends on whether the observed variables change during the period of observation or experiment. The distinction of the two states is time-scale dependent.

4.2.1   Jacob Solution

Let the drawdown at r and a reference radial distance R be Δh and Δh_R, respectively. Then, according to Jacob's semilog solution, the difference between the two drawdowns at the same time is

$$\Delta h - \Delta h_R = \frac{Q}{2\pi T} \ln \frac{R}{r}. \tag{4.4}$$

Since drawdowns $\Delta h = h_o - h$ and $\Delta h_R = h_o - h_R$, the above relation can be expressed in terms of hydraulic head

$$h - h_R = \frac{Q}{2\pi T} \ln \frac{r}{R}, \tag{4.5}$$

where h_o is the prepumping static hydraulic head, and h and h_R are the hydraulic heads at r and R, respectively.

4.2.2   Thiem's Formula

An alternative to the above solution can be obtained directly from solving the steady radial-dependent hydraulic equation

$$\frac{1}{r} \frac{\partial}{\partial r} \left(r \frac{\partial h}{\partial r} \right) = 0. \tag{4.6}$$

This has the general solution

$$h = A \ln r + C. \tag{4.7}$$

Coefficients A and C are determined from two boundary conditions. First, the total steady flow across any cylindrical surface equals the well discharge rate

$$-2\pi r b k \frac{\partial h}{\partial r} = -Q, \tag{4.8}$$

where b is the thickness of aquifer and k is the hydraulic conductivity [ms^{-1}]. This gives

$$A = \frac{Q}{2\pi T} \qquad (4.9)$$

with the transmissivity $T = kb$.

Application of another boundary condition (i.e., given head h_R at radial distance R) gives

$$C = h_R - A \ln R. \qquad (4.10)$$

The solution for the relative change in hydraulic head is accordingly

$$h - h_R = \frac{Q}{2\pi T} \ln \frac{r}{R}, \qquad (4.11)$$

which is *Dupuit*'s or *Thiem*'s formula.

Note that the above two solutions for head difference between different locations ($h - h_R$) are identical, although one is derived at steady state while the other is valid for $u \leq 0.01$ (large time, but h is not at steady state).

4.3 Full-Penetration Pumping Well, $r_w \neq 0$

4.3.1 Formulation

Consider a pumping well of which the radius r_w is not zero (Figure 2.1). Otherwise, all assumptions regarding the Theis well function are applicable. The governing differential equation for the drawdown is

$$\frac{\partial^2 \Delta h}{\partial r^2} + \frac{1}{r} \frac{\partial \Delta h}{\partial r} = \frac{1}{\kappa} \frac{\partial \Delta h}{\partial t}, \qquad (4.12)$$

where κ is the hydraulic diffusivity ($= T/S$), [$m^2 s^{-1}$].

The first-order time derivative requires one boundary condition, i.e., the initial condition,

$$\text{Condition 1:} \quad \Delta h[r,0] = 0. \qquad (4.13)$$

The second-order spatial derivative requires two boundary conditions, one at infinity and the other at the pumping well,

$$\text{Condition 2:} \quad \Delta h[\infty,t] = 0 \qquad (4.14)$$

and

$$\text{Condition 3:} \quad 2\pi r_{\text{w}} bk \left.\frac{\partial \Delta h}{\partial r}\right|_{r=r_{\text{w}}} = -Q, \qquad (4.15)$$

where b is the aquifer thickness $[m]$ and k is the hydraulic conductivity $[m\ s^{-1}]$. Note that $\partial \Delta h/\partial r = -\partial h/\partial r$. This is a mass balance relation for well discharge. Condition 2 will hereafter be referred to as far-field condition (of zero drawdown) and Condition 3 as discharge or well-face condition.

4.3.2 Laplace-Domain Solution

The procedures for solving equation (4.12) are given below.

☐ Step 1. Convert a partial differential equation into an ordinary one.

Apply the Laplace transform to equation (4.12) to generate the following modified Bessel equation

$$\frac{\partial^2 \overline{\Delta h}}{\partial r^2} + \frac{1}{r}\frac{\partial \overline{\Delta h}}{\partial r} - q^2 \overline{\Delta h} = 0, \quad q^2 = \frac{p}{\kappa}, \qquad (4.16)$$

where p is the Laplace transform parameter $[s^{-1}]$, and the symbol with an overbar designates a transformed variable. In so doing, the initial condition has also been incorporated (see Section 3.1.2).

☐ Step 2. Find the general solution.

The general solution to the modified Bessel equation is

$$\overline{\Delta h}[r,p] = A K_0[qr] + C I_0[qr], \qquad (4.17)$$

where $I_0[qr]$ and $K_0[qr]$ are the zeroth-order modified Bessel functions of the first and second kind, respectively (see Section 3.4.1). A and C are coefficients to be determined.

☐ Step 3. Impose boundary conditions.

The far-field condition of drawdown requires that $C = 0$; otherwise, the behavior that $I_0[\infty] \to \infty$ will violate the boundary condition at infinity.

In the transform domain, the discharge condition at the well face becomes

$$\frac{\partial \overline{\Delta h}}{\partial r} = -\frac{Q}{2\pi r_{\text{w}} T p}, \qquad (4.18)$$

which gives

$$A \left.\frac{\partial K_0[qr]}{\partial r}\right|_{r_{\text{w}}} = -A q K_1[qr_{\text{w}}] = -\frac{Q}{2\pi r_{\text{w}} T p}. \qquad (4.19)$$

Once A is determined, the solution in the transform domain is

$$\overline{\Delta h}[r, p] = \frac{Q}{2\pi T p} \frac{K_0[qr]}{q r_{\mathrm{w}} K_1[q r_{\mathrm{w}}]}.$$ (4.20)

☐ Step 4. Validate the solution.

Rather than proving directly that we have a correct answer, we will show that as $r_{\mathrm{w}} \to 0$, the solution approaches the Theis well function. This is an indirect means to validate the derivations. At small argument of the modified Bessel function,

$$K_\nu[z] \approx \frac{1}{2} \Gamma[\nu] \left(\frac{z}{2}\right)^{-\nu} \quad \text{as } z \to 0,$$ (4.21)

where $\Gamma[\nu]$ is the Gamma function. Accordingly,

$$q r_{\mathrm{w}} K_1[q r_{\mathrm{w}}] \to 1 \qquad \text{as} \qquad q r_{\mathrm{w}} \to 0$$ (4.22)

and equation (4.20) approaches

$$\overline{\Delta h}[r, p] = \frac{Q}{2\pi T} \frac{K_0[qr]}{p},$$ (4.23)

which is the Theis well function in the Laplace domain. See Section 3.2.12 for the inverse Laplace transform of $K_0[qr]$, then use the convolution theorem to obtain the Theis well function in the time domain.

4.3.3   Time-Domain Solution

The inversion of equation (4.20) is available in Carslaw and Jaeger (1959, p. 338). Nevertheless, let us recreate the solution by inverting $K_0[qr]/q K_1[q r_{\mathrm{w}}]$ and convolving the result with 1 to account for the factor of $1/p$.

☐ Step 1. Start the inversion by using the Bromwich integral.
Let

$$\begin{aligned} g[r, t] &= \mathcal{L}^{-1} \left\{ \frac{K_0[qr]}{q K_1[q r_{\mathrm{w}}]} \right\} \\ &= \frac{1}{i2\pi} \int_{c-i\infty}^{c+i\infty} \frac{K_0[qr] e^{pt}}{q K_1[q r_{\mathrm{w}}]} dp, \quad c > 0. \end{aligned}$$ (4.24)

The integrand has a branch point and singular point at $p = 0$ because $q = \sqrt{p/\kappa}$. So make a branch cut along the negative real axis and take the integration path that excludes the singular point (Figure 3.7).

Since the residue is zero at $p = 0$, the contour integration vanishes (Cauchy theorem). Accordingly,

$$g[r,t] = \frac{1}{i2\pi} \int_{c-i\infty}^{c+i\infty} \cdots = -\frac{1}{i2\pi} \left(\int_C^D \cdots + \int_E^F \cdots \right). \qquad (4.25)$$

Integration along the branch cut is pursued as follows.

□ Step 1a. Along the upper path CD of the branch cut, let $p = \kappa x^2 e^{i\pi}$ to obtain

$$-\frac{1}{i2\pi} \int_C^D \cdots = \frac{-1}{i2\pi} \int_\infty^0 \frac{K_0[rxe^{i\pi/2}]2\kappa x e^{i\pi} e^{-x^2\kappa t}}{xe^{i\pi/2} K_1[r_w x e^{i\pi/2}]} dx. \qquad (4.26)$$

Then, use the relation that

$$K_\nu[z] = \frac{-i\pi}{2} e^{-i\nu\pi/2} H_\nu^{(2)}[ze^{-i\pi/2}] \qquad (4.27)$$

(see Section 3.4.1) to yield

$$\begin{aligned}
-\frac{1}{i2\pi} \int_C^D \cdots &= -\frac{\kappa}{\pi} \int_0^\infty \frac{e^{-x^2\kappa t} H_0^{(2)}[xr]}{e^{-i\pi/2} H_1^{(2)}[xr_w]} dx \\
&= \frac{\kappa}{i\pi} \int_0^\infty \frac{e^{-x^2\kappa t} \left(J_0[xr] - iY_0[xr] \right)}{J_1[xr_w] - iY_1[xr_w]} dx. \qquad (4.28)
\end{aligned}$$

□ Step 1b. Similarly, let $p = \kappa x^2 e^{-i\pi}$ along the lower path EF and use the relation that

$$K_\nu[z] = \frac{i\pi}{2} e^{i\nu\pi/2} H_\nu^{(1)}[ze^{i\pi/2}] \qquad (4.29)$$

to obtain

$$-\frac{1}{i2\pi} \int_E^F \cdots = -\frac{\kappa}{i\pi} \int_0^\infty \frac{e^{-x^2\kappa t} \left(J_0[xr] + iY_0[xr] \right)}{J_1[xr_w] + iY_1[xr_w]} dx. \qquad (4.30)$$

□ Step 1c. Add the last two integrals together and rationalize the denominator to yield

$$g[r,t] = -\frac{2\kappa}{\pi} \int_0^\infty \frac{e^{-x^2\kappa t} \left(J_0[xr]Y_1[xr_w] - Y_0[xr]J_1[xr_w] \right)}{J_1^2[xr_w] + Y_1^2[xr_w]} dx. \qquad (4.31)$$

This represents the impulse response of the aquifer to an instantaneous pumping from a well of radius r_w.

□ Step 2. Incorporate the input function.

Convolving the input function with the impulse response yields the final solution for a well discharged at a steady rate of Q (Hantush, 1964),

$$
\begin{aligned}
\Delta h[r,t] &= \frac{Q}{2\pi r_w T} \int_0^t 1 \cdot g[r,\tau] d\tau \\
&= \frac{Q}{4\pi T} \frac{4}{\pi r_w} \int_0^\infty \left(1 - e^{-x^2 \kappa t}\right) \\
&\quad \cdot \frac{Y_0[xr] J_1[xr_w] - J_0[xr] Y_1[xr_w]}{\left(J_1^2[xr_w] + Y_1^2[xr_w]\right) x^2} dx.
\end{aligned} \tag{4.32}
$$

□ Step 3. Obtain the drawdown at the pumping well.

At $r = r_w$, the numerator in equation (4.32) can be simplified by means of the Wronskian

$$
Y_0[xr_w] J_1[xr_w] - J_0[xr_w] Y_1[xr_w] = \frac{2}{\pi x r_w} \tag{4.33}
$$

(see Section 3.4.1). Therefore the drawdown at the pumping well is

$$
\boxed{\Delta h[r_w, t] = \frac{Q}{4\pi T} \frac{8}{\pi^2 r_w^2} \int_0^\infty \frac{1 - \exp\left[-x^2 \kappa t\right]}{\left(J_1^2[xr_w] + Y_1^2[xr_w]\right) x^3} dx} \tag{4.34}
$$

4.4 Hantush's Leaky Aquifer

When the drawdown at an observation well or piezometer does not increase at a rate expected from the Theis function (i.e., drawdown curve flattens out with time), the drawdown may have been influenced by lateral inhomogeneity such as sediments with higher hydraulic conductivity or greater thickness at distance, or by leakage from river, lake, recharge pond or through the confining beds.

Here we deal with one of the plausible causes: leakage through the upper confining layer (Hantush and Jacob, 1955; Hantush, 1964).

4.4.1   Assumptions

With the exception of the leakage of water from an overlying water-table aquifer through the upper confining aquitard into the confined aquifer (Figure 4.1), all assumptions and requirements for the Theis well function are applicable in the present case. Additionally we make the following assumptions for the aquitard and the water-table aquifer.

☐ The confining aquitard is assumed to be incompressible and has no storage. The leakage through the aquitard is vertical, Darcian, and instantaneous.

For a very small conductivity ratio k'/k (k' is the conductivity of aquitard), the tangent law of flow-line refraction at the aquifer-aquitard interface,

$$\frac{\tan \theta'}{\tan \theta} = \frac{k'}{k},\tag{4.35}$$

indicates that the angle θ' between the flowline and the normal to the interface is near zero in the aquitard but the corresponding θ in the aquifer is near 90°. Therefore, the flowline is nearly vertical in the aquiard but is nearly horizontal in the aquifer.

☐ The unconfined aquifer is horizontal and of infinite lateral extent. Its hydraulic head (water table) stays constant during pumping. This condition can be met approximately if the recharge rate to the water-table aquifer is high or the transmissivity is much greater in the water-table aquifer than in the confined aquifer, or if the pumping duration is short.

4.4.2   Mass Balance Equation

Hydraulic response to pumping of a Theis well in a nonleaky confined aquifer (Figure 4.1) is governed by the mass balance equation,

$$\frac{\partial}{\partial r}\left(-2\pi r k b \frac{\partial h}{\partial r}\right)\Delta r = -\left[\pi(r+\Delta r)^2 - \pi r^2\right]bS_s\frac{\partial h}{\partial t}\tag{4.36}$$

in the cylindrical shell between r and $r + \Delta r$, where b is the thickness of the confined aquifer $[m]$, and S_s is the specific storage $[m^{-1}]$. This equation says that the net efflux of water is accompanied by a decline in storage.

As the hydraulic head in the confined aquifer declines, water will flow from the water-table aquifer into the confined aquifer if the confining aquitard is somewhat permeable. The leakage flux (per unit surface area of the interface)

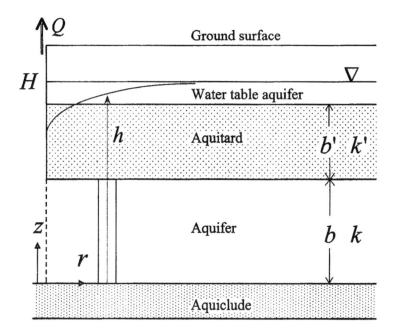

Figure 4.1: Leakage from the confining aquifer is distributed uniformly in the control volume of a cylindrical shell. Piezometric surface represented by h is the hydraulic head at the aquitard-aquifer interface.

into the confined aquifer is $-k'(H - h)/b'$ for the coordinate shown in Figure 4.1, where H is the prepumping head in the water table aquifer.

The vertical flux term is added to the radial efflux term to form the mass balance equation

$$\frac{\partial}{\partial r}\left(-2\pi rkb\frac{\partial h}{\partial r}\right)\Delta r - k'\frac{H - h}{b'}\left[\pi(r + \Delta r)^2 - \pi r^2\right]$$
$$= -\left[\pi(r + \Delta r)^2 - \pi r^2\right]bS_\mathrm{s}\frac{\partial h}{\partial t}. \qquad (4.37)$$

Thus, the vertical flux term is treated here as a volumetric source term that represents mass production per unit volume. As will be pursued later, a more appropriate formulation would treat it as a boundary flux and the drawdown would also depend on z.

Simplifying the mass balance relation and using the definition that drawdown $\Delta h = H - h$ yields

$$\frac{\partial^2 \Delta h}{\partial r^2} + \frac{1}{r}\frac{\partial \Delta h}{\partial r} - \frac{\Delta h}{B^2} = \frac{S}{T}\frac{\partial \Delta h}{\partial t}, \quad B^2 = \frac{Tb'}{k'}, \tag{4.38}$$

where $T = kb$ is the transmissivity $[m^2 s^{-1}]$, $S = S_s b$ is the storativity, and B is the leakage factor $[m]$. Note that $h[r, 0] = H$ initially, otherwise pre-pumping leakage would have occurred.

The initial and boundary conditions are described in Section 4.3.1 except that the discharge condition is now modified to

$$\text{Condition 3a:} \quad \lim_{r \to 0}\left(2\pi r b k \frac{\partial \Delta h}{\partial r}\right) = -Q. \tag{4.39}$$

Equation (4.38) together with the associated boundary conditions describes Hantush's leaky aquifer. A lengthy derivation was made originally by Hantush and Jacob (1955) and Hantush (1956). A simple solution using the Laplace transform method was given by Hantush (1964, p. 309). What follows is a rederivation.

4.4.3   Full-Penetration Pumping Well, $r_w = 0$

● **Laplace-Domain Solution**

Take the Laplace transform of equation (4.38) and incorporate the initial condition to obtain

$$\frac{\partial^2 \overline{\Delta h}}{\partial r^2} + \frac{1}{r}\frac{\partial \overline{\Delta h}}{\partial r} - \sigma^2 \overline{\Delta h} = 0, \tag{4.40}$$

$$\sigma^2 = \frac{1}{B^2} + \frac{p}{\kappa}, \quad \kappa = \frac{T}{S}.$$

Its general solution is

$$\overline{\Delta h}[r, p] = A K_0[\sigma r] + C I_0[\sigma r]. \tag{4.41}$$

Because $I_0[\sigma r] \to \infty$ as $r \to \infty$, we must set coefficient $C = 0$ to satisfy the far-field condition of zero drawdown.

Coefficient A is then determined from the remaining condition of well discharge,

$$\text{Condition 3b:} \quad \lim_{r \to 0}\left(r\frac{d\overline{\Delta h}}{dr}\right) = -\frac{Q}{2\pi T p}. \tag{4.42}$$

Substitution of the following relations

$$K_0[\sigma r] \approx -0.5772 - \ln \frac{\sigma r}{2} \qquad (4.43)$$

and

$$r\frac{\partial K_0[\sigma r]}{\partial r} \to -1 \quad \text{as} \quad r \to 0, \qquad (4.44)$$

(Abramowitz and Stegun, 1964, p. 379) yields $A = Q/2\pi Tp$ and

$$\boxed{\overline{\Delta h}[r, p] = \frac{Q}{2\pi Tp} K_0[\sigma r]}. \qquad (4.45)$$

See Problem 4 for an alternative means of obtaining this result.

• Time-Domain Solution

☐ Step 1. Find the inverse of $K_0[\sigma r]$ according to the parameter-shifting rule and the result of Example 9 in Chapter 3:

$$
\begin{aligned}
\mathcal{L}^{-1}\{K_0[\sigma r]\} &= \mathcal{L}^{-1}\left\{K_0\left[r\sqrt{(p+\kappa/B^2)/\kappa}\right]\right\} \\
&= \exp\left[\frac{-\kappa t}{B^2}\right] \cdot \mathcal{L}^{-1}\left\{K_0\left[r\sqrt{\frac{p}{\kappa}}\right]\right\} \\
&= \exp\left[-\frac{\kappa t}{B^2}\right]\frac{\exp(-r^2/4\kappa t)}{2t}. \qquad (4.46)
\end{aligned}
$$

☐ Step 2. Apply the convolution theorem to obtain the drawdown

$$\Delta h[r, t] = \frac{Q}{2\pi T}\int_0^t 1 \cdot \exp\left[-\frac{\kappa\tau}{B^2}\right]\frac{\exp[-r^2/4\kappa\tau]}{2\tau}d\tau. \qquad (4.47)$$

Let $y = r^2/4\kappa\tau$, then

$$\Delta h[r, t] = \frac{Q}{4\pi T}W[u, r/B], \qquad (4.48)$$

where

$$W[u, \beta] = \int_u^\infty \frac{\exp\left(-y - \beta^2/4y\right)}{y}dy, \quad u = \frac{r^2 S}{4Tt}. \qquad (4.49)$$

The integral $W[u, r/B]$ is Hantush's well function for a leaky aquifer. As $B \to \infty$ when $k' \to 0$ (when leakage is insignificant), the integral becomes the Theis well function. The Hantush well function is illustrated in Figure (4.2) for different values of r/B.

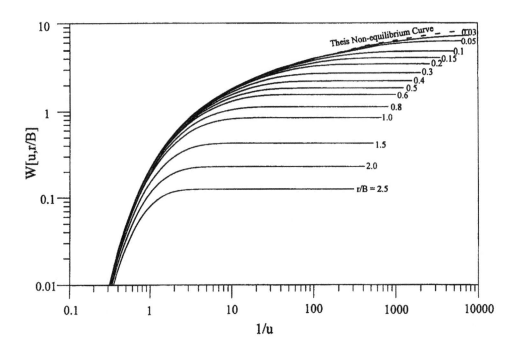

Figure 4.2: Drawdown for Hantush's leaky aquifer, $W[u, r/B]$, in comparision with the Theis well function, $W[u]$.

• A Series Representation

Hantush's leaky well function is not expressible in terms of common functions. It can be integrated numerically according to the procedures described in Section 2.5 or evaluated by the technique of numerical inverse Laplace transform.

A solution that is expressible in terms of the exponential integral

$$E_1[u] = \int_u^\infty \frac{e^{-y}}{y} dy = \int_1^\infty \frac{e^{-uz}}{z} dz \qquad (4.50)$$

is described here (Hunt, 1983). Substitute the series representation of the exponential function

$$\exp\left[-\frac{r^2}{4B^2 y}\right] = \sum_{n=0}^\infty \frac{(-1)^n}{n!} \left(\frac{r^2}{4B^2 y}\right)^n \qquad (4.51)$$

into equation (4.48) and use the method of integration by parts to yield

$$\Delta h[r,t] = \frac{Q}{4\pi T} \sum_{n=0}^{\infty} \frac{(-1)^n}{n!} \left(\frac{r^2}{4B^2} \right)^n \frac{E_{n+1}[u]}{u^{n+1}}, \tag{4.52}$$

where $E_{n+1}[u]$ is obtainable from the recurrence relation

$$\begin{aligned} E_{m+1}[u] &= \int_1^{\infty} \frac{\exp(-uz)}{z^{m+1}} dz \\ &= \frac{1}{m} \left(e^{-u} - uE_m[u] \right), \quad m = 1, 2, \dots \end{aligned} \tag{4.53}$$

- **Remark**

This alternative could be troublesome for small u during numerical computation even though $E_{m+1}[u]$ is well behaved.

It is noted that Bownds and Rizk (1992) have used Green function and series representation to obtain a closed form solution.

4.4.4 Full-Penetration Pumping Well, $r_w \neq 0$

- **Laplace-Domain Solution**

If the radius r_w of the pumping well is not zero, the discharge condition is modified to

$$\text{Condition 3c:} \quad \lim_{r \to r_w} \left(r \frac{d\overline{\Delta h}}{dr} \right) = -\frac{Q}{2\pi T p} \tag{4.54}$$

in the Laplace domain. Use this condition and the relation that

$$\partial K_0[\sigma r]/\partial r = -\sigma K_1[\sigma r] \tag{4.55}$$

to determine coefficient A in equation (4.41).

As a result,

$$\boxed{\overline{\Delta h}[r,p] = \frac{Q}{2\pi T p} \frac{K_0[\sigma r]}{\sigma r_w K_1[\sigma r_w]}}, \tag{4.56}$$

which is reduced to equation (4.20) for a nonleaky aquifer when $B \to \infty$.

- **Time-Domain Solution**

Since

$$\sigma = \sqrt{(p + \kappa/B^2)/\kappa}, \tag{4.57}$$

one can use the rule of parameter shifting to obtain the inverse Laplace transform

$$\mathcal{L}^{-1}\left\{\frac{K_0[\sigma r]}{\sigma K_1[\sigma r_{\rm w}]}\right\} = \exp\left[-\frac{\kappa t}{B^2}\right] \cdot \mathcal{L}^{-1}\left\{\frac{K_0[qr]}{qK_1[qr_{\rm w}]}\right\}, \quad q^2 = \frac{p}{\kappa}. \tag{4.58}$$

The inversion of $\{K_0[qr]/qK_1[qr_{\rm w}]\}$ is given in equation (4.31). Hence, the impulse response is

$$\mathcal{L}^{-1}\left\{\frac{K_0[\sigma r]}{\sigma K_1[\sigma r_{\rm w}]}\right\} \tag{4.59}$$

$$= -\frac{2\kappa}{\pi} \int_0^\infty e^{-(x^2 + 1/B^2)\kappa t} \frac{J_0[xr]Y_1[xr_{\rm w}] - Y_0[xr]J_1[xr_{\rm w}]}{J_1^2[xr_{\rm w}] + Y_1^2[xr_{\rm w}]} dx. \tag{4.60}$$

Now, convolve the steady well discharge Q with the impulse response to give

$$\Delta h[r, t] = \frac{Q}{2\pi r_{\rm w}T} \int_0^t 1 \cdot \mathcal{L}^{-1}\left\{\frac{K_0[\sigma r]}{\sigma K_1[\sigma r_{\rm w}]}\right\} d\tau. \tag{4.61}$$

Simplification leads to

$$\boxed{\Delta h[r, t] = \frac{Q}{4\pi T} \frac{4}{\pi r_{\rm w}} \int_0^\infty \left(1 - \exp[-(x^2 + 1/B^2)\kappa t]\right) F[x] dx}, \tag{4.62}$$

where

$$F[x] = \frac{Y_0[xr]J_1[xr_{\rm w}] - J_0[xr]Y_1[xr_{\rm w}]}{\left(J_1^2[xr_{\rm w}] + Y_1^2[xr_{\rm w}]\right)\left(x^2 + 1/B^2\right)}. \tag{4.63}$$

This result has been given in dimensionless form by Kabala (1993).

At the well face $r = r_{\rm w}$ of the pumping well, using the Wronskian in equation (4.33) yields

$$\Delta h[r_{\rm w}, t] = \frac{Q}{4\pi T} \cdot \frac{8}{\pi^2 r_{\rm w}^2} \int_0^\infty \frac{1 - \exp\left[-(x^2 + 1/B^2)\kappa t\right]}{\left(J_1^2[xr_{\rm w}] + Y_1^2[xr_{\rm w}]\right)\left(x^2 + 1/B^2\right)x} dx. \tag{4.64}$$

4.4.5 Partial Penetration Pumping Well, $r_w = 0$

• Formulation

For a partial-penetrating well in Hantush's leaky aquifer (Hantush, 1964, Example 9), the governing differential equation requires the addition of a z-dependency term to equation (4.38)

$$\frac{\partial^2 \Delta h}{\partial r^2} + \frac{1}{r}\frac{\partial \Delta h}{\partial r} + \frac{\partial^2 \Delta h}{\partial z^2} - \frac{\Delta h}{B^2} = \frac{1}{\kappa}\frac{\partial \Delta h}{\partial t}. \tag{4.65}$$

We need two additional boundary conditions to constrain the second-order z-differential. Let z be upward positive and the well discharge be limited between $z = Z_{\text{bot}}$ and $z = Z_{\text{top}}$ (Figure 4.3). Then, the new conditions are

$$\text{Condition 3d:} \quad 2\pi k r \int_{Z_{\text{bot}}}^{Z_{\text{top}}} \frac{\partial \Delta h}{\partial r} dz = -Q,$$

$$\text{Condition 4:} \quad \left.\frac{\partial \Delta h}{\partial z}\right|_{r,0,t} = 0,$$

$$\text{Condition 5:} \quad \left.\frac{\partial \Delta h}{\partial z}\right|_{r,b,t} = 0, \tag{4.66}$$

where b is the aquifer thickness. Condition 4 will be hereafter referred to as the lower-interface condition and Condition 5 as the upper-interface condition. Both are conditioned at zero flux for Hantush's problems of leaky aquifer.

Assuming that $\partial \Delta h/\partial r$ is independent of z at the pumping well ($r = 0$), the discharge condition becomes

$$\text{Condition 3e} \quad : \quad \lim_{r \to 0} (Z_{\text{top}} - Z_{\text{bot}}) r \frac{\partial \Delta h}{\partial r} = -\frac{Q}{2\pi k}\delta, \tag{4.67}$$

$$\delta = \begin{cases} 1, & Z_{\text{bot}} < z < Z_{\text{top}}; \\ 0, & \text{otherwise.} \end{cases}$$

• Laplace Transform

Applying the Laplace transform to the governing differential equation and incorporating the initial condition yields

$$\frac{\partial^2 \overline{\Delta h}}{\partial r^2} + \frac{1}{r}\frac{\partial \overline{\Delta h}}{\partial r} + \frac{\partial^2 \overline{\Delta h}}{\partial z^2} - \left(\frac{p}{\kappa} + \frac{1}{B^2}\right)\overline{\Delta h} = 0. \tag{4.68}$$

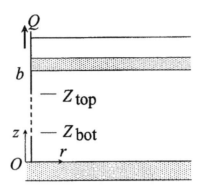

Figure 4.3: Sketch of a partially screened or perforated well in a confined aquifer.

The discharge condition for uniform $\partial \Delta h / \partial r$ along the pumping screen is

$$\text{Condition 3f:} \quad \lim_{r \to 0} r \frac{\partial \overline{\Delta h}}{\partial r} = -\frac{Q}{2\pi k p \left(Z_{\text{top}} - Z_{\text{bot}} \right)} \delta. \qquad (4.69)$$

- **Cosine Series**

Recall the half-range cosine series expansion of a function to obtain

$$\overline{\Delta h} \left[r, z, p \right] = \frac{\overline{\Delta H_0}}{2} + \sum_{n=1}^{\infty} \overline{\Delta H_n} \cos \frac{n \pi z}{b}, \qquad (4.70)$$

$$\overline{\Delta H_n} \left[r, p \right] = \frac{2}{b} \int_0^b \overline{\Delta h} \cos \frac{n \pi z}{b} dz, \quad n = 0, 1, 2, \dots$$

Then, replace $\overline{\Delta h}$ in equation (4.68) by its series expansion and let the coefficient of each $\cos[n \pi z / b]$ term be zero to obtain a modified Bessel equation

$$\frac{\partial^2 \overline{\Delta H_n}}{\partial r^2} + \frac{1}{r} \frac{\partial \overline{\Delta H_n}}{\partial r} - \sigma_n^2 \overline{\Delta H_n} = 0, \qquad (4.71)$$

where

$$\sigma_n^2 = \frac{p}{\kappa} + \left(\frac{n \pi}{b} \right)^2 + \frac{1}{B^2}, \quad n = 0, 1, 2, 3, \dots \qquad (4.72)$$

- **Laplace-Domain Solution**

The solution to the modified Bessel equation is

$$\overline{\Delta H_n}[r, p] = A_n K_0 \left[\sigma_n r \right]. \qquad (4.73)$$

Coefficient A_n is determined from discharge condition, rewritten as

$$\text{Condition 3g:} \qquad \lim_{r \to 0} r \frac{\partial}{\partial r} \left(\frac{\overline{\Delta H_0}}{2} + \sum_{n=1}^{\infty} \overline{\Delta H_n} \cos \frac{n\pi z}{b} \right)$$

$$= - \frac{Q}{2\pi k p \left(Z_{\text{top}} - Z_{\text{bot}} \right)} \delta, \qquad (4.74)$$

in the transformed domain. Noting again that

$$\lim \left[r \partial K_0[\sigma_n r] / \partial r \right] = - \lim \left[\sigma_n r K_1[\sigma_n r] \right] = -1 \quad \text{as} \quad r \to 0, \qquad (4.75)$$

we obtain

$$\frac{A_0}{2} + \sum_{n=1}^{\infty} A_n \cos \frac{n\pi z}{b} = \frac{Q}{2\pi k \left(Z_{\text{top}} - Z_{\text{bot}} \right) p} \delta. \qquad (4.76)$$

Multiplying both sides by $\cos[m\pi z/b]$ (m is a positive integer) and integrating z from 0 to b on the left-hand side and from Z_{bot} to Z_{top} on the right-hand side yields

$$A_n = \frac{Q}{\pi T \left(Z_{\text{top}} - Z_{\text{bot}} \right)} \frac{b}{n\pi p} \left(\sin \frac{n\pi Z_{\text{top}}}{b} - \sin \frac{n\pi Z_{\text{bot}}}{b} \right). \qquad (4.77)$$

Putting A_n into the inverse cosine transform results in

$$\boxed{\overline{\Delta h}[r, z, p] = \frac{Q}{2\pi T p} \left\{ K_0 \left[\sigma_0 r \right] + \sum_{n=1}^{\infty} A_n' K_0 \left[\sigma_n r \right] \cos \frac{n\pi z}{b} \right\}}, \qquad (4.78)$$

where

$$A_n' = \frac{2b}{n\pi \left(Z_{\text{top}} - Z_{\text{bot}} \right)} \left(\sin \frac{n\pi Z_{\text{top}}}{b} - \sin \frac{n\pi Z_{\text{bot}}}{b} \right). \qquad (4.79)$$

Note that an alternative solution can be obtained by taking the cosine transform of the well-face condition (Problem 7).

• **Time-Domain Solution**

Each of the series terms in equation (4.78) contains a factor of

$$\frac{Q}{2\pi T p} K_0 \left[\sigma_n r \right] = \frac{Q}{2\pi T p} K_0 \left[r \sqrt{\frac{p}{\kappa} + \frac{1}{B_n^2}} \right], \qquad (4.80)$$

$$B_n = \frac{B}{\sqrt{1 + B^2 (n\pi/b)^2}},$$

which resembles the solution for a full-penetration well in Hantush's leaky aquifer. Thus for each n, the time-domain solution is

$$\Delta h_n[r, z, t] = \frac{Q}{4\pi T} W[u, r/B_n],$$ (4.81)

$$W[u, \beta] = \int_u^\infty \frac{\exp\left(-y - \beta^2/4y\right)}{y} dy.$$

The final time-domain solution is accordingly

$$\Delta h[r, z, t] = \frac{Q}{4\pi T} \left\{ W\left[u, \frac{r}{B}\right] + \sum_{n=1}^\infty A_n' W\left[u, \frac{r}{B_n}\right] \cos \frac{n\pi z}{b} \right\}.$$ (4.82)

- **Remark**

  □ If the conductivity is anisotropic $[k_r, k_z]$, then σ_n and B_n in the above relations are replaced, respectively, by

$$\sigma_n^2 = \frac{p}{\kappa} + k_D \left(\frac{n\pi}{b}\right)^2 + \frac{1}{B_r^2}, \quad k_D = \frac{k_z}{k_r},$$

$$B_n = \frac{B_r}{\sqrt{1 + B_r^2 k_D (n\pi/b)^2}}, \quad B_r^2 = \frac{T_r b'}{k'}, \quad T_r = k_r b. \quad (4.83)$$

  □ The assumption of uniform $\partial \Delta h/\partial r$ in the screened section and zero $\partial \Delta h/\partial r$ in the cased section implies a discontinuity of radial flow at the junction of the screened and cased sections along the well face. See problem 1 for some consequences of the assumption.

  □ Illustrated in Figure 4.4 are type curves for average drawdown over the screen depth interval from 0 to 0.1 due to pumping through the screen depths between 0 and 0.25, in which $\beta = k_D r^2/b^2$. The type curves are similar in shape to those for a full-penetration pumping well, Figure 4.2. However, the two sets of type curves are distinctive in reference to the Theis curve.

4.4.6   Partial Penetration Pumping Well, $r_w \neq 0$

Let the radius of a partial penetration well be r_w. Then, the discharge condition in equation (4.67) is modified to

$$\text{Condition 3h:} \quad r\frac{\partial \Delta h}{\partial r}\bigg|_{r=r_w} = -\frac{Q}{2\pi k \left(Z_{\text{top}} - Z_{\text{bot}}\right)} \delta, \quad (4.84)$$

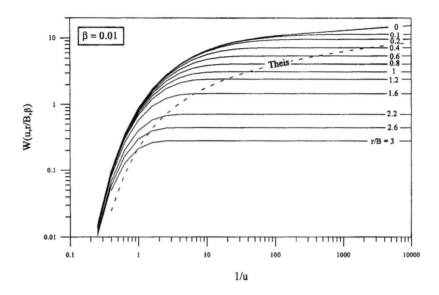

Figure 4.4: Average drawdown in an observation well screened between depths 0 and 0.1 (elevations between 0.9 and 1.0) due to pumping through the screen interval between 0 and 0.25 in depth.

but other conditions stay the same. In the Laplace-cosine transform domain, the discharge condition becomes

Condition 3i:
$$\frac{\partial \overline{\Delta H_0}}{2} + \sum_{n=1}^{\infty} \overline{\Delta H_n} \cos \frac{n\pi z}{b}$$

$$= -\frac{A_0}{2} \sigma_0 K_1[\sigma_0 r_{\mathrm{w}}] - \sum_{n=1}^{\infty} A_n \sigma_n K_1[\sigma_n r_{\mathrm{w}}] \overline{\Delta H_n} \cos \frac{n\pi z}{b}$$

$$= -\frac{Q}{2\pi k p \left(Z_{\mathrm{top}} - Z_{\mathrm{bot}}\right) r_{\mathrm{w}}} \delta. \qquad (4.85)$$

Multiplying both sides by $\cos(n\pi z/b)$ and integrating z from 0 to b for the left-hand side and from Z_{bot} to Z_{top} for the right-hand side yields

$$A_n = \frac{Q}{\pi T p \left(Z_{\mathrm{top}} - Z_{\mathrm{bot}}\right) \sigma_n r_{\mathrm{w}} K_1\left[\sigma_n r_{\mathrm{w}}\right]}$$
$$\cdot \frac{b}{n\pi} \left(\sin \frac{n\pi Z_{\mathrm{top}}}{b} - \sin \frac{n\pi Z_{\mathrm{bot}}}{b} \right), \qquad (4.86)$$

which also gives

$$A_0 = \frac{Q}{\pi T p q r_{\rm w} K_1 [q r_{\rm w}]}.$$

(4.87)

The solution in the Laplace domain is

$$\overline{\Delta h}[r, z, p] = \frac{\overline{\Delta H_0}}{2} + \sum_{n=0}^{\infty} \overline{\Delta H_n} \cos \frac{n\pi z}{b}$$

(4.88)

or

$$\boxed{\overline{\Delta h}[r, z, p] = \frac{A_0}{2} K_0 [\sigma_0 r] + \sum_{n=1}^{\infty} A_n K_0 [\sigma_n r] \cos \frac{n\pi z}{b}}.$$

(4.89)

By analogy to equation (4.56) for a full-penetration pumping well, the time-domain solution can be obtained by inverting $K_0[\sigma_n r]/p\sigma_n K_1 [\sigma_n r_{\rm w}]$ but it is omitted here because it can be easily evaluated by means of the Stehfest algorithm.

4.5   Leakage as Boundary Flux

As stated earlier, the solutions for the Hantush leaky aquifer were obtained by lumping the leakage as volumetric water production, although the solutions are intended for leakage from an overlying unconfined aquifer through an intervening aquitard. This section treats the leakage term as a boundary flux (Figure 4.5). Hence, the solution is z-dependent, even for a full-penetration pumping well.

The governing differential equation for the drawdown is

$$\frac{\partial^2 \Delta h}{\partial r^2} + \frac{1}{r} \frac{\partial \Delta h}{\partial r} + k_{\rm D} \frac{\partial^2 \Delta h}{\partial z^2} = \frac{1}{\alpha_{\rm s}} \frac{\partial \Delta h}{\partial t},$$

$$k_{\rm D} = k_z/k_r, \quad \alpha_{\rm s} = k_r/S_{\rm s},$$

(4.90)

where k_r and k_z are the horizontal and vertical hydraulic conductivities, $[m\, s^{-1}]$, respectively, and $\alpha_{\rm s}$ is the hydraulic diffusivity, $[m^2 s^{-1}]$ (a ratio of horizontal transmissivity to storativity, $\alpha_{\rm s} = T_{\rm r}/S$, $T_{\rm r} = k_r b$).

We will use two techniques to solve the boundary flux problem, one by means of the Hankel transform (Section 4.5.1) and the other by the method of separation of variables (Section 4.5.2).

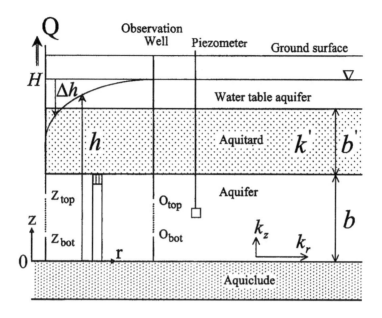

Figure 4.5: Leakage from the water table aquifer through the confining aquitard into the confined aquifer is treated as a boundary flux, as symbolized by the hatched spot atop a shell control volume. Hydraulic head in the water table aquifer stays at H during pumping. Δh is the piezometric drawdown.

4.5.1 Full-Penetration Pumping Well, $r_w = 0$

The boundary conditions are

Condition 1a: $\quad \Delta h[r, z, 0] = 0,$

Condition 2a: $\quad \Delta h[\infty, z, t] = 0,$

Condition 3j: $\quad 2\pi k_r r \int_0^b \dfrac{\partial \Delta h}{\partial r} dz = -Q \quad$ as $r \to 0,$

Condition 5b: $\quad k_z \dfrac{\partial \Delta h}{\partial z}\bigg|_{r,b,t} = -k' \dfrac{\Delta h}{b'}.$ $\qquad$ (4.91)

See Figure (4.5) for notations. The hydraulic conductivity in the confined aquifer is anisotropic. The equation and the conditions are equivalent to those of Neuman's water-table model (Chapter 5) except Condition 5b, which does not contain a time-derivative term. Note that the upper-interface condition (Condition 5b) is now a nonzero flux condition and the lower-interface condition stays at zero flux.

- **Laplace-Hankel Domain Solution**

Applying the Laplace transform to equation (4.90) yields

$$\frac{\partial^2 \overline{\Delta h}}{\partial r^2} + \frac{1}{r}\frac{\partial \overline{\Delta h}}{\partial r} + k_D \frac{\partial^2 \overline{\Delta h}}{\partial z^2} = q^2 \overline{\Delta h}, \quad q^2 = p/\alpha_s. \qquad (4.92)$$

Then, employ the Hankel transform and incorporate the discharge condition (using the result in Problem 3) to obtain

$$\frac{\partial^2 \overline{\Delta h}^*}{\partial z^2} - \eta^2 \overline{\Delta h}^* = -\frac{Q}{2\pi T_r p k_D}, \quad \eta^2 = \left(a^2 + q^2\right)/k_D, \qquad (4.93)$$

where a is the Hankel transform parameters. An overbar designates a Laplace-transformed variable and an asterisk designates a Hankel-transformed variable.

The general solution to the above ordinary differential equation is

$$\overline{\Delta h}^*[a, z, p] = C \cosh[\eta z] + \frac{Q}{2\pi T_r p(a^2 + q^2)}. \qquad (4.94)$$

This solution satisfies all conditions but the upper-interface condition. Thus, impose Condition 5b in the Laplace-Hankel domain to determine coefficient C and obtain

$$\overline{\Delta h}^*[a, z, p] = \frac{Q}{2\pi T_r p(a^2 + q^2)} \left(1 - \frac{\cosh[\eta z]}{c\eta \sinh[\eta b] + \cosh[\eta b]}\right), \qquad (4.95)$$

where $c = k_z b'/k'$.

- **Laplace-Domain Solution**

The inverse Hankel transform of the first term is the Theis solution in the Laplace domain

$$\begin{aligned}
\overline{\Delta h}^I[r, z, p] &= \frac{Q}{2\pi T_r p} \int_0^\infty \frac{a J_0[ar]}{a^2 + q^2} da \\
&= \frac{Q}{2\pi T_r} \frac{K_0[qr]}{p}
\end{aligned} \qquad (4.96)$$

(see Example 2H in Chapter 3). The inverse Hankel transform of the second term is

$$\overline{\Delta h}^{II}[r,z,p] = \frac{-Q}{2\pi T_r p} \int_0^\infty \frac{a \cosh[\eta z] J_0[ar]}{(a^2+q^2)(c\eta \sinh[\eta b] + \cosh[\eta b])} da. \qquad (4.97)$$

Its result is available from Example 5H in Chapter 3.

Combining the two yields

$$\boxed{\overline{\Delta h}[r,z,p] = \frac{Q}{\pi T p} \sum_{n=0}^\infty \lambda_n K_0[\xi_n r] \cos[\zeta_n z]}, \qquad (4.98)$$

where

$$\begin{aligned} \lambda_n &= \frac{c}{(c+b+bc^2\zeta_n^2)\cos[\zeta_n b]} \\ &= \frac{\sin[\zeta_n b]}{\zeta_n b + 0.5\sin[2\zeta_n b]} = \frac{(-1)^n \sin[\theta_n]}{\zeta_n b + 0.5\sin[2\theta_n]}, \end{aligned} \qquad (4.99)$$

$$\xi_n = \sqrt{\zeta_n^2 k_D + p/\alpha_s}, \qquad (4.100)$$

and $\zeta_n b$ is the root of

$$\tan[\zeta_n b] = \frac{1}{c\zeta_n}. \qquad (4.101)$$

For ease of root-finding, the root $\zeta_n b$ can be redefined in terms of θ_n (Figure 3.12)

$$\begin{aligned} \zeta_n b &= n\pi + \theta_n, \quad 0 \leq \theta_n \leq \pi/2, \\ &\theta_{n+1} < \theta_n, \quad n = 0, 1, 2, \ldots \end{aligned} \qquad (4.102)$$

The average drawdown is

$$\overline{\Delta h}^{ave}[r,p] = \frac{1}{b} \int_0^b \overline{\Delta h}[r,z,p] dz \qquad (4.103)$$

or

$$\boxed{\overline{\Delta h}^{ave}[r,p] = \frac{Q}{\pi T p} \sum_{n=0}^\infty \lambda_n^{ave} K_0[\xi_n r]}, \qquad (4.104)$$

where

$$\begin{aligned} \lambda_n^{ave} &= \frac{1}{(b+c+bc^2\zeta_n^2)b\zeta_n^2} \\ &= \frac{\sin^2[\theta_n]}{(\zeta_n b + 0.5\sin[2\theta_n])\zeta_n b}. \end{aligned} \qquad (4.105)$$

• Time-Domain Solution

The inversion of $\overline{\Delta h}[r, z, p]$ to the time-domain solution can be obtained numerically through the Stehfest inversion scheme

$$\boxed{\Delta h[r, z, t] = \frac{\ln 2}{t} \sum_{j=1}^{M} v_j \overline{\Delta h}[r, z, p_j]} , \qquad (4.106)$$

where $p_j = (j \ln 2)/t$, M is an optimal number for maximum accuracy, and v_j is the weighting coefficient (see Section 3.3 for details).

• Remark

Defining a vertical leakage factor

$$B_z^2 = \frac{T_z b'}{k'} \qquad (4.107)$$

leads to

$$\lambda_n = \frac{1}{\left(1 + b^2/B_z^2 + \zeta_n^2 b^2 B_z^2/b^2\right) \cos(\zeta_n b)} \qquad (4.108)$$

and

$$\lambda_n^{\mathrm{ave}} = \frac{1}{\left(\zeta_n^2 B_z^2 + 1\right) \zeta_n^2 B_z^2 + \zeta_n^2 b^2}, \qquad (4.109)$$

where $T_z = k_z b$ is the vertical transmissivity. If $k_z = k_r$, the vertical leakage factor B_z is the same as Hantush's horizontal leakage factor B_r.

Note that in terms of the leakage factor, the root equation is

$$\tan[\zeta_n b] = \frac{b^2}{B_z^2} \frac{1}{\zeta_n b}. \qquad (4.110)$$

Introducing another variable

$$\beta = k_D r^2/b^2 \qquad (4.111)$$

and recalling $p = j \ln 2/t$ for the Stehfest scheme of the inverse Laplace transform results in

$$(\xi_n r)_j = \sqrt{\zeta_n^2 b^2 \beta + p r^2/\alpha_s} = \sqrt{\zeta_n^2 b^2 \beta + 4 j u \ln 2}. \qquad (4.112)$$

Thus, the normalized well function is described completely by three parameters: b/B_z, β and u. Since

$$\frac{b^2}{B_z^2} = \frac{k'}{b'} \Big/ \frac{k_z}{b} \tag{4.113}$$

is a ratio of vertical Darcy's velocities in the aquitard and the confined aquifer if the same hydraulic head difference is imposed respectively across the two zones, b/B_z is thus a measure of the significance of leakage. As the conductivity of the aquitard $k' \to 0$, the leakage factor $B_z \to \infty$ and all well functions approach the Theis well function (see Problem 6).

• **Early-Time Functional Behavior**

The root equation indicates that roots $\zeta_n b$ are independent of the Laplace transform parameter p and hence time, but $\xi_n r$ is time dependent. If $p \to \infty$ as time $t \to 0$, then

$$\begin{aligned} K_0[\xi_n r] &= K_0\left[r\sqrt{\zeta_n^2 k_D + p/\alpha_s}\right] \\ &\to K_0\left[qr\right]. \end{aligned} \tag{4.114}$$

At this limit, the average drawdown becomes

$$\overline{\Delta h}^{\text{ave}}[r, p] \to \frac{Q}{2\pi T} \frac{K_0[qr]}{p} 2 \sum_{n=0}^{\infty} \lambda_n^{\text{ave}}. \tag{4.115}$$

Resembling equation (4.23), this $\overline{\Delta h}^{\text{ave}}[r, z, p]$ behaves like a Theis well function but not exactly. Since λ_n is independent of time, the early-time curve shifts by $2\sum_{n=0}^{\infty} \lambda_n^{\text{ave}}$ relative to Theis curve unless $\sum_{n=0}^{\infty} \lambda_n^{\text{ave}}$ happens to be 0.5 (as remarked later, it is 0.5 for special cases only).

• **Late-Time Functional Behavior**

If $p \to 0$ as time $t \to \infty$, then

$$\begin{aligned} K_0[\xi_n r] &= K_0\left[r\sqrt{\zeta_n^2 k_D + p/\alpha_s}\right] \\ &\to K_0\left[\zeta_n r\sqrt{k_D}\right], \end{aligned} \tag{4.116}$$

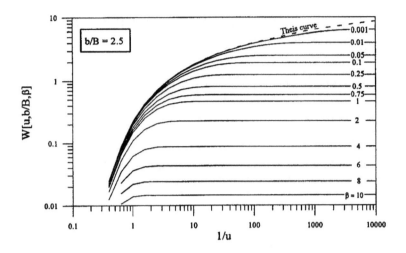

Figure 4.6: Type curves for a full-penetration pumping well in a confined aquifer with boundary leakage.

which is independent of time. At this limit, the average drawdown becomes

$$\overline{\Delta h}^{\mathrm{ave}}[r, p] \rightarrow \frac{Q}{\pi T_r p} \sum_{n=0}^{\infty} \lambda_n^{\mathrm{ave}} K_0 \left[\zeta_n r \sqrt{k_{\mathrm{D}}} \right]. \tag{4.117}$$

Since the inverse Laplace transform of $1/p$ is one, the average drawdown is a time-independent constant

$$\Delta h^{\mathrm{ave}}[r, t] \rightarrow \frac{Q}{4\pi T_r} \sum_{n=0}^{\infty} 4\lambda_n^{\mathrm{ave}} K_0 \left[\zeta_n b \sqrt{\beta} \right]. \tag{4.118}$$

● **Remark**

Figure 4.6 depicts the type curves for a full-penetration pumping well in a confined aquifer with boundary leakage

$$W[u, b/B_z, \beta] = \Delta h^{\mathrm{ave}}[r, t] / (Q/4\pi T). \tag{4.119}$$

The predicted early- and late-time behaviors are clearly demonstrated. It also shows that there is less drawdown at higher β values, in accordance with equation (4.118). An increase in $\beta = k_z r^2 / k_r b^2$ may mean an increase in k_z/k_r to allow more downward leakage flow, or it may mean that drawdown is less at greater r if other parameters are held constant.

If $\beta \to 0$ (implying $k_z \to 0$), then $\xi_n r \to qr$ according to equations (4.100) and (4.112). Consequently, the leakage diminishes and the well function should behave like a Theis function. At this limit, $B_z \to 0$, $\zeta_n b \to (n + 1/2)\pi$ and

$$\sum_{n=0}^{\infty} \lambda_n^{\text{ave}} \to \sum_{n=0}^{\infty} \frac{1}{(n + 1/2)^2 \pi^2} = \frac{1}{2}. \tag{4.120}$$

Therefore equation (4.104) becomes

$$\overline{\Delta h}^{\text{ave}}[r, p] = \frac{Q}{\pi T_r p} K_0[qr] \sum_{n=0}^{\infty} \lambda_n^{\text{ave}} = \frac{Q}{2\pi T_r p} K_0[qr]. \tag{4.121}$$

4.5.2 Partial-Penetration Pumping Well, $r_w = 0$

For a pumping well screened between Z_{bot} and Z_{top}, the governing equation and the boundary conditions are the same as described for a full-penetration pumping well, equation (4.91), except that the well-face condition is now modified to

$$\text{Condition 3k:} \quad \lim_{r \to 0} 2\pi k_r r \int_{Z_{\text{bot}}}^{Z_{\text{top}}} \frac{\partial \Delta h}{\partial r} dz = -Q\delta. \tag{4.122}$$

In Section 4.4.5 a cosine-series representation of $\overline{\Delta h}$ was used for solving Hantush's partial-penetration problem. Here we will solve the partial-penetration problem formally by using the technique of separation of variables.

• **Separation of Variables**

Assume that drawdown $\overline{\Delta h}$ in the Laplace domain is separable in r and z, i.e.,

$$\overline{\Delta h}[r, z] = \Phi[r]\Psi[z]. \tag{4.123}$$

Substituting it into equation (4.92) and dividing the result by $\Phi\Psi$ yields

$$\frac{1}{\Phi} \frac{\partial^2 \Phi}{\partial r^2} + \frac{1}{r\Phi} \frac{\partial \Phi}{\partial r} + \frac{k_D}{\Psi} \frac{\partial^2 \Psi}{\partial z^2} - q^2 = 0. \tag{4.124}$$

Because each term of the differentials depends on either r or z and q is independent of either, Φ and Ψ are completely separated.

Let the separation constant be $-k_D\zeta^2$. Then equation (4.124) is split into two equations:

$$\frac{k_D}{\Psi}\frac{\partial^2\Psi}{\partial z^2} = -k_D\zeta^2, \tag{4.125}$$

$$\frac{\partial^2\Phi}{\partial r^2} + \frac{1}{r}\frac{\partial\Phi}{\partial r} - \xi^2\Phi = 0, \quad \xi^2 = \zeta^2 k_D + q^2. \tag{4.126}$$

The solution to the Ψ equation is a linear combination of $\sin[\zeta z]$ and $\cos[\zeta z]$ terms. The no-flow condition at the lower interface (Condition 4 at $z = 0$) necessitates that the $\sin[\zeta z]$ term be zero. The Φ equation (a modified Bessel equation) has a solution which is composed of $I_0[\xi r]$ and $K_0[\xi r]$ terms. The far-field condition of zero drawdown (Condition 2) precludes $I_0[\xi r]$ term from consideration. Therefore the solution is of the form

$$\overline{\Delta h} = \Phi\Psi = AK_0[\xi r]\cos[\zeta z], \tag{4.127}$$

where A and ζ are to be determined from the remaining two boundary conditions.

Imposition of the upper-interface condition (Condition 5b) at $z = b$ yields

$$k_z\zeta\sin[\zeta b] = (k'/b')\cos[\zeta b]. \tag{4.128}$$

Since the trigonometric functions are multivalued, this constraint for ζ is identical to the root equation (4.101). The multivalued ζ_n is thus the root of

$$\tan[\zeta_n b] = \frac{1}{c\zeta_n}, \quad n = 0, 1, 2, \ldots \tag{4.129}$$

All linear combination of $\Psi_n = \cos[\zeta_n z]$ is a solution. To each n, there corresponds a $\Phi_n = K_0[\xi_n r]$. Therefore

$$\boxed{\overline{\Delta h}[r, z] = \sum_{n=0}^{\infty} A_n K_0[\xi_n r]\cos[\zeta_n z]}. \tag{4.130}$$

• **Discharge Condition**

Assume that the mass flux through the screened interval is independent of z. The discharge condition becomes

$$\text{Condition 3l:} \qquad \lim_{r\to 0}\left(r\frac{\partial\overline{\Delta h}}{\partial r}\right) = -\Omega\delta,$$

$$\Omega = \frac{Q}{2\pi k_r p \left(Z_{\text{top}} - Z_{\text{bot}}\right)}, \qquad (4.131)$$

in the Laplace domain. Imposing this discharge condition and recalling that

$$\lim_{r\to 0}\left(r\frac{\partial K_0[\xi_n r]}{\partial r}\right) = -\lim_{r\to 0}\xi_n r K_1[\xi_n r] = -1 \qquad (4.132)$$

yields

$$\sum_{n=0}^{\infty} A_n \cos[\zeta_n z] = \Omega\delta. \qquad (4.133)$$

Multiply both sides of the above relation by $\cos[\zeta_m z]$ (m is an integer) and integrate the result from 0 to b to yield

$$A_n = \frac{2\Omega\left(\sin[\zeta_n Z_{\text{top}}] - \sin[\zeta_n Z_{\text{bot}}]\right)}{\zeta_n b + 0.5\sin[2\zeta_n b]}. \qquad (4.134)$$

This coefficient A_n together with equation (4.130) in various functional forms will be used for other functions of aquifer responses to pumping. For this particular problem, we use

$$A_n = \frac{Qb}{\pi T p\left(Z_{\text{top}} - Z_{\text{bot}}\right)}\frac{\sin[\zeta_n Z_{\text{top}}] - \sin[\zeta_n Z_{\text{bot}}]}{\left(\zeta_n b + 0.5\sin[2\zeta_n b]\right)}. \qquad (4.135)$$

Conforming with the notations used in the last section, the final solution for a partial-penetration pumping well with infinitesimal diameter is

$$\boxed{\overline{\Delta h}[r, z, p] = \frac{Q}{\pi T_r p}\sum_{n=0}^{\infty}\lambda_n^{\text{part}} K_0[\xi_n r]\cos[\zeta_n z]}, \qquad (4.136)$$

where

$$\lambda_n^{\text{part}} = \frac{b}{Z_{\text{top}} - Z_{\text{bot}}}\frac{\sin[\zeta_n Z_{\text{top}}] - \sin[\zeta_n Z_{\text{bot}}]}{\sin[\zeta_n b]}\lambda_n. \qquad (4.137)$$

• **Remark**

□ It is obvious that equation (4.136) becomes the drawdown for a full-penetration pumping well as $Z_{\text{top}} \to b$ and $Z_{\text{bot}} \to 0$.

□ If the aquitard is not leaky, i.e., $k' = 0$, then $1/B_z \to 0$ and $\tan[\zeta_n b] \to 0$. As a result, $\lambda_0^{\text{part}} = 1/2$, $\lambda_n^{\text{part}} = A_n'$, and equation (4.136) is in agreement with equation (4.78) for $1/B_r \to 0$.

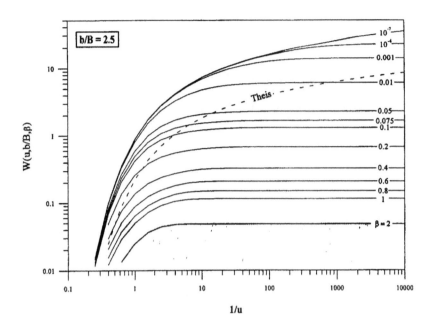

Figure 4.7: Type curves for average drawdown in observation wells screened at elevations between 0.9 and 1.0 due to pumping through screen elevations between 0.75 and 1.0 – a boundary flux model.

☐ It is noted that equation (4.136) is identical with the hydraulic response to pumping of a partial penetration well in unconfined aquifers (Section 5.2.2) except that the c here is independent of p and time t.

☐ Illustrated in Figure 4.7 are type curves for average drawdown in an observation well screened between elevations 0.9 and 1.0 above the bottom surface of the confined aquifer (depths 0.0 and 0.1 below the top surface of the confined aquifer) due to pumping through screen elevations from 0.75 to 1.0 in a confined aquifer with boundary flux.

4.5.3   Partial-Penetration Pumping Well, $r_w \neq 0$

For a well of finite diameter, the drawdown is governed by the same differential equation (4.90) for a well of infinitesimal diameter. All boundary conditions remain the same except that the discharge condition is now modified to

$$\text{Condition 3m:} \qquad \lim_{r \to r_w} \left(r \frac{\partial \overline{\Delta h}}{\partial r} \right) = -\Omega \delta. \tag{4.138}$$

Substituting equation (4.130) into this modified condition, recalling that $\partial K_0[\xi_n r]/\partial r = -\xi_n r K_1[\xi_n r]$, and using the orthogonal property of cosine functions yields the drawdown due to a partial-penetration pumping well of finite diameter

$$\overline{\Delta h}[r, z, p] = \sum_{n=0}^{\infty} A_n K_0[\xi_n r] \cos[\zeta_n z], \tag{4.139}$$

where

$$A_n = \frac{2\Omega \left(\sin[\zeta_n Z_{\text{top}}] - \sin[\zeta_n Z_{\text{bot}}] \right)}{\xi_n r_w K_1[\xi_n r_w] \left(\zeta_n b + 0.5 \sin[2\zeta_n b] \right)}. \tag{4.140}$$

Expressing in terms of λ yields

$$\boxed{\overline{\Delta h}[r, z, p] = \frac{Q}{\pi T_r p} \sum_{n=0}^{\infty} \lambda_n'^{\text{part}} K_0[\xi_n r] \cos[\zeta_n z]}, \tag{4.141}$$

where

$$\lambda_n'^{\text{part}} = \frac{\lambda_n^{\text{part}}}{\xi_n r_w K_1[\xi_n r_w]}. \tag{4.142}$$

As $r_w \to 0$ and $\xi_n r_w K_1[\xi_n r_w] \to 1$, then the solution is reduced to that of a pumping well with infinitesimal diameter.

4.6 Slug Test

Conventional testing for aquifer response requires long-term pumping. In testing contaminated aquifers, however, the disposal of extracted water is costly. Hence, it is desirable to pump as little as possible. A slug test is appealing because it uses a small volume of water. The test is initiated by either injection or extraction of a given volume of water instantaneously. In recent years, compressed air (or partial vacuum) has been used to displace a slug of water in the well in lieu of extraction (or injection) of water. A slug of solid instead of liquid can also be used to displace the water; the removal of the solid slug can be used for conducting a recovery slug test. In many cases, the slug displacement of water is so small that the measured aquifer parameters reflect only the properties near the test well. This drawback can be compensated to some extent by conducting more slug tests at many sites.

Slug tests are applicable only to aquifers with relatively low hydraulic conductivity; a high dissipation rate of a slug disturbance yields poor data resolution. Because the test results reflect aquifer properties near the test wells, one should be aware that irregularities in well construction may affect the test results. See Butler's (1998) text for a full discussion.

4.6.1　Full-Penetration Well

• Formulation

Consider a well of finite diameter with full penetration and screening in a confined aquifer. At time zero, a slug of water is injected into the well. What is the hydraulic head variation? Cooper, Bredehoeft, and Papadopulos (1967) formulated the problem as follows.

The governing differential equation is

$$\frac{\partial^2 \Delta h}{\partial r^2} + \frac{1}{r}\frac{\partial \Delta h}{\partial r} = \frac{S}{T}\frac{\partial \Delta h}{\partial t}, \quad r \geq r_s, \tag{4.143}$$

where Δh is the hydraulic head (inverse drawdown) above the static head, S is the storativity $[-]$, T is the transmissivity $[m^2 s^{-1}]$, r_s is the radius of the screened section of the well, and r_c is the radius of the cased well (Figure 4.8).

The initial and boundary conditions are

$$\text{Condition 3n:} \quad -2\pi T r_s \frac{\partial \Delta h}{\partial r}\bigg|_{r=r_s} = -\pi r_c^2 \frac{\partial H}{\partial t},$$

$$\text{Condition 6:} \quad H[0] = H_o = \frac{V}{\pi r_c^2},$$

$$\text{Condition 7:} \quad \Delta h[r_s, t] = H[t], \tag{4.144}$$

where V is the volume of injected water, and $H[t]$ is the hydraulic head inside the well. Conditions 4 and 5 are not applicable for the slug-test problems. Condition 6 will be hereafter referred as the initial slug-height condition and Condition 7 as the skin-effect condition although skin effect is not considered in this section (it is used in Section 5.4).

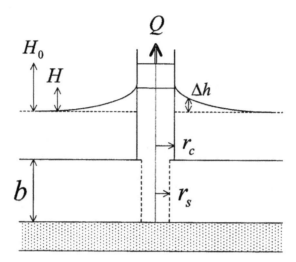

Figure 4.8: Sketch for a slug test. H_0 is the initial height of the slug. H is the height of water in the well. r_c = cased well radius. r_s = screened well radius.

• Solution

Take the Laplace transform of the governing differential equation and implement the initial condition to yield a modified Bessel equation

$$\frac{\partial^2 \overline{\Delta h}}{\partial r^2} + \frac{1}{r}\frac{\partial \overline{\Delta h}}{\partial r} - \frac{S}{T}p\Delta \bar{h} = 0. \tag{4.145}$$

In the Laplace domain, the discharge condition becomes

Condition 3o: $\qquad \left.\frac{\partial \overline{\Delta h}}{\partial r}\right|_{r=r_s} = \frac{r_c^2}{2r_sT}\left(p\overline{H} - H_o\right)$

$$= \frac{r_c^2}{2r_sT}\left(p\overline{\Delta h}[r_s, p] - H_o\right) \tag{4.146}$$

in which Conditions 6 and 7 have also been implemented.

Equation (4.145) has a solution consisting of two independent terms: $I_0[qr]$ and $K_0[qr]$ with $q = \sqrt{p/\kappa}$. The far-field condition is satisfied if the $I_0[qr]$ term is set to zero. So, we obtain

$$\boxed{\overline{\Delta h}[r, p] = AK_0[qr]}, \tag{4.147}$$

where coefficient A is determined from the discharge condition (Condition 3o)

$$A[p] = \frac{r_s S H_o}{Tq\left(r_s q K_0[qr_s] + 2\alpha K_1[qr_s]\right)}, \quad \alpha = \frac{r_s^2 S}{r_c^2}. \tag{4.148}$$

The inverse for $\Delta h[r, t]$ has been obtained by Cooper, Bredehoeft, and Papadopulos (1967) via a relation for heat conduction in Carslaw and Jaeger (1959). The time-domain solution requires numerical integration of Bessel functions $J_0[u]$ and $Y_0[u]$ and is difficult to implement. For the convenience of programming, we recommend using the Stehfest inverse-Laplace method to obtain $h[r, t]$. Nevertheless, see Problems 5 for the derivation of the real-time solution. See also Section 5.4 for slug test in the water-table aquifer.

4.7   Problems, Keys, and Suggested Readings

• **Problems**

1. Consider Hantush's solution (equation 4.78) for a partial-penetration pumping well. (a) Find the drawdown at $z = b/2$. (b) Find the drawdown averaged over the aquifer thickness at radial distance r. (c) Discuss the implications of the results in comparison to the Theis well function for different screening intervals in the pumping well.

2. Use the following series representation

$$J_\nu[x] = \left(\frac{x}{2}\right)^\nu \sum_{k=0}^\infty \frac{(-x^2/4)^k}{k!\Gamma[\nu + k + 1]} \tag{4.149}$$

to show that

$$x\frac{\partial J_1[x]}{\partial x} = xJ_0[x] - J_1[x], \tag{4.150}$$

where $\Gamma[z]$ is the Gamma function (see equation 10.51).

3. Show that

$$\mathcal{H}\left\{\frac{1}{r}\frac{\partial}{\partial r}\left(r\frac{\partial \overline{\Delta h}}{\partial r}\right)\right\} = -\lim_{r\to 0}\left[r\frac{\partial \overline{\Delta h}}{\partial r}\right] - a^2\overline{\Delta h}^*. \tag{4.151}$$

4. Use the Hankel transform technique to solve the equation

$$\frac{\partial^2 \Delta h}{\partial r^2} + \frac{1}{r}\frac{\partial \Delta h}{\partial r} - \frac{\Delta h}{B^2} = \frac{S}{T}\frac{\partial \Delta h}{\partial t} \tag{4.152}$$

for Hantush's leaky confined aquifer. Show that in the Laplace domain, the drawdown is

$$\overline{\Delta h}[r,p] = \frac{Q}{2\pi T}\frac{K_0[r\sqrt{1/B^2 + p/\kappa}]}{p}, \quad \kappa = T/S. \qquad (4.153)$$

5. Find the inverse Laplace transform (Carslaw and Jaeger, 1959, p. 342)

$$\frac{K_0[qr]}{q\left(\beta q K_0[\beta q] + \alpha K_1[\beta q]\right)}, \quad q^2 = \frac{p}{\kappa} \qquad (4.154)$$

where α and β are constants unrelated to the notations used elsewhere.

6. Show that equation (4.104) behaves like a Theis well function as the conductivity of the aquitard k' approaches zero.

7. Use a cosine series to represent $r\partial\Delta h/\partial r$ at the pumping well and derive a well function for partial-penetration pumping wells in confined aquifers with boundary leakage.

8. Prove equation (4.134).

• **Keys**

Key 2

The solution for this problem is straightforward. First, recall that the Gamma function of an integer argument is a factorial

$$\Gamma[k+1] = k\Gamma[k] = k!. \qquad (4.155)$$

Then, proceed as follows

$$
\begin{aligned}
x\frac{\partial J_1[x]}{\partial x} &= \frac{x}{2}\sum_{k=0}^{\infty}\frac{(2k+1)(-x^2/4)^k}{k!\Gamma[k+2]} \\
&= x\sum_{k=0}^{\infty}\frac{(k+1-1/2)(-x^2/4)^k}{k!(k+1)\Gamma[k+1]} \\
&= x\sum_{k=0}^{\infty}\frac{(-x^2/4)^k}{k!\Gamma[k+1]} - \frac{x}{2}\sum_{k=0}^{\infty}\frac{(-x^2/4)^k}{k!(k+1)\Gamma[k+1]} \\
&= xJ_0[x] - J_1[x].
\end{aligned} \qquad (4.156)
$$

Key 3

Take the Hankel transform, then perform integration by parts twice to get

$$\int_0^\infty J_0[ar]\frac{\partial}{\partial r}\left(r\frac{\partial\overline{\Delta h}}{\partial r}\right)dr$$

$$= J_0[ar]\left(r\frac{\partial\overline{\Delta h}}{\partial r}\right)\Big|_0^\infty - \int_0^\infty r\frac{\partial\overline{\Delta h}}{\partial r}\frac{\partial J_0[ar]}{\partial r}dr \qquad (4.157)$$

$$= -\left(r\frac{\partial\overline{\Delta h}}{\partial r}\right)_{r=0} + a\int_0^\infty rJ_1[ar]\frac{\partial\overline{\Delta h}}{\partial r}dr$$

$$= -\left(r\frac{\partial\overline{\Delta h}}{\partial r}\right)_{r=0} + arJ_1[ar]\overline{\Delta h}\Big|_0^\infty - a\int_0^\infty \overline{\Delta h}\frac{\partial}{\partial r}\left(rJ_1[ar]\right)dr$$

$$= -\left(r\frac{\partial\overline{\Delta h}}{\partial r}\right)_{r=0} - a\int_0^\infty \overline{\Delta h}\left(J_1[ar] + r\frac{\partial}{\partial r}J_1[ar]\right)dr. \qquad (4.158)$$

Now use the relation in Problem 2 to yield

$$\int_0^\infty J_0[ar]\frac{\partial}{\partial r}\left(r\frac{\partial\overline{\Delta h}}{\partial r}\right)dr = -\left(r\frac{\partial\overline{\Delta h}}{\partial r}\right)_{r=0} - a^2\int_0^\infty r\overline{\Delta h}J_0[ar]dr$$

$$= -\left(r\frac{\partial\overline{\Delta h}}{\partial r}\right)_{r=0} - a^2\overline{\Delta h}^*. \qquad (4.159)$$

Key 4

Rewrite the governing differential equation

$$\frac{1}{r}\frac{\partial}{\partial r}\left(r\frac{\partial\Delta h}{\partial r}\right) - \frac{\Delta h}{B^2} = \frac{1}{\kappa}\frac{\partial\Delta h}{\partial t}. \qquad (4.160)$$

Take the Laplace transform to obtain

$$\frac{1}{r}\frac{\partial}{\partial r}\left(r\frac{\partial\overline{\Delta h}}{\partial r}\right) - \left(\frac{1}{B^2} + \frac{p}{\kappa}\right)\overline{\Delta h} = 0. \qquad (4.161)$$

Applying the Hankel transform to the above equation yields

$$\int_0^\infty \frac{\partial}{\partial r}\left(r\frac{\partial\overline{\Delta h}}{\partial r}\right)J_0[ar]dr - \left(\frac{1}{B^2} + \frac{p}{\kappa}\right)\overline{\Delta h}^* = 0. \qquad (4.162)$$

According to Problem 3, this relation becomes

$$-\lim_{r \to 0}\left[r\frac{\partial \overline{\Delta h}}{\partial r}\right] - \left(a^2 + \frac{1}{B^2} + \frac{p}{\kappa}\right)\overline{\Delta h}^* = 0, \qquad (4.163)$$

and

$$\overline{\Delta h}^*[a, p] = \frac{Q}{2\pi T p}\frac{1}{(a^2 + \sigma^2)}, \qquad \sigma^2 = \frac{1}{B^2} + \frac{p}{\kappa}. \qquad (4.164)$$

Use the result of Example 2H in Chapter 3 to get the inverse Hankel transform

$$\overline{\Delta h}[r, p] = \frac{Q}{2\pi T}\frac{K_0[\sigma r]}{p}. \qquad (4.165)$$

This is the alternative way to obtain the drawdown for a full-penetration well in a leaky confined aquifer (equation 4.45).

Key 5

Let the desired inverse Laplace transform be

$$A = \frac{1}{i2\pi}\int_{b-i\infty}^{b+i\infty}\frac{\exp[pt]K_0[qr]}{q\left(\beta q K_0[\beta q] + \alpha K_1[\beta q]\right)}dp. \qquad (4.166)$$

The integrand has one branch point at $p = 0$ but no singular point (Figure 3.5).

Draw a branch cut along the negative real axis. According to the Cauchy theorem, a closed contour integration vanishes if the path encloses no singularity. Hence, A is equal to the integration along the upper and lower paths along the branch cut

$$A = -\frac{1}{i2\pi}\underbrace{\int \cdots}_{\text{upper}} - \frac{1}{i2\pi}\underbrace{\int \cdots}_{\text{lower}}. \qquad (4.167)$$

Let $p = e^{i\pi}\kappa x^2/\beta^2$ along the upper integration path, then

$$\text{upper} = \frac{-1}{i2\pi}\int_{\infty}^{0}\frac{\exp[-\kappa x^2 t/\beta^2]K_0[e^{i\pi/2}xr/\beta]2\kappa x e^{i\pi}/\beta^2}{e^{i\pi/2}x/\beta\left(xe^{i\pi/2}K_0[xe^{i\pi/2}] + \alpha K_1[xe^{i\pi/2}]\right)}dx. \qquad (4.168)$$

Recalling

$$\begin{aligned}
K_\nu[z] &= -\frac{i\pi}{2}e^{-i\nu\pi/2}H_\nu^{(2)}\left[ze^{-i\pi/2}\right] \quad \text{and} \\
H_\nu^{(2)}[z] &= J_\nu[z] - iY_\nu[z],
\end{aligned} \qquad (4.169)$$

one gets

$$
\begin{aligned}
\text{upper} &= \frac{-\kappa}{i\pi\beta} \int_0^\infty \frac{\exp[-\kappa x^2 t/\beta^2] H_0^{(2)}[xr/\beta]}{x H_0^{(2)}[x] + \alpha H_1^{(2)}[x]} dx \\
&= \frac{-\kappa}{i\pi\beta} \int_0^\infty \frac{\exp[-\kappa x^2 t/\beta^2] \left(J_0[xr/\beta] - iY_0[xr/\beta]\right)}{x\left(J_0[x] - iY_0[x]\right) - \alpha\left(J_1[x] - iY_1[x]\right)} dx \\
&= \frac{-\kappa}{\pi\beta} \int_0^\infty \frac{\exp[-\kappa x^2 t/\beta^2]\left(iJ_0[xr/\beta] + Y_0[xr/\beta]\right)}{\left(xJ_0[x] - \alpha J_1[x]\right) - i\left(xY_0[x] - \alpha Y_1[x]\right)} dx. \quad (4.170)
\end{aligned}
$$

The integration along the lower path is the complex conjugate of the upper-path integration

$$
\text{lower} = \frac{-\kappa}{\pi\beta} \int_0^\infty \frac{\exp[-\kappa x^2 t/\beta^2]\left(-iJ_0[xr/\beta] + Y_0[xr/\beta]\right)}{\left(xJ_0[x] - \alpha J_1[x]\right) + i\left(xY_0[x] - \alpha Y_1[x]\right)} dx. \quad (4.171)
$$

Add the "upper" and "lower" solutions together and rationalize their denominators to finalize

$$
\begin{aligned}
A &= \frac{2\kappa}{\pi\beta} \int_0^\infty \frac{dx}{F[x]} \exp\left[-\frac{\kappa x^2 t}{\beta^2}\right] \left\{ J_0\left[rx/\beta\right]\left(xY[x] - \alpha Y_1[x]\right)\right. \\
&\quad \left. - Y_0\left[rx/\beta\right]\left(xJ_0[x] - \alpha J_1[x]\right)\right\}, \quad\quad (4.172)
\end{aligned}
$$

where

$$
F[x] = \left(xJ_0[x] - \alpha J_1[x]\right)^2 + \left(xY_0[x] - \alpha Y_1[x]\right)^2. \quad (4.173)
$$

Key 6

As $k' \to 0$,

$$
B_z = \frac{T_z b'}{k'} \to \infty \quad \text{and} \quad \frac{b}{B_z} \to 0. \quad (4.174)
$$

As a result, the root equation (4.110) yields

$$
\zeta_n b \to n\pi, \quad \theta_n \to 0 \quad \text{for all } n \quad (4.175)
$$

and

$$
\begin{aligned}
\lambda_n &\to 0 \quad \text{for} \quad n \neq 0 \\
\lambda_0 &\to \frac{1}{2}.
\end{aligned} \quad (4.176)
$$

Equation (4.104) therefore approaches the Theis well function.

Key 7

Applying the result of Problem 3 to the governing differential equation in the Hankel-Laplace domain yields

$$\frac{\partial^2 \overline{\Delta h}^*}{\partial z^2} - \eta^2 \overline{\Delta h}^* = \frac{1}{k_D} \lim_{r \to 0} r \frac{\partial \overline{\Delta h}}{\partial r}. \tag{4.177}$$

We can represent $\lim_{r \to 0} r \frac{\partial \overline{\Delta h}}{\partial r}$ by the cosine series

$$\lim_{r \to 0} r \frac{\partial \overline{\Delta h}}{\partial r} = \frac{F_0}{2} + \sum_{m=1}^{\infty} F_m \cos \left[\frac{m \pi z}{b} \right], \tag{4.178}$$

where

$$\begin{aligned}
F_m &= \frac{2}{b} \int_0^b \lim_{r \to 0} r \frac{\partial \overline{\Delta h}}{\partial r} \cos \left[\frac{m \pi z}{b} \right] dz \\
&= \frac{-2}{b} \int_0^b \Omega \delta \cos \left[\frac{m \pi z}{b} \right] dz = \frac{-Q}{\pi T_r p} F_m', \tag{4.179}
\end{aligned}$$

$$F_m' = \frac{b}{Z_{\text{top}} - Z_{\text{bot}}} \frac{1}{m\pi} \left\{ \sin \left[\frac{m \pi Z_{\text{top}}}{b} \right] - \sin \left[\frac{m \pi Z_{\text{bot}}}{b} \right] \right\} \tag{4.180}$$

in which $F_0' = 1$.

Substitute equations (4.178) into (4.177) to get the general solution

$$\overline{\Delta h}^*[a, z, p] = C \cosh[\eta z] - \frac{1}{k_D} \left\{ \frac{F_0}{2\eta^2} + \sum_{m=1}^{\infty} \frac{F_m \cos[m\pi z/b]}{\eta^2 + (m\pi/b)^2} \right\}. \tag{4.181}$$

Then, impose the upper-interface condition (Condition 5b) to determine

$$C = \frac{1}{c\eta \sinh[\eta b] + \cosh[\eta b]} \frac{1}{k_D} \left\{ \frac{F_0}{2\eta^2} + \sum_{m=1}^{\infty} \frac{(-1)^m F_m}{\eta^2 + (m\pi/b)^2} \right\}. \tag{4.182}$$

Explicitly we get

$$\begin{aligned}
\overline{\Delta h}^*[a, z, p] &= \frac{-F_0}{2\eta^2 k_D} \left\{ 1 - \frac{\cosh[\eta z]}{c\eta \sinh[\eta b] + \cosh[\eta b]} \right\} \\
&\quad - \frac{1}{k_D} \sum_{m=1}^{\infty} \frac{F_m}{\eta^2 + (m\pi/b)^2} \left\{ \cos \left[\frac{m\pi z}{b} \right] \right. \\
&\quad \left. - \frac{(-1)^m \cosh[\eta z]}{c\eta \sinh[\eta b] + \cosh[\eta b]} \right\}. \tag{4.183}
\end{aligned}$$

Now we examine the individual terms. The term bearing F_0,

$$\overline{\Delta h_1^*}[a, z, p] = \frac{Q}{2\pi T_r p} \frac{1}{a^2 + q^2} \left\{ 1 - \frac{\cosh[\eta z]}{c\eta \sinh[\eta b] + \cosh[\eta b]} \right\}, \qquad (4.184)$$

is equivalent to the drawdown due to pumping in a full-penetration well. The term bearing $F_m \cos[m\pi z/b]$ is

$$\overline{\Delta h_2^*}[a, z, p] = \frac{Q}{\pi T_r p} \sum_{m=1}^{\infty} \frac{F_m'}{a^2 + q_m^2} \cos\left[\frac{m\pi z}{b}\right], \qquad (4.185)$$

$$q_m^2 = q^2 + (m\pi/b)^2 k_D.$$

Its inverse Hankel transform (according to Example 2H in Chapter 3) is

$$\overline{\Delta h_2}[r, z, p] = \frac{Q}{\pi T_r p} \sum_{m=1}^{\infty} F_m' K_0[q_m r] \cos\left[\frac{m\pi z}{b}\right]. \qquad (4.186)$$

The term bearing $F_m \cosh[\eta z]$ is

$$\overline{\Delta h_3^*}[a, z, p] = \frac{Q}{\pi T_r p} \sum_{m=1}^{\infty} \frac{(-1)^m F_m'}{a^2 + q_m^2} \frac{\cosh[\eta z]}{c\eta \sinh[\eta b] + \cosh[\eta b]}. \qquad (4.187)$$

Following the procedures for Example 5H in Chapter 3, we get

$$\overline{\Delta h_3}[r, z, p] = \frac{Q}{\pi T_r p} \sum_{m=1}^{\infty} (-1)^m F_m'$$

$$\cdot \frac{i}{\pi} \int_{-i\infty}^{i\infty} \frac{\xi K_0[\xi r] \cos[\zeta z]}{\left(\xi^2 - q_m^2\right)\left(c\zeta \sin[\zeta b] - \cos[\zeta b]\right)} d\xi. \qquad (4.188)$$

The contribution from the residue at $\xi = q_m$ is

$$\overline{\Delta h_{3a}}[r, z, p] = \frac{-Q}{\pi T_r p} \sum_{m=1}^{\infty} F_m' K_0[q_m r] \cos\left[\frac{m\pi z}{b}\right], \qquad (4.189)$$

which nullifies the $\overline{\Delta h_2}[r, z, p]$ term. The contribution by the residues at the roots of

$$c\zeta \sin[\zeta b] - \cos[\zeta b] = 0 \qquad (4.190)$$

is

$$\overline{\Delta h_{3b}}[r,z,p] = \frac{Q}{\pi T_r p} \sum_{m=1}^{\infty} (-1)^m F'_m \cdot (-i2\pi)\frac{i}{\pi}$$

$$\cdot \sum_{n=0}^{\infty} \frac{\xi K_0[\xi r] \cos[\zeta z]}{(\xi^2 - q_m^2) \frac{\partial}{\partial \zeta} (c\zeta \sin[\zeta b] - \cos[\zeta b]) \frac{\partial \zeta}{\partial \xi}} \bigg|_{\xi=\xi_n} \quad (4.191)$$

$$= \frac{Q}{\pi T_r p} \sum_{n=0}^{\infty} \sum_{m=1}^{\infty} (-1)^m F'_m \frac{\xi_n^2 - q^2}{\xi_n^2 - q_m^2}$$

$$\cdot \lambda_n K_0[\xi_n r] \cos[\zeta_n z]. \quad (4.192)$$

Finally the drawdown in the Laplace domain is

$$\overline{\Delta h}[r,z,p] = \overline{\Delta h_1}[r,z,p] + \overline{\Delta h_{3b}}[r,z,p]$$

$$= \frac{Q}{\pi T_r p} \sum_{n=0}^{\infty} \sum_{m=0}^{\infty} (-1)^m F'_m \frac{\xi_n^2 - q^2}{\xi_n^2 - q_m^2}$$

$$\cdot \lambda_n K_0[\xi_n r] \cos[\zeta_n z]. \quad (4.193)$$

Note that in this final expression, the m-summation starts from $m = 0$. Thus the result obtained by a cosine series representation of the discharge at the well face is more complicated in functional form than that obtained with the technique of separation of variables, equation (4.136).

Key 8

Multiplying both sides of

$$\sum_{n=0}^{\infty} A_n \cos[\zeta_n z] = \Omega \delta \quad (4.194)$$

by $\cos[\zeta_m z]$ and integrating the result with respect to z from 0 to b yields

$$\frac{1}{2} \sum_{n=0}^{\infty} A_n \left(\frac{\sin[(\zeta_n + \zeta_m)b]}{\zeta_n + \zeta_m} + \frac{\sin[(\zeta_n - \zeta_m)b]}{\zeta_n - \zeta_m} \right)$$

$$= \frac{\Omega \left(\sin[\zeta_m Z_{\text{top}}] - \sin[\zeta_m Z_{\text{bot}}] \right)}{\zeta_m}. \quad (4.195)$$

For $n \neq m$, the left-hand side can be simplified

$$\frac{1}{2} \sum_{n=0}^{\infty} A_n \left\{ \frac{\zeta_n \left(\sin[(\zeta_n + \zeta_m)b + \sin[(\zeta_n - \zeta_m)b]\right)}{\zeta_n^2 - \zeta_m^2} \right.$$
$$\left. - \frac{\zeta_m \left(\sin[(\zeta_n + \zeta_m)b - \sin[(\zeta_n - \zeta_m)b]\right)}{\zeta_n^2 - \zeta_m^2} \right\}$$
$$= \frac{2\zeta_n \sin[\zeta_n b] \cos[\zeta_m b]}{\zeta_n^2 - \zeta_m^2} \left(1 - \frac{\zeta_m}{\zeta_n} \frac{\tan[\zeta_n b]}{\tan[\zeta_m b]} \right) = 0 \qquad (4.196)$$

in which the root equation is used.

For $m = n$,

$$\frac{1}{2} A_n \left(\frac{\sin[2\zeta_n b]}{2\zeta_n} + b \right) = \frac{\Omega \left(\sin[\zeta_m Z_{\text{top}}] - \sin[\zeta_m Z_{\text{bot}}]\right)}{\zeta_m}. \qquad (4.197)$$

This concludes the derivation.

• Suggested Readings

Several textbooks are available for further reading on well hydraulics or flow toward pumping wells. For practitioners, we recommend Walton's (1987) *Groundwater Pumping Tests* and Driscoll's (1986) *Groundwater and Wells*. Some programs are available in Dawson and Istok's (1991) *Aquifer Testing*. For more practice on applied mathematics, read de Marsily's (1986) *Quantitative Hydrogeology*.

For readings on groundwater flow other than its relation to wells, we suggest Strack's (1989) *Groundwater Mechanics* (which is an excellent source for analytic solutions) and Zaradny's (1993) *Groundwater Flow in Saturated and Unsaturated Soil*. In the order of increasing sophistication, the following are recommended: Fetter's (1994) *Applied Hydrogeology*, Freeze and Cherry's (1979) *Groundwater*, and Domenico and Schwartz's (1998) *Physical and Chemical Hydrogeology*. For slug tests, see *The Design, Performance, and Analysis of Slug Tests* by Butler (1998).

4.8   Notations

Symbol	Definition	SI Unit	Dimension
a	Hankel transform parameter	m^{-1}	L^{-1}
b, b'	Aquifer, aquitard thickness	m	L
B, B_r, B_z	Leakage factor, Eq. (4.38), (4.107)	m	L
c	$c = k_z b'/k'$, eq.(4.95)	m	L
$E_1[u]$	Exponential integral	-	-
h, H	Hydraulic head	m	L
I_0, I_1	Modified Bessel function, first kind	-	-
J_0, J_1	Bessel function, first kind	-	-
k, k'	Hydraulic conductivity	$m\,s^{-1}$	LT^{-1}
k_D	k_r/k_z	-	-
k_r, k_z	Hydraulic conductivity	$m\,s^{-1}$	LT^{-1}
K_0, K_1	Modified Bessel function, second kind	-	-
p	Laplace parameter	s^{-1}	T^{-1}
q	$q^2 = p/\alpha_s$, or p/κ	m^{-1}	L^{-1}
q_m	$(m\pi/b)^2 k_D + q^2$	m^{-1}	L^{-1}
Q	Well discharge rate	$m^3 s^{-1}$	$L^3 T^{-1}$
r, R	Radial distance	m	L
r_c, r_s	Casing radius, screen radius	m	L
r_w	Well radius	m	L
S	Storativity, $S = S_s b$	-	-
S_s	Specific storage	m^{-1}	L^{-1}
t	Time	s	T
t_s	Dimensionless time, $t_s = Tt/r^2 S$	-	-
T	Transmissivity, $T = kb$	$m^2 s^{-1}$	$L^2 T^{-1}$
T_r, T_z	$T_r = k_r b, \quad T_z = k_z b$	$m^2 s^{-1}$	$L^2 T^{-1}$
u	$u = r^2 S/4Tt = 1/4t_s$	-	-
W	Well function	-	-
Y_0, Y_1	Bessel function, second kind	-	-
$Z_{\text{bot}}, Z_{\text{top}}$	z coordinate to bottom or top screen	m	L

Symbol	Definition	SI Unit	Dimension
$\mathcal{H}$	Hankel transform operator		
$\mathcal{L}$	Laplace transform operator		
α	$r_s^2 S/r_c^2$, eq. (4.148)	-	-
α_s	Hydraulic diffusivity, $= k_r/S_s$	$m^2 s^{-1}$	$L^2 T$
β	$k_D r^2/b^2$, eq. (4.111)	-	-
γ	Euler constant	-	-
Γ	Gamma function	-	-
Δh	Drawdown	m	L
$\overline{\Delta h}$	Laplace transform of Δh	$m\, s$	LT
$\overline{\Delta h^*}$	Hankel transform of $\overline{\Delta h}$	$m^3 s$	$L^3 T$
$\overline{\Delta H}$	Fourier transform of $\overline{\Delta h}$	$m^2 s$	$L^2 T$
$\zeta_n b$	Root of $\tan[\zeta b] = 1/c\zeta$	-	-
η	$\eta^2 = (a^2 + q^2)/k_D$	m^{-1}	L^{-1}
θ_n	$\theta_n = \zeta_n b - n\pi$	-	-
κ	Hydraulic diffusivity, $\kappa = T/S$	$m^2 s^{-1}$	$L^2 T^{-1}$
$\lambda_n, \lambda_n^{ave}$	Eq. (4.99), (4.105)	-	-
λ_n^{part}	Eq. (4.137)	-	-
$\lambda_n^{\prime part}$	Eq. (4.142)	-	-
ξ_n	$\xi_n^2 = \zeta_n^2 k_D + p/\alpha_s$	-	-
σ	$\sigma^2 = q^2 + 1/B^2$	m^{-1}	L^{-1}
σ_n	$\sigma_n^2 = q^2 + 1/B^2 + (n\pi/b)^2$	m^{-1}	L^{-1}
Φ	Function of r		
Ψ	Function of z		
ω_n	$\omega_n^2 = q^2 + (n\pi/b)^2 k_D$	m^{-1}	L^{-1}
Ω	Discharge-related parameter	m	L

Chapter 5

DRAWDOWN IN UNCONFINED AQUIFERS

By definition, an unconfined aquifer has a water table. The water table is not an interface separating saturated from unsaturated zones. It is the surface where the pressure is atmospheric (i.e., gauge pressure is zero). Depending on the height of capillary fringe, the saturated-unsaturated interface can be a few centimeters to a couple of meters above the water table. The water in the capillary-saturated pores is not directly extractable by pumping; however, the capillary-saturated zone can fall with the lowering of the water table. Generally, water in an unconfined aquifer is extracted through gravity drainage of pores while water in a confined aquifer is extracted through aquifer compression and pore water expansion such that the confined pores remain saturated during pumping. Extensive pumping can lead to a locally unconfined condition around the pumped well in a confined aquifer.

When a water-table aquifer is pumped, the drawdown behaves like a Theis well function at early time. The drawdown-time curve flattens at intermediate time as if there is recharge or leakage through the underlying or overlying aquitard, then resumes a Theis-like well-function behavior at late time. Neuman's solutions (1972, 1974, and 1975) can generally explain such observational behavior. Although some questions have been raised about whether Neuman's theory on delayed gravity response can adequately predict field observations (Nwankwor et al., 1992; Endres et al., 1997), his solutions remain the most viable analytical well functions for estimating aquifer properties for unconfined aquifers.

Various attempts have been made to enhance the efficiency of computing

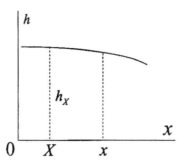

Figure 5.1: Two-dimensional, depth-independent steady flow.

Neuman's solutions (Moench, 1993, 1996). Here we will develop alternative solutions to Neuman's type of unconfined aquifers. Then, the method is extended for a well of finite diameter, well storage, and well head loss.

5.1   Dupuit-Forchheimer Theory

5.1.1   Lateral Flow

The Dupuit-Forchheimer theory of steady flow in an unconfined aquifer bounded by a free surface (water table) is based on two assumptions: 1) The flowlines are horizontal and the equipotential surfaces are vertical. 2) The hydraulic gradient equals the slope of the water table.

Let the flow be in the x direction. Then, the two assumptions essentially say that flow Q per unit area of the section perpendicular to the flow direction is constant. Mathematically

$$-kh\frac{\partial h}{\partial x} = Q, \qquad (5.1)$$

where $h[x]$ is the elevation (hydraulic head) of the free surface above the reference elevation at the bottom of the unconfined aquifer (Figure 5.1).

This nonlinear differential equation can be easily solved by rewriting it as

$$-\frac{k}{2}\frac{\partial h^2}{\partial x} = Q \qquad (5.2)$$

and treating h^2 as a new dependent variable. Given the water table h_X at

X, Dupuit-Forchheimer formula is

$$h^2 - h_{\mathrm{X}}^2 = -\frac{2Q}{k}(x - X).$$ (5.3)

This formula is not related to any well functions but serves as a prelude to the following well function.

5.1.2 Steady Radial Flow

If a well fully penetrates an unconfined aquifer and if Dupuit's flow is applicable, the mass balance relation is

$$-k2\pi r h \frac{\partial h}{\partial r} = -Q.$$ (5.4)

If the hydraulic head is h_{R} at $r = R$, then Thiem's formula at constant pumping rate Q is

$$h^2 - h_{\mathrm{R}}^2 = \frac{Q}{\pi k} \ln \frac{r}{R}.$$ (5.5)

Take the saturated thickness b as the average of the two hydraulic head values, i.e., $b \approx (h + h_{\mathrm{R}})/2$. The above relation becomes

$$h - h_{\mathrm{R}} = \frac{Q}{\pi k(h + h_{\mathrm{R}})} \ln \frac{r}{R} \approx \frac{Q}{2\pi T} \ln \frac{r}{R},$$ (5.6)

where $T = bk$ is the transmissivity. This relation is identical to the solution for steady flow in a confined aquifer (Section 4.2).

5.2 Pumping Wells of Infinitesimal Diameter

This section presents two alternative solutions to Neuman's well functions. One employs the Laplace-Hankel transform to solve the governing differential equation, and the other uses the technique of separation of variables. Both repeat the methods used for boundary-flux problems in leaky confined aquifers and utilize the Stehfest algorithm to obtain time-domain solutions.

5.2.1   Full-Penetration Pumping Well, $r_w = 0$

• **Assumptions**

☐ All assumptions and requirements about the aquifer and the pumping well for the Theis well function are applicable in an unconfined aquifer except that, in addition to the instantaneous elastic response of aquifer compression and water expansion, gravity drainage of the pores plays a major role in the aquifer response.

☐ Two additional assumptions are essential for Neuman's solution: 1) Unsaturated water flow is negligible. 2) The drawdown is small compared to the saturated thickness so that the boundary conditions can be imposed on the prepumping, horizontal water table, instead of a migrating free surface.

☐ The effect of gravity drainage is delayed with respect to the instantaneous elastic behavior (Theis well response). Its effect is greatest when the drawdown cone is moderate in size and diminishes when the drawdown cone flattens out after a well is pumped for a long time.

• **Formulation**

According to Neuman (1972), drawdown $\Delta h[r, z, t]$ in an unconfined aquifer (Figure 5.2) is governed by

$$\frac{\partial^2 \Delta h}{\partial r^2} + \frac{1}{r}\frac{\partial \Delta h}{\partial r} + k_{\mathrm{D}}\frac{\partial^2 \Delta h}{\partial z^2} = \frac{1}{\alpha_{\mathrm{s}}}\frac{\partial \Delta h}{\partial t}, \tag{5.7}$$

where $k_{\mathrm{D}} = k_z/k_r$ is the ratio of vertical (k_z) to horizontal (k_r) hydraulic conductivity, $\alpha_{\mathrm{s}} = k_r/S_{\mathrm{s}}$ is the hydraulic diffusivity, and S_{s} is the specific storage. The drawdown is subject to the following initial and boundary conditions

$$\text{Condition 1:} \quad \Delta h[r, z, 0] = 0,$$

$$\text{Condition 2:} \quad \left.\frac{\partial \Delta h}{\partial r}\right|_{r\to\infty} = 0,$$

$$\text{Condition 3:} \quad \lim_{r\to 0}\int_0^b r\frac{\partial \Delta h}{\partial r}dz = -\frac{Q}{2\pi k_r},$$

$$\text{Condition 4:} \quad \left.\frac{\partial \Delta h}{\partial z}\right|_{z=0} = 0,$$

$$\text{Condition 5:} \quad \left. k_z\frac{\partial \Delta h}{\partial z}\right|_{z=b} = -S_{\mathrm{y}}\left.\frac{\partial \Delta h}{\partial t}\right|_{z=b}, \tag{5.8}$$

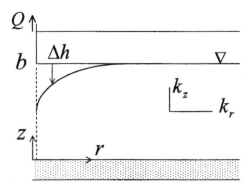

Figure 5.2: Sketch of a full-penetration pumping well in an unconfined aquifer.

where Q is the well discharge rate $[m^3 s^{-1}]$, b is the prepumping thickness of the unconfined aquifer $[m]$, and S_y is the specific yield [-]. These conditions will be referred to, respectively, as initial, far-field, discharge (or well-face), lower-interface, and upper-interface conditions.

Note that the upper integration limit b for the discharge or well-face condition (Condition 3) and the upper-interface condition (Condition 5) at $z = b$ reflect the assumption that the water table is stationary during pumping for the purpose of obtaining an analytic solution. The far-field condition (Condition 2) can be alternatively set as $\Delta h[\infty, z, t] = 0$ without incurring any consequence on the resulting formulas.

We note that, in the Laplace domain, the governing differential equation and the initial and boundary conditions are identical to those used for the boundary flux problems in Chapter 4, equation (4.91), except different parameter c for the upper-interface condition (Condition 5). For the water table aquifers, the rate of drawdown instead of the drawdown itself is imposed.

• Solution

Neuman (1972) has solved the problem by splitting the original differential equation and boundary conditions into two new sets. After applying the Laplace transform for time t and the Hankel transform for radial distance r, he obtained a closed-form inverse Laplace transform but left the inverse Hankel transform as a semidefinite integral, which constitutes his real-time solution. That semidefinite integral contains an infinite series, of which each term bears a slowly converging, oscillatory Bessel function.

Instead of using oscillatory Bessel functions, the alternative solutions presented here employ rapidly converging modified Bessel functions and use the Stehfest algorithm to numerically invert the Laplace transform for time-domain solutions. The first alternative solution is obtained through analytical inverse Hankel transform.

Applying the Laplace transform to the governing differential equation and incorporating the initial condition yields

$$\frac{1}{r}\frac{\partial}{\partial r}\left(r\frac{\partial\overline{\Delta h}}{\partial r}\right) + k_\mathrm{D}\frac{\partial^2\overline{\Delta h}}{\partial z^2} - \frac{p}{\alpha_\mathrm{s}}\overline{\Delta h} = 0, \tag{5.9}$$

where p is the Laplace transform parameter $[s^{-1}]$ and $\overline{\Delta h}$ is the drawdown in the Laplace-transform domain, $[ms]$. The discharge and upper-interface conditions in the Laplace domain are, respectively,

$$\text{Condition 3a:}\quad \lim_{r\to0}\int_0^b r\frac{\partial\overline{\Delta h}}{\partial r}dz = -\frac{Q}{2\pi k_r p},$$

$$\text{Condition 5a:}\quad \left.\frac{\partial\overline{\Delta h}}{\partial z}\right|_{z=b} = -\left.\frac{p\overline{\Delta h}}{\alpha_y}\right|_{r=b}, \quad \alpha_y = \frac{k_z}{S_y}. \tag{5.10}$$

Apply the Hankel transform to equation (5.9) and use the relation that

$$\int_0^\infty J_0[ar]\frac{\partial}{\partial r}\left(r\frac{\partial\overline{\Delta h}}{\partial r}\right)dr = -\left.J_0[ar]\left(r\frac{\partial\overline{\Delta h}}{\partial r}\right)\right|_{r=0} - a^2\overline{\Delta h}^* \tag{5.11}$$

(see Problem 3 in Chapter 4) to incorporate the far-field and discharge conditions (Conditions 2 and 3a), where a is the Hankel transform parameter, $[m]$, and $\overline{\Delta h}^*$ is the drawdown in the Laplace-Hankel transform domain, $[m^3s]$. The resulting equation is

$$\frac{\partial^2\overline{\Delta h}^*}{\partial z^2} - \eta^2\overline{\Delta h}^* = -\frac{Q}{2\pi Tpk_\mathrm{D}}, \tag{5.12}$$

$$\eta^2 = \left(a^2 + q^2\right)/k_\mathrm{D}, \quad q^2 = p/\alpha_\mathrm{s},$$

where q and η have the units of $[m^{-1}]$.

The solution that satisfies the lower-interface condition (Condition 4) is

$$\overline{\Delta h}^*[a, z, p] = C\cosh[\eta z] + \frac{Q}{2\pi Tp\left(a^2 + q^2\right)}. \tag{5.13}$$

Impose the upper-interface condition (Condition 5a) to determine coefficient C and

$$\overline{\Delta h}^*[a, z, p] = \frac{Q}{2\pi Tp(a^2 + q^2)} \left\{ 1 - \frac{\cosh[\eta z]}{c\eta \sinh[\eta b] + \cosh[\eta b]} \right\}, \quad (5.14)$$

$$c = \alpha_y / p.$$

• Inverse Hankel Transform

Except for the different meaning of c, equation (5.14) is identical to the result for the case of a leaky confined aquifer (see Section 4.5.1). Therefore, the solution in the Laplace domain is

$$\overline{\Delta h}[r, z, p] = \frac{Q}{\pi Tp} \sum_{n=0}^{\infty} \lambda_n K_0[\xi_n r] \cos[\zeta_n z], \quad (5.15)$$

where

$$\lambda_n = \frac{c}{(b + c + bc^2\zeta_n^2)\cos[\zeta_n b]} = \frac{\sin[\zeta_n b]}{\zeta_n b + 0.5\sin[2\zeta_n b]}$$

$$= \frac{(-1)^n \sin[\theta_n]}{n\pi + \theta_n + 0.5\sin[2\theta_n]}, \quad (5.16)$$

$$\xi_n = \sqrt{\zeta_n^2 k_D + q^2}, \quad (5.17)$$

and ζ_n satisfies the root equation

$$\tan[\zeta_n b] = \frac{1}{c\zeta_n}, \quad (5.18)$$

and

$$\zeta_n b = n\pi + \theta_n. \quad (5.19)$$

The drawdown averaged over the aquifer thickness is

$$\overline{\Delta h}^{\text{ave}}[r, p] = \frac{Q}{\pi Tp} \sum_{n=0}^{\infty} \lambda_n^{\text{ave}} K_0[\xi_n r], \quad (5.20)$$

where

$$\lambda_n^{\text{ave}} = \frac{1}{(b + c + bc^2\zeta_n^2)b\zeta_n^2}$$

$$= \frac{\sin^2[\theta_n]}{\zeta_n b(\zeta_n b + 0.5\sin[2\theta_n])}. \quad (5.21)$$

• Time-Domain Solution

The time-domain solution can be obtained from $\overline{\Delta h}[r, z, p]$ numerically through the Stehfest inversion scheme

$$\Delta h[r, z, t] = \frac{\ln 2}{t} \sum\nolimits_{j=1}^{M} v_j \overline{\Delta h}[r, z, p_j] \,, \tag{5.22}$$

where $p_j = (j \ln 2)/t$, M is an optimal number for maximum accuracy, and v_j is the weighting coefficient (see Section 3.3 for details).

For the purpose of combining parameters to form a set with the least number of independent parameters, let us introduce the dimensionless variables (Neuman, 1975)

$$\sigma = \frac{S}{S_y}, \quad \beta = \left(\frac{r}{b}\right)^2 k_D, \quad t_s = \frac{Tt}{r^2 S} = \frac{1}{4u}, \tag{5.23}$$

where $S = S_s b$ is the storativity [-] and u is the parameter of the Theis well function. Using $p_j = (j \ln 2)/t$, the root equation becomes

$$\tan[\zeta_{n,j} b] = \frac{j}{\sigma \beta t_s} \frac{\ln 2}{\zeta_{n,j} b} = \frac{ju}{\sigma \beta} \frac{\ln 16}{\zeta_{n,j} b}. \tag{5.24}$$

Thus $\zeta_{n,j}$, $\lambda_{n,j}$, and

$$\xi_n r = \sqrt{\beta(n\pi + \theta_{n,j})^2 + \frac{j \ln 2}{t_s}} \tag{5.25}$$

can be obtained from the root-finding procedures described in Section 3.4.6.

The drawdown is accordingly

$$\Delta h[r, z, t] = \frac{Q}{4\pi T} W[u, \sigma, \beta] \,, \tag{5.26}$$

$$W[u, \sigma, \beta] = \sum_{j=1}^{M} \frac{4v_j}{j} \sum_{n=0}^{\infty} \lambda_{n,j} \cos[\zeta_{n,j} z] K_0[\xi_{n,j} r].$$

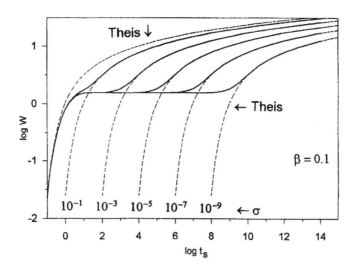

Figure 5.3: Average drawdown $W[u, \sigma, \beta]$ (solid lines) versus time t_s. W behaves like a Theis curve (dashed curves) at early or late time. The delay time for the Theis-like behavior increases with decreasing σ at a given β.

• Early-Time Functional Behavior

As the Laplace transform parameter $p \to \infty$ when time $t \to 0$, one sees $\theta_n \to \pi/2$ and $\zeta_n b \to (n + 1/2)\pi$ according to Example 5H in Chapter 3. In this extreme case,

$$\lambda_n^{\text{ave}} \to \frac{4}{(2n + 1)^2\pi^2}, \quad \text{and} \quad \xi_n r \to qr. \tag{5.27}$$

Because $\xi_n r$ is independent of n, the average drawdown shown in equation (5.20) becomes

$$\overline{\Delta h}^{\text{ave}}[r, p] \to \frac{4Q}{\pi^3 T p} K_0[qr] \sum_{n=0}^{\infty} \frac{1}{(2n + 1)^2}. \tag{5.28}$$

Since

$$\sum_{n=0}^{\infty} \frac{1}{(2n + 1)^2} = \frac{\pi^2}{8}, \tag{5.29}$$

one obtains

$$\overline{\Delta h}^{\text{ave}}[r, p] \to \frac{Q}{2\pi T p} K_0[qr], \tag{5.30}$$

which is the Theis solution in the Laplace domain for a confined aquifer.

• Late-Time Functional Behavior

As $t \to \infty$ and parameter $p \to 0$, we have $\theta_n \to 0$ and $\zeta_n b \to n\pi$. Thus

$$\lambda_0^{\text{ave}} \to 1/2 \quad \text{and} \quad \lambda_n^{\text{ave}} \to 0 \quad \text{for} \quad n \neq 0 \tag{5.31}$$

and equation (5.20) also approaches a Theis well function.

• Intermediate-Time Functional Behavior

Rewrite equation (5.20)

$$\overline{\Delta h}^{\text{ave}}[r, p] = \frac{Q}{2\pi T p} \sum_{n=0}^{\infty} 2\lambda_n^{\text{ave}} K_0 \left[r \sqrt{\frac{p}{\alpha_s} + \zeta_n^2 k_{\text{D}}} \right]. \tag{5.32}$$

If λ_n^{ave} and ζ_n were independent of p, each term in the series would represent a Hantush's leaky well function in a confined aquifer with leakage factor $B_n = 1/\zeta_n \sqrt{k_{\text{D}}}$ (see Section 4.4).

According to equation (5.24), root $\zeta_n b$ depends solely on the values of $\beta \sigma t_s$ or $u/\beta\sigma$. If $\beta\sigma << 1$, root $\zeta_n b$ is insensitive to variations of time t_s over a certain range. The smaller $\beta\sigma$ is, the wider the time range over which $\zeta_n b$ is nearly constant. As shown in Figure 5.3, this time range over which the drawdown appears steady (near flat on the drawdown-time curve) indicates that the response is delayed with respect to the resumption of Theis-like behavior. The delay period increases with decreasing σ for a given β.

Figure 5.4 illustrates the drawdown for various β at a given σ. As compared to the effect of σ variation, the graph indicates that the delay period is insensitive to variation of β. Both Figures 5.3 and 5.4 indeed show that the drawdown at early or late time behaves like a Theis well function.

• Remark

Drawdown as used in this book is defined as $\Delta h[r, z, t] = h[r, z, 0] - h[r, z, t]$. At a given distance from the pumping well, Δh varies with depth. For the purpose of obtaining an analytic solution, the boundary at the prepumping water table was treated as stationary. Strictly speaking, the migrating water table is indeterminable from the analytical solution. Here we consider the water table to lie approximately at $\Delta h[r, b, t]$ below the prepumping water level at $z = b$.

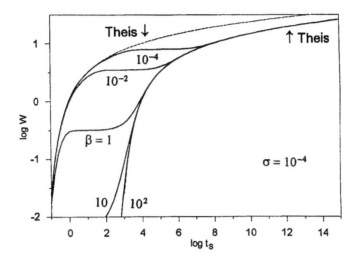

Figure 5.4: Average drawdown $W[u, \sigma, \beta]$ versus time t_s for different values of β at a given value of σ. W behaves like a Theis curve (dashed) at early or late time. The delay time for the Theis behavior is insensitive to the β variation.

5.2.2 Partial-Penetration Pumping Well, $r_w = 0$

• Formulation

For a pumping well perforated or screened between elevations $z = Z_{\mathrm{bot}}$ and $z = Z_{\mathrm{top}}$ in an unconfined aquifer with prepumping thickness b (Figure 5.5), drawdown $\Delta h[r, z, t]$ satisfies all the boundary conditions described in equation (5.8) except that

$$\text{Condition 3b:} \qquad \lim_{r \to 0} \int_{Z_{\mathrm{bot}}}^{Z_{\mathrm{top}}} r \frac{\partial \Delta h}{\partial r} dz = -\frac{Q\delta}{2\pi k_r},$$

$$\delta = \begin{cases} 1, & Z_{\mathrm{bot}} < z < Z_{\mathrm{top}} \\ 0, & \text{elsewhere} \end{cases}. \qquad (5.33)$$

If $\partial \Delta h / \partial r$ is constant over the screen interval, this discharge condition becomes

$$\text{Condition 3c:} \qquad \lim_{r \to 0} \left(r \frac{\partial \overline{\Delta h}}{\partial r} \right) = -\Omega \delta, \qquad (5.34)$$

$$\Omega = \frac{Q}{2\pi k_r p \left(Z_{\mathrm{top}} - Z_{\mathrm{bot}} \right)}$$

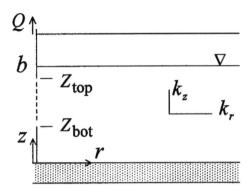

Figure 5.5: Sketch of a partial-penetration pumping well in an unconfined aquifer.

in the Laplace domain.

Since the governing equation and boundary conditions in the Laplace domain are the same for the boundary-flux leaky aquifers and Neuman's type of unconfined aquifers (except the different meaning of c), the solution in Section 4.5.2 can be transferred here,

$$\overline{\Delta h}[r, z] = \sum_{n=0}^{\infty} A_n K_0[\xi_n r] \cos[\zeta_n z] , \tag{5.35}$$

where

$$A_n = \frac{2\Omega \left(\sin[\zeta_n Z_{\text{top}}] - \sin[\zeta_n Z_{\text{bot}}] \right)}{\zeta_n b + 0.5 \sin[2\zeta_n b]} . \tag{5.36}$$

In terms of λ, the drawdown can be rewritten as

$$\overline{\Delta h}[r, z, p] = \frac{Q}{\pi T p} \sum_{n=0}^{\infty} \lambda_n^{\text{part}} \cos[\zeta_n z] K_0[\xi_n r] , \tag{5.37}$$

where

$$\lambda_n^{\text{part}} = \frac{b}{Z_{\text{top}} - Z_{\text{bot}}} \frac{\sin[\zeta_n Z_{\text{top}}] - \sin[\zeta_n Z_{\text{bot}}]}{\sin[\zeta_n b]} \lambda_n. \tag{5.38}$$

For a full-penetration pumping well ($Z_{\text{top}} = b$ and $Z_{\text{bot}} = 0$), λ_n^{part} is reduced to λ_n.

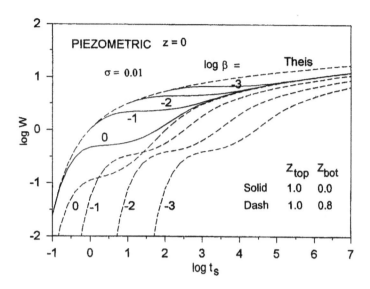

Figure 5.6: Piezometric drawdown versus time at the base of an aquifer for different pumping screen intervals (modified after Neuman, 1974, by extending t_s from 10^4 to 10^7).

• **Type Curves**

Figure 5.6 depicts the piezometric drawdown at the bottom of a water table aquifer. Plots for both full and partial penetration wells extend Neuman's (1974) curves from time 10^4 to 10^7. The Theis curve is shifted by $-\log 4$ in $1/u$ to conform with plotting in t_s unit. Note that type curves can vary greatly in shape, depending on screening intervals.

Figure 5.7 illustrates average drawdowns at observation wells screened near the top of the aquifer, and Figure 5.8 depicts the responses screened near the lower interface. Both are induced by a pumping well screened near the water table. Solid curves duplicate those of Moench (1993) for comparison; others are curves for interpolated parameters.

5.3 Pumping Well of Nonzero Diameter

We will derive the drawdown responses to pumping in a well of nonzero diameter by two methods. The first one continues the Hankel-Laplace transform

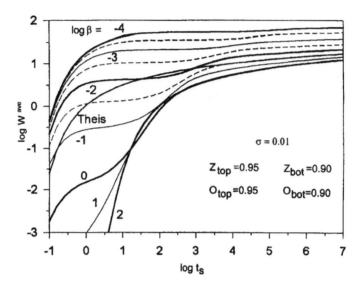

Figure 5.7: Average drawdown versus time near the water table due to pumping near the water table. Heavy solid curves repeat those of Moench (1993). Thin solid and dashed curves are drawdowns for interpolated β values.

technique, and the second extends the solution obtained with the technique of the separation of variables. The first one is mathematically more involved but we intend to state that the Hankel transform does not have to be in the range from $r = 0$ to ∞.

5.3.1   Full-Penetration Pumping Well, $r_w \neq 0$

• Formulation

For a pumping well of nonzero diameter, the hydraulic response follows the same governing differential equation and boundary conditions for a pumping well of infinitesimal diameter except the discharge condition, which is now modified to

$$\text{Condition 3d:} \quad \lim_{r \to r_w} \int_0^b r \frac{\partial \Delta h}{\partial r} dz = -\frac{Q}{2\pi k_r}. \tag{5.39}$$

In order to work in the Laplace-Hankel domain, let us modify the Hankel

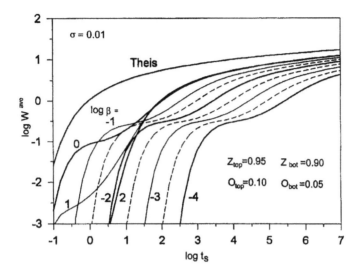

Figure 5.8: Average drawdown versus time near the bottom of the unconfined aquifer due to pumping near the water table. Heavy solid curves repeat those of Moench (1993). Thin solid and dashed curves are drawdowns for interpolated β values.

transform pair to

$$\overline{\Delta h}^*[a, z, p] = \int_{r_w}^{\infty} r J_0[ar]\overline{\Delta h}[r, z, p]\,dr,$$

$$\overline{\Delta h}[r, z, p] = \int_0^{\infty} a J_0[ar]\overline{\Delta h}^*[a, z, p]\,da. \tag{5.40}$$

The usage of r_w as the lower integration limit for the Hankel transform instead of the customary 0 does not matter if we still follow the rules of inverse Hankel transforms for a from 0 to ∞.

Take the Hankel transform of equation (5.9) and use the relation that

$$\int_{r_w}^{\infty} J_0[ar]\frac{\partial}{\partial r}\left(r\frac{\overline{\Delta h}}{\partial r}\right)dr = -J_0[ar_w]r_w\frac{\partial\overline{\Delta h}}{\partial r_w} - a^2\overline{\Delta h}^*[a, z, p]$$

$$-ar_w J_1[ar_w]\overline{\Delta h}[r_w, z, p] \tag{5.41}$$

to accommodate the discharge condition (see Problem 1). The resulting equation is

$$\frac{\partial^2 \overline{\Delta h}^*}{\partial z^2} - \eta^2 \overline{\Delta h}^* = -\frac{Q J_0[ar_w]}{2\pi Tpk_D} + \frac{ar_w J_1[ar_w]}{k_D}\overline{\Delta h}[r_w, z, p], \tag{5.42}$$

where

$$\eta^2 = (a^2 + q^2)/k_D, \quad q^2 = p/\alpha_s. \tag{5.43}$$

Assuming that $\overline{\Delta h}[r_w, z, p]$ at the well face is independent of z (as assumed for using the discharge condition, i.e., no vertical flow at the well face, $\overline{\Delta h}[r_w, z, p] = \overline{\Delta h}[r_w, b, p]$), the general solution is

$$\begin{aligned}
\overline{\Delta h}^*[a, z, p] &= C \cosh[\eta z] + \frac{Q J_0[ar_w]}{2\pi Tp(a^2 + q^2)} \\
&\quad - \frac{ar_w J_1[ar_w]}{a^2 + q^2}\overline{\Delta h}[r_w, b, p],
\end{aligned} \tag{5.44}$$

which satisfies all conditions in equation (5.8) but the upper-interface condition.

Imposing the upper-interface condition in the Laplace-Hankel domain yields

$$\begin{aligned}
C &= \left\{ \frac{-Q J_0[ar_w]}{2\pi Tp(a^2 + q^2)} + \frac{ar_w J_1[ar_w]}{a^2 + q^2}\overline{\Delta h}[r_w, b, p] \right\} \\
&\quad \cdot \frac{1}{c\eta \sinh[\eta b] + \cosh[\eta b]}.
\end{aligned} \tag{5.45}$$

Therefore, the solution in the Laplace-Hankel domain is

$$\overline{\Delta h}^*[a, z, p] = \overline{\Delta h}_I^*[a, z, p] + \overline{\Delta h}_{II}^*[a, z, p], \tag{5.46}$$

where

$$\overline{\Delta h}_I^*[a, z, p] = \frac{Q}{2\pi Tp}\frac{J_0[ar_w]}{a^2 + q^2}\left\{ 1 - \frac{\cosh[\eta z]}{c\eta \sinh[\eta b] + \cosh[\eta b]} \right\} \tag{5.47}$$

and

$$\begin{aligned}
\overline{\Delta h}_{II}^*[a, z, p] &= -\overline{\Delta h}[r_w, b, p]\frac{ar_w J_1[ar_w]}{a^2 + q^2} \\
&\quad \cdot \left\{ 1 - \frac{\cosh[\eta z]}{c\eta \sinh[\eta b] + \cosh[\eta b]} \right\}.
\end{aligned} \tag{5.48}$$

• Inverse Hankel Transform

With the exception of the additional factor of $J_0[ar_w]$, component $\overline{\Delta h}_I^*[a, z, p]$ is the same as $\overline{\Delta h}^*[a, z, p]$ in equation (5.14) for a pumping well of infinitesimal diameter. Using the inversion result in Problem 2 by analogy to Example 3H (Section 3.4.4), the $J_0[ar_w]$ factor is simply transformed into the factor $I_0[\xi_n r_w]$ which should be attached to equation (5.15) for the present case. The result is

$$\overline{\Delta h}_I[r, z, p] = \frac{Q}{\pi T p} \sum_{n=0}^{\infty} \lambda_n I_0[\xi_n r_w] K_0[\xi_n r] \cos[\zeta_n z]. \tag{5.49}$$

By comparison with the first component $\overline{\Delta h}_I^*$ and by analogy with Example 4H in Chapter 3, the inversion of the second component (see Problem 3) is

$$\overline{\Delta h}_{II}[r, z, p] = 2\overline{\Delta h}[r_w, b, p] \sum_{n=0}^{\infty} \lambda_n \xi_n r_w I_1[\xi_n r_w] K_0[\xi_n r] \cos[\zeta_n z]. \tag{5.50}$$

Now, add the two components together to get

$$\overline{\Delta h}[r, z, p] = \sum_{n=0}^{\infty} \lambda_n \left\{ \frac{Q}{\pi T p} I_0[\xi_n r_w] + E_n \right\} K_0[\xi_n r] \cos[\zeta_n z], \tag{5.51}$$

where

$$E_n = 2\xi_n r_w I_1[\xi_n r_w] \overline{\Delta h}[r_w, b, p]. \tag{5.52}$$

• Unknown Drawdown at Well Face

Drawdown $\overline{\Delta h}[r, z, p]$ contains an unknown factor $\overline{\Delta h}[r_w, b, p]$ on the right-hand side of the above equation. This unknown is determined as follows.

The average drawdown over the aquifer thickness is

$$\overline{\Delta h}^{ave}[r_w, p] = \sum_{n=0}^{\infty} \lambda_n^{ave} K_0[\xi_n r] \left\{ \frac{Q}{\pi T p} I_0[\xi_n r_w] + E_n \right\}, \tag{5.53}$$

where λ_n^{ave} is defined in equation (5.21). Because the average drawdown at the well face $\overline{\Delta h}^{ave}[r_w, p]$ is identical with $\overline{\Delta h}[r_w, b, p]$ (a consequence of the assumption that the drawdown at the well face is independent of z), one obtains

$$\overline{\Delta h}[r_{\mathrm{w}}, b, p] = \frac{Q}{\pi T p D} G, \tag{5.54}$$

where

$$G = \sum_{m=0}^{\infty} \lambda_m^{\mathrm{ave}} I_0[\xi_m r_{\mathrm{w}}] K_0[\xi_m r_{\mathrm{w}}] \tag{5.55}$$

$$D = 1 - 2 \sum_{k=0}^{\infty} \lambda_k^{\mathrm{ave}} \xi_k r_{\mathrm{w}} I_1[\xi_k r_{\mathrm{w}}] K_0[\xi_k r_{\mathrm{w}}]. \tag{5.56}$$

The final solution in the Laplace domain is

$$\boxed{\overline{\Delta h}[r, z, p] = \frac{Q}{\pi T p D} \sum_{n=0}^{\infty} \gamma_n \lambda_n K_0[\xi_n r] \cos[\zeta_n z]}, \tag{5.57}$$

where

$$\gamma_n = D I_0[\xi_n r_{\mathrm{w}}] + 2 G \xi_n r_{\mathrm{w}} I_1[\xi_n r_{\mathrm{w}}] \tag{5.58}$$

is the modifier for the finite radius, with reference to the result for a pumping well of infinitesimal diameter.

The average drawdown over the aquifer thickness is

$$\boxed{\overline{\Delta h}^{\mathrm{ave}}[r, p] = \frac{Q}{\pi T p} \sum_{n=0}^{\infty} \gamma_n \lambda_n^{\mathrm{ave}} K_0[\xi_n r]}. \tag{5.59}$$

• Functional Behavior as Radius Goes to Zero

As $r_{\mathrm{w}} \to 0$, equations (5.57) and (5.59) approach the solutions for a pumping well of infinitesimal diameter because $I_0[\xi_n r_{\mathrm{w}}] \to 1$, $\xi_n r_{\mathrm{w}} I_1[\xi_n r_{\mathrm{w}}] \to 0$, $D \to 1$, and $\gamma_n \to 1$.

• Early-Time Functional Behavior

At early time when $t \to 0$, then $p \to \infty$. As a result

$$\begin{aligned}
\theta_n &\to \pi/2, \quad \zeta_n b \to n\pi + \pi/2, \\
\lambda_n^{\mathrm{ave}} &= 1/(\zeta_n b)^2 = 4/(2n+1)^2 \pi^2, \quad \xi_n r \to qr.
\end{aligned} \tag{5.60}$$

The average drawdown in the limit is

$$\overline{\Delta h}^{\text{ave}}[r,p] \rightarrow \frac{Q}{\pi T p} K_0[qr] \sum_{n=0}^{\infty} \lambda_n^{\text{ave}}$$

$$\cdot \left\{ I_0[qr_{\text{w}}] + \frac{2qr_{\text{w}} I_1[qr_{\text{w}}] I_0[qr_{\text{w}}] K_0[qr_{\text{w}}] \sum_{m=0}^{\infty} \lambda_m^{\text{ave}}}{1 - 2qr_{\text{w}} I_1[qr_{\text{w}}] K_0[qr_{\text{w}}] \sum_{k=0}^{\infty} \lambda_k^{\text{ave}}} \right\}. \qquad (5.61)$$

Because

$$\sum_{n=0}^{\infty} \lambda_n^{\text{ave}} = \frac{4}{\pi^2} \sum_{n=0}^{\infty} \frac{1}{(2n+1)^2} = \frac{1}{2}, \qquad (5.62)$$

the average drawdown is

$$\overline{\Delta h}^{\text{ave}}[r,p] = \frac{Q}{2\pi T p} \frac{I_0[qr_{\text{w}}] K_0[qr]}{1 - qr_{\text{w}} I_1[qr_{\text{w}}] K_0[qr_{\text{w}}]}. \qquad (5.63)$$

Using the Wronskian relation (Section 3.4.1 in Chapter 3),

$$I_0[x] K_1[x] + I_1[x] K_0[x] = 1/x, \qquad (5.64)$$

yields

$$\overline{\Delta h}^{\text{ave}}[r,p] \rightarrow \frac{Q}{2\pi T p} \frac{K_0[qr]}{qr_{\text{w}} K_1[qr_{\text{w}}]} \quad \text{as} \quad t \rightarrow 0. \qquad (5.65)$$

This approaches the drawdown in the Laplace domain for a full-penetration pumping well in a nonleaky confined aquifer, equation (4.20).

• Late-Time Functional Behavior

At large time when $t \rightarrow \infty$, then $\theta_n \rightarrow 0$ and

$$\lambda_0^{\text{ave}} = 1/2, \quad \xi_0 r \rightarrow qr; \quad \lambda_n = 0 \quad \text{for} \quad n \geq 1. \qquad (5.66)$$

Hence,

$$\overline{\Delta h}^{\text{ave}}[r,p] \rightarrow \frac{Q}{2\pi T p} \frac{K_0[qr]}{qr_{\text{w}} K_1[qr_{\text{w}}]} \quad \text{as} \quad t \rightarrow \infty. \qquad (5.67)$$

Again, this response approaches equation 4.20.

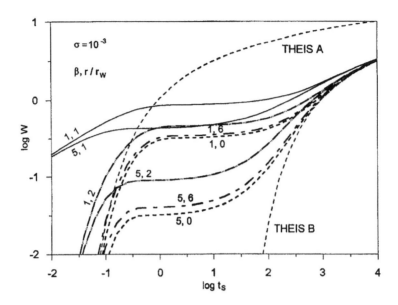

Figure 5.9: Well-size effect on drawdown at different distances from the pumping well. The paired numbers are β and r/r_w except that the curves of $r/r_w = 0$ designate the drawdown due to a pumping well of infinitesimal diameter. Note that as r/r_w increases, type curves drift toward the curves for $r/r_w = 0$.

5.3.2   Partial-Penetration Pumping Well, $r_w \neq 0$

For a pumping well of finite diameter, the response function is the same if A_n in equation (5.36) is now modified in accordance with the revised discharge condition

$$\text{Condition 3e:} \qquad \lim_{r \to r_w} \left(r \frac{\partial \overline{\Delta h}}{\partial r} \right) = -\Omega \delta, \qquad (5.68)$$

$$\Omega = \frac{Q}{2\pi k_r p \left(Z_{\text{top}} - Z_{\text{bot}} \right)}.$$

In this case

$$A_n = \frac{2\Omega \left(\sin[\zeta_n Z_{\text{top}}] - \sin[\zeta_n Z_{\text{bot}}] \right)}{\xi_n r_w K_1[\xi_n r_w] \left(\zeta_n b + 0.5 \sin[2\zeta_n b] \right)}. \qquad (5.69)$$

Substituting this A_n into equation (5.35) yields the desired solution, which is equivalent to equation (4.141), the solution for the boundary-flux problems,

$$\boxed{\overline{\Delta h}[r, z, p] = \frac{Q}{\pi T_r p} \sum_{n=0}^{\infty} \lambda_n'^{\text{part}} K_0[\xi_n r] \cos[\zeta_n z]}, \tag{5.70}$$

where

$$\lambda_n'^{\text{part}} = \frac{\lambda_n^{\text{part}}}{\xi_n r_{\text{w}} K_1[\xi_n r_{\text{w}}]}. \tag{5.71}$$

Note that for early- and late-time behaviors, it can be easily shown that this $\overline{\Delta h}[r, z, p]$ agrees to the result obtained by the Hankel transform method.

5.3.3 Effect of Finite Diameter

To evaluate the effect of pumping-well size on drawdown responses, consider a full-penetration pumping well ($\lambda_n^{\text{part}} \to \lambda_n$). According to equation (5.70), the drawdown averaged over a full-screened observation well is

$$\overline{\Delta h}^{\text{ave}}[r, z, p] = \frac{Q}{\pi T_r p} \sum_{n=0}^{\infty} \frac{\lambda_n^{\text{ave}}}{\xi_n r_{\text{w}} K_1[\xi_n r_{\text{w}}]} K_0[\xi_n r]. \tag{5.72}$$

Figure 5.9 illustrates the effect of well size on the type curves, which are computed from equation (5.59) instead of the above simple, alternative solution. As referenced to the case of infinitesimal well diameter, the effect is significant in the vicinity of the pumping well but negligible at distances beyond 10 times the well radius.

The early- and late-time responses to pumping of a nonzero diameter well in the water table aquifer approach the corresponding behavior for a pumping well of nonzero diameter in a confined aquifer.

As $q r_{\text{w}} \to 0$, the average drawdowns at early and late time approach that of the Theis well function. The Theis-like behavior is not likely as $t \to 0$ unless $r_{\text{w}} \to 0$ also; however, it is likely at large time for $r_{\text{w}} \neq 0$ since $q \to 0$ as $t \to \infty$ (because $q^2 = p/\alpha_{\text{s}}$ and $p \propto 1/t$). In other words, the type curves deviate from the Theis curve as $t \to 0$ if $r_{\text{w}} \neq 0$ but behave Theis-like as $t \to \infty$ irrespective of well size.

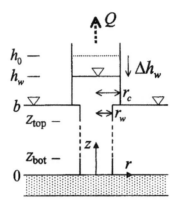

Figure 5.10: Sketch of a parital-penetration and -screening pumping well in an unconfined aquifer. The well has nonzero radius (r_w) and its casing radius is r_c. h_0 is the prepumping head and h_w is the water level in the well.

5.4   Well Storage and Skin Effect

5.4.1   Full-Penetration Well

• Formulation

In this section we will consider the effect of well bore storage and hydraulic head loss across the well face in a pumping well (Figure 5.10). The desired solution is subject to the same initial and boundary conditions as described in equation (5.8), except that the discharge condition is now modified to include a well bore storage term,

$$\text{Condition 3f:} \quad C_w \frac{\partial \Delta h_w}{\partial t} = Q + 2\pi r_w b k_r \left. \frac{\partial \Delta h}{\partial r} \right|_{r_w}, \qquad (5.73)$$

where Δh_w is the change of hydraulic head in the well bore with respect to the prepumping hydraulic head h_0 in the aquifer (i.e., $\Delta h_w = h_0 - h_w$). Condition 3f expresses mass balance among storage rate in the well bore, discharge rate, and flow rate to the well. Here the discharge rate Q can be time dependent and storage coefficient C_w is an appropriate cross-sectional area of the well bore; for example, πr_c^2 for a well with a casing radius of r_c, which may differ from the screen radius r_w.

Condition 3f also serves as a differential equation for hydraulic head in the well bore. Hence, $\Delta h_{\mathrm{w}}[t]$ needs an initial slug-height condition

$$\text{Condition 6:} \quad \Delta h_{\mathrm{w}}[0] = -h_{\mathrm{w}0}. \tag{5.74}$$

Parameter $h_{\mathrm{w}0}$ represents water level (hydraulic head) above the prepumping hydraulic head h_0 in the aquifer. For an injection slug test, $h_{\mathrm{w}0}$ is the positive slug height and $\Delta h_{\mathrm{w}}[0]$ is negative; and for an extraction slug, $h_{\mathrm{w}0} < 0$ and $\Delta h_{\mathrm{w}}[0]$ is positive. Conditions 1 through 6 completely define the problem for the well bore storage.

Being an indicator of well efficiency, the discontinuity in drawdown across the well face is approximated by

$$\text{Condition 7:} \quad \Delta h_{\mathrm{w}}[t] = \Delta h[r_{\mathrm{w}}, z, t] - \epsilon k_r \left. \frac{\partial \Delta h}{\partial r} \right|_{r_{\mathrm{w}}} \tag{5.75}$$

(Dougherty and Babu, 1984, and references therein used similar forms for hydraulic head loss across the well face at r_{w}). Head loss, $(h - h_{\mathrm{w}})$ or $(\Delta h_{\mathrm{w}} - \Delta h)$, usually increases with discharge rate. The functional form is probably nonlinear because Darcy's law may not be applicable around the well bore where the flow is turbulent. Being an approximation, Condition 7 is referred to as the skin-effect condition.

In step-drawdown tests (Driscoll, 1986), the drawdown is frequently expressed with a quadratic flow term such as

$$\Delta h = EQ + FQ^2, \tag{5.76}$$

where EQ represents the theoretical drawdown as customarily approximated by the Theis well function, and the nonlinear term FQ^2 represents an additional head loss. Here, for simplicity, the loss is presumed to increase linearly with the flow rate $(-k_r \partial \Delta h / \partial r)$ into the well bore. Thus coefficient ϵ represents the skin effect – and we have one more unknown parameter to describe a well function. This effect can be observationally estimated from head loss by extrapolating drawdowns in monitoring wells near the pumping well. It is noted that the effect can be negative for enhanced permeability around the well (e.g., hydraulic fracturing in oil or gas reservoir).

Condition 7 can also be expressed in a form like

$$k_{\mathrm{s}} \frac{\Delta h[r_{\mathrm{w}}, z, t] - \Delta h_{\mathrm{w}}[t]}{r_{\mathrm{s}}} = k_r \left. \frac{\partial \Delta h}{\partial r} \right|, \tag{5.77}$$

where k_s and r_s are, respectively, the hydraulic conductivity and the thickness of the well bore skin. The two forms are equivalent if $\epsilon = r_s/k_s$. Moench (1997) has used this alternative skin-effect condition to formulate the flow toward a well.

- **Combining Conditions**

The solution to Δh and Δh_w can be obtained by eliminating Δh_w. In the Laplace domain, the discharge and initial slug-height conditions together yield

$$\text{Condition 3g:} \quad C_w \left(p\overline{\Delta h_w} + h_{w0} \right) = \overline{Q} + 2\pi r_w b k_r \left. \frac{\partial \overline{\Delta h}}{\partial r} \right|_{r_w} \qquad (5.78)$$

and the skin-effect condition is

$$\text{Condition 7b:} \quad \overline{\Delta h_w}[p] = \overline{\Delta h}[r_w, z, p] - \epsilon k_r \left. \frac{\partial \overline{\Delta h}}{\partial r} \right|_{r_w}. \qquad (5.79)$$

Substituting Conditions 7b into 3g yields

$$\text{Condition 3h:} \quad r_w \left. \frac{\partial \overline{\Delta h}}{\partial r} \right|_{r_w} = -\Omega, \qquad (5.80)$$

where

$$\begin{aligned} \Omega &= \left\{ \overline{Q} - C_w \left(p\overline{\Delta h}[r_w, b, p] + h_{w0} \right) \right\} / \omega, \\ \omega &= \epsilon k_r p C_w / r_w + 2\pi T, \end{aligned} \qquad (5.81)$$

in which the drawdown $\overline{\Delta h}[r_w, z, p]$ at the well face is assumed to be independent of z, i.e.,

$$\overline{\Delta h}[r_w, z, p] = \overline{\Delta h}[r_w, b, p]. \qquad (5.82)$$

- **Laplace-Hankel Domain**

Take the Laplace-Hankel transform of the governing differential equation, follow the procedure in Section 5.3, and use the result of Problem 1. By analogy to equation (5.44), the solution in the Laplace-Hankel transform domain is

$$\overline{\Delta h}^*[a, z, p] = C \cosh[\eta z] + \Omega \frac{J_0[a r_w]}{a^2 + q^2} - \frac{a r_w J_1[a r]}{a^2 + q^2} \overline{\Delta h}[r_w, b, p]. \qquad (5.83)$$

Now, impose Condition 5a in equation (5.10) to determine coefficient C. As a result,

$$\overline{\Delta h}^*[a, z, p] = \left\{ \Omega \frac{J_0[ar_w]}{a^2 + q^2} - \frac{ar_w J_1[ar]}{a^2 + q^2} \overline{\Delta h}[r_w, b, p] \right\}$$
$$\cdot \left\{ 1 - \frac{\cosh[\eta z]}{c\eta \sinh[\eta b] + \cosh[\eta b]} \right\}. \tag{5.84}$$

• Inverse Hankel Transform

Function $\overline{\Delta h}^*[a, z, p]$ is identical in functional form to equation (5.46). Therefore, the solution in the Laplace domain is

$$\overline{\Delta h}[r, z, p] = \sum_{n=0}^{\infty} \lambda_n \left\{ 2I_0[\xi_n r_w]\Omega + E_n \right\} K_0[\xi_n r] \cos[\zeta_n z], \tag{5.85}$$

where E_n is defined in equation (5.52) with a different $\overline{\Delta h}[r_w, b, p]$ given below.

The unknown drawdown $\overline{\Delta h}[r_w, b, p]$ at the well face is obtained by taking the mean drawdown at r_w. Recall the assumption that hydraulic head at the well face is independent of z to get

$$\overline{\Delta h}[r_w, b, p] = \overline{\Delta h}^{\text{ave}}[r_w, p] = \overline{\mathcal{Q}}G, \tag{5.86}$$

where

$$\overline{\mathcal{Q}} = 2 \left(\overline{Q} - C_w h_{w0} \right) / D_s, \tag{5.87}$$
$$D_s = D\omega + 2pC_w G, \tag{5.88}$$

and D and G are respectively given in equations (5.56) and (5.55).

Putting $\overline{\Delta h}[r_w, b, p]$ into equation (5.85) yields

$$\boxed{\overline{\Delta h}[r, z, p] = \overline{\mathcal{Q}} \sum_{n=0}^{\infty} \lambda_n \gamma_n K_0[\xi_n r] \cos[\zeta_n z]} \tag{5.89}$$

where γ_n is given in equation (5.58).

• Special Case 1: No well bore storage

If well bore storage $C_w = 0$, then

$$\omega = 2\pi T, \quad D_s = 2\pi TD, \quad \overline{\mathcal{Q}} = \overline{Q}/\pi TD. \tag{5.90}$$

If discharge rate Q is constant, $\overline{Q} = Q/p$ and equations (5.86) and (5.89) are reduced to equations (5.54) and (5.57), respectively.

- **Special Case 2: Slug test**

For a slug test,

$$Q = 0, \quad \epsilon = 0. \tag{5.91}$$

Therefore

$$D_{\mathrm{s}} = 2\pi T D + 2p C_{\mathrm{w}} G. \tag{5.92}$$

5.4.2   Partial-Penetration Well

- **Condition at the Screened Well Face**

Now we consider the effect of screening between Z_{bot} and Z_{top} in the pumping well. All conditions are the same as described in the last section except that the radial gradient for the well-face condition in equation (5.73) is modified here to

$$\text{Condition 3i:} \qquad 2\pi r_{\mathrm{w}} k_r (Z_{\mathrm{top}} - Z_{\mathrm{bot}}) \left.\frac{\partial \Delta h}{\partial r}\right|_{r_{\mathrm{w}}}$$

$$= -\left(Q - C_{\mathrm{w}}\frac{\partial \Delta h_{\mathrm{w}}}{\partial t}\right)\delta, \tag{5.93}$$

where

$$\delta = 1 \quad \text{if} \quad Z_{\mathrm{bot}} \le z \le Z_{\mathrm{top}}; \quad \text{else} \quad \delta = 0. \tag{5.94}$$

In the Laplace domain, this condition becomes

$$\text{Condition 3j:} \qquad 2\pi k_r (Z_{\mathrm{top}} - Z_{\mathrm{bot}})\, r\left.\frac{\partial \overline{\Delta h}}{\partial r}\right|_{r_{\mathrm{w}}}$$

$$= -\left[\overline{Q} - C_{\mathrm{w}}(p\overline{\Delta h_{\mathrm{w}}} + h_{\mathrm{w}0})\right]\delta. \tag{5.95}$$

Using the skin-effect condition,

$$\text{Condition 7c:} \quad \epsilon k_r \left.\frac{\partial \overline{\Delta h}}{\partial r}\right|_{r_{\mathrm{w}}} = \left(\overline{\Delta h}[r_{\mathrm{w}}, z, p] - \overline{\Delta h_{\mathrm{w}}}[p]\right)\delta, \tag{5.96}$$

to eliminate $\overline{\Delta h_{\mathrm{w}}}$ from Condition 3j yields

$$\text{Condition 3k:} \quad r\left.\frac{\partial \overline{\Delta h}}{\partial r}\right|_{r_{\mathrm{w}}} = -\Omega^{\mathrm{part}}\delta, \tag{5.97}$$

where

$$\Omega^{\text{part}} = \frac{1}{\omega^{\text{part}}} \left\{ \overline{Q} - C_{\text{w}}(p\overline{\Delta h}[r_{\text{w}}, z, p] + h_{\text{w}0}) \right\}, \tag{5.98}$$

$$\omega^{\text{part}} = \epsilon k_r p C_{\text{w}}/r_{\text{w}} + 2\pi T(Z_{\text{top}} - Z_{\text{bot}})/b. $$

• Cosine-Series Representation

As shown in Problem 4, the presence of $\overline{\Delta h}[r_{\text{w}}, z, p]\delta$ in Condition 3k prevents the problem from being solved by using the technique of separation of variables, as used for other cases of partial-penetration pumping wells. Hence, we will solve the problem by means of the Hankel-Fourier transform.

Assuming no vertical flow at r_{w}, we use a continuous cosine series to represent the discontinuous $r\partial\Delta h/\partial r$ (discontinuous at the junctions of the screened and unscreened sections along the well face),

$$\left(r\frac{\partial\overline{\Delta h}}{\partial r} \right)_{r_{\text{w}}} = \frac{F_0}{2} + \sum_{m=1}^{\infty} F_m \cos\frac{m\pi z}{b}, \tag{5.99}$$

where

$$\begin{aligned} F_m &= \frac{2}{b}\int_0^b \left(r\frac{\partial\overline{\Delta h}}{\partial r} \right)_{r_{\text{w}}} \cos\left[\frac{m\pi z}{b}\right] dz \\ &= -\frac{2\Omega^{\text{part}}}{m\pi} \left\{ \sin\left[\frac{m\pi Z_{\text{top}}}{b}\right] - \sin\left[\frac{m\pi Z_{\text{bot}}}{b}\right] \right\}. \end{aligned} \tag{5.100}$$

• Laplace-Hankel Domain Solution

Take the Laplace and Hankel transforms of the governing differential equation and use the result of Problem 1 to obtain

$$\begin{aligned} \frac{\partial^2\overline{\Delta h}^*}{\partial z^2} - \eta^2\overline{\Delta h}^* &= \frac{J_0[ar_{\text{w}}]}{k_{\text{D}}} \left(r\frac{\partial\overline{\Delta h}}{\partial r} \right)_{r_{\text{w}}} + \frac{ar_{\text{w}}J_1[ar_{\text{w}}]}{k_{\text{D}}}\overline{\Delta h}[r_{\text{w}}, b, p] \\ &= \frac{J_0[ar_{\text{w}}]}{k_{\text{D}}} \left(\frac{F_0}{2} + \sum_{m=1}^{\infty} F_m \cos\left[\frac{m\pi z}{b}\right] \right) \\ &\quad + \frac{ar_{\text{w}}J_1[ar_{\text{w}}]}{k_{\text{D}}}\overline{\Delta h}[r_{\text{w}}, b, p]. \end{aligned} \tag{5.101}$$

Its general solution is

$$\overline{\Delta h}^*[a, z, p] = C \cosh[\eta z] - \frac{a r_w J_1[a r_w]}{a^2 + q^2} \overline{\Delta h}[r_w, b, p] - \frac{J_0[a r_w] F_0}{2 (a^2 + q^2)}$$

$$- J_0[a r_w] \sum_{m=1}^{\infty} \frac{F_m}{a^2 + q_m^2} \cos \left[\frac{m \pi z}{b} \right], \tag{5.102}$$

where

$$q_m^2 = q^2 + k_D (m\pi/b)^2. \tag{5.103}$$

Now, recall the upper-interface condition in equation (5.10) to determine

$$C = \frac{1}{c\eta \sinh[\eta b] + \cosh[\eta b]} \left\{ \frac{a r_w J_1[a r_w]}{a^2 + q^2} \overline{\Delta h}[r_w, b, p] \right.$$

$$\left. + \frac{J_0[a r_w] F_0}{2 (a^2 + q^2)} + J_0[a r_w] \sum_{m=1}^{\infty} (-1)^m \frac{F_m}{a^2 + q_m^2} \right\}. \tag{5.104}$$

The final solution in the Laplace-Hankel transform domain is

$$\overline{\Delta h}^*[a, z, p] = \overline{\Delta h_I}^*[a, z, p] + \overline{\Delta h_{II}}^*[a, z, p] + \overline{\Delta h_{III}}^*[a, z, p], \tag{5.105}$$

where

$$\overline{\Delta h_I}^*[a, z, p] = - \left\{ \frac{J_0[a r_w] F_0}{2 (a^2 + q^2)} + \frac{a r_w J_1[a r_w]}{a^2 + q^2} \overline{\Delta h}[r_w, b, p] \right\}$$

$$\cdot \left\{ 1 - \frac{\cosh[\eta z]}{c\eta \sinh[\eta b] + \cosh[\eta b]} \right\}, \tag{5.106}$$

$$\overline{\Delta h_{II}}^*[a, z, p] = - \sum_{m=1}^{\infty} \frac{F_m J_0[a r_w] \cos[m\pi z/b]}{a^2 + q_m^2}, \tag{5.107}$$

and

$$\overline{\Delta h_{III}}^*[a, z, p] = \sum_{m=1}^{\infty} \frac{(-1)^m F_m}{a^2 + q_m^2} \frac{\cosh[\eta z] J_0[a r_w]}{c\eta \sinh[\eta b] + \cosh[\eta b]}. \tag{5.108}$$

- ## Laplace Domain Solution

Since equation (5.106) is similar to equation (5.84), its inverse Hankel transform can be written immediately

$$\overline{\Delta h_I}[r, z, p] = -F_0 \sum_{n=0}^{\infty} \lambda_n I_0[\xi_n r_w] K_0[\xi_n r] \cos[\zeta_n z] + 2\overline{\Delta h}[r_w, b, p]$$

$$\cdot \sum_{n=0}^{\infty} \lambda_n \xi_n r_w I_1[\xi_n r_w] K_0[\xi_n r] \cos[\zeta_n z]. \tag{5.109}$$

The inverse Hankel transform of the second part, $\overline{\Delta h_{II}}^*[a, z, p]$, is

$$\overline{\Delta h_{II}}[r, z, p] = -\sum_{m=1}^{\infty} F_m I_0[q_m r_w] K_0[q_m r] \cos\left[\frac{m\pi z}{b}\right], \tag{5.110}$$

according to Example 3H in Section 3.4.3.

The third part has the following inverse Hankel transform

$$\overline{\Delta h_{III}}[r, z, p] = \sum_{m=1}^{\infty} F_m \left\{ I_0[q_m r_w] K_0[q_m r] \cos\left[\frac{m\pi z}{b}\right] - (-1)^m \right.$$

$$\left. \cdot 2 \sum_{n=0}^{\infty} \frac{\xi_n^2 - q^2}{\xi_n^2 - q_m^2} \lambda_n I_0[\xi_n r_w] K_0[\xi_n r] \cos[\zeta_n z] \right\}, \tag{5.111}$$

where ξ_n, $\zeta_n b$, and λ_n are defined in Section 5.2.1.

Noting that $\overline{\Delta h_{II}}[r, z, p]$ is cancelled by the first series term in $\overline{\Delta h_{III}}[r, z, p]$, one obtains the drawdown in the Laplace domain

$$\overline{\Delta h}[r, z, p] = \sum_{n=0}^{\infty} \lambda_n \left\{ 2 I_0[\xi_n r_w] \Omega_n^F + E_n \right\} K_0[\xi_n r] \cos[\zeta_n z], \tag{5.112}$$

where

$$E_n = 2\xi_n r_w I_1[\xi_n r_w] \overline{\Delta h}[r_w, b, p] \tag{5.113}$$

and

$$\Omega_n^F = -\frac{F_0}{2} - \sum_{m=1}^{\infty} (-1)^m F_m \frac{\xi_n^2 - q^2}{\xi_n^2 - q_m^2} = \Omega^{part} P_n \tag{5.114}$$

with

$$
P_n = \frac{Z_{\text{top}} - Z_{\text{bot}}}{b} + 2 \sum_{m=1}^{\infty} \frac{(-1)^m}{m\pi} \frac{\xi_n^2 - q^2}{\xi_n^2 - q_m^2}
$$
$$
\cdot \left\{ \sin\left[\frac{m\pi Z_{\text{top}}}{b}\right] - \sin\left[\frac{m\pi Z_{\text{bot}}}{b}\right] \right\}. \qquad (5.115)
$$

Both E_n and Ω^{part} contain the unknown $\overline{\Delta h}[r_{\text{w}}, b, p]$. It can be determined from the assumption that $\overline{\Delta h}[r_{\text{w}}, z, p]$ is independent of z at the well face by taking the average value

$$
\overline{\Delta h}^{\text{ave}}[r_{\text{w}}, p] = \overline{\Delta h}[r_{\text{w}}, b, p] = \overline{Q}^{\text{part}} G^{\text{part}}, \qquad (5.116)
$$

where

$$
\overline{Q}^{\text{part}} = 2\left(\overline{Q} - C_{\text{w}} h_{\text{w}0}\right) / D_{\text{s}}^{\text{part}}, \qquad (5.117)
$$

$$
D_{\text{s}}^{\text{part}} = D\omega^{\text{part}} + 2p C_{\text{w}} G^{\text{part}}, \qquad (5.118)
$$

$$
G^{\text{part}} = \sum_{m=0}^{\infty} \lambda_m^{\text{ave}} P_m I_0[\xi_m r_{\text{w}}] K_0[\xi_m r_{\text{w}}], \qquad (5.119)
$$

and D is defined in equation (5.56).

Putting $\overline{\Delta h}^{\text{ave}}[r_{\text{w}}, p]$ into equation (5.112) yields the final form of the *generalized well function*

$$
\boxed{\overline{\Delta h}[r, z, p] = \overline{Q}^{\text{part}} \sum_{n=0}^{\infty} \lambda_n \gamma_n^{\text{part}} K_0[\xi_n r] \cos[\zeta_n z]}, \qquad (5.120)
$$

where

$$
\gamma_n^{\text{part}} = D P_n I_0[\xi_n r_{\text{w}}] + 2\xi_n r_{\text{w}} I_1[\xi_n r_{\text{w}}] G^{\text{part}}. \qquad (5.121)
$$

The series in equation (5.120) is a dimensionless function of $[u, \sigma, \beta]$ but

$$
\overline{Q}^{\text{part}} = \frac{\left(\overline{Q} - C_{\text{w}} h_{\text{w}0}\right) b}{\pi T \left(Z_{\text{top}} - Z_{\text{bot}}\right)} \left\{ D + \frac{p C_{\text{w}} b \left(D\epsilon k_r / 2r_{\text{w}} + G^{\text{part}}\right)}{\pi T \left(Z_{\text{top}} - Z_{\text{bot}}\right)} \right\}^{-1} \qquad (5.122)
$$

requires more information about the pumping well. $\overline{Q}^{\text{part}}$ is also time dependent through p and $\overline{Q}$ for nonzero well storage C_{w} and nonsteady discharge rate $Q[t]$, respectively.

• **Remark**

For a full-penetration pumping well, $F_m = 0$ for $m \neq 0$. Consequently, $P_n = 1$ and all parameters with superscript "part" are reduced to the corresponding ones without it and equation (5.120) for partial-penetration pumping wells is reduced to equation (5.89) for full-penetration pumping wells.

Following the discussion of special cases in the last section, one can render all well functions listed in Chapters 4 and 5 (except Hantush's solutions for leaky aquifers) as the special cases of the general solution, equation (5.120).

5.5 Problems, Keys, and Suggested Readings

• **Problems**

    1. Find

$$\int_{r_\mathrm{w}}^{\infty} J_o[ar] \frac{\partial}{\partial r} \left(r \frac{\partial \overline{\Delta h}}{\partial r} \right) dr. \tag{5.123}$$

    2. Find the inverse Hankel transform of

$$\frac{J_0[ar_\mathrm{w}]}{a^2 + q^2} \frac{\cosh[\eta z]}{c\eta \sinh[\eta b] + \cosh[\eta b]}, \quad c = \alpha_\mathrm{y}/p. \tag{5.124}$$

    3. Find the inverse Hankel transform of

$$\frac{ar_\mathrm{w} J_1[ar_\mathrm{w}]}{a^2 + q^2} \frac{\cosh[\eta z]}{c\eta \sinh[\eta b] + \cosh[\eta b]}. \tag{5.125}$$

    4. Assume that the solution to the problem of well storage and skin effect for a partial-penetration pumping well can be represented by

$$\overline{\Delta h}[r, z, p] = \sum_{n=0}^{\infty} A_n K_0[\xi_n r] \cos [\zeta_n z]. \tag{5.126}$$

Use Condition 3k in equation (5.97) to determine A_n.

• **Keys**

Key 1

The desired integral is a Hankel transform. Let the transform be

$$A = \int_{r_\mathrm{w}}^{\infty} r J_0[ar] \frac{1}{r} \frac{\partial}{\partial r} \left(r \frac{\partial \overline{\Delta h}}{\partial r} \right) dr. \tag{5.127}$$

Using integration by parts yields

$$
\begin{aligned}
A &= J_0[ar]\left(r\frac{\partial\overline{\Delta h}}{\partial r}\right)\bigg|_{r_w}^{\infty} - \int_{r_w}^{\infty} r\frac{\partial\overline{\Delta h}}{\partial r}\frac{\partial J_0[ar]}{\partial r}dr \\
&= -J_0[ar_w]\left(r\frac{\partial\overline{\Delta h}}{\partial r}\right)\bigg|_{r=r_w} + a\int_{r_w}^{\infty} rJ_1[ar]\frac{\partial\overline{\Delta h}}{\partial r}dr. \quad (5.128)
\end{aligned}
$$

Use integration by parts again to yield

$$
\begin{aligned}
A &= -J_0[ar_w]\left(r\frac{\partial\overline{\Delta h}}{\partial r}\right)\bigg|_{r=r_w} \\
&\quad + \left(arJ_1[ar]\overline{\Delta h}\right)\big|_{r_w}^{\infty} - a\int_{r_w}^{\infty}\overline{\Delta h}\frac{\partial\left(rJ_1[ar]\right)}{\partial r}dr. \quad (5.129)
\end{aligned}
$$

Simplify the above integral by using the relation that

$$
x\partial J_1[x]/\partial x = xJ_0[x] - J_1[x] \quad (5.130)
$$

and noting that $J_1[x]$ is proportional to $x^{-3/2}$ as $x \to \infty$ to obtain

$$
\begin{aligned}
A &= -J_0[ar_w]\left(r\frac{\partial\overline{\Delta h}}{\partial r}\right)\bigg|_{r=r_w} \\
&\quad -ar_wJ_1[ar_w]\overline{\Delta h}[r_w, z, p] - a^2\int_{r_w}^{\infty}\overline{\Delta h}J_0[ar]rdr. \quad (5.131)
\end{aligned}
$$

Now, as an example, use the discharge condition at r_w in equation (5.39) to get

$$
\boxed{A = \frac{QJ_0[ar_w]}{2\pi Tp} - ar_wJ_1[ar_w]\overline{\Delta h}[r_w, z, p] - a^2\overline{\Delta h}^*[a, z, p]}. \quad (5.132)
$$

Key 2

Let the inverse Hankel transform be

$$
A = \int_0^{\infty} aJ_0[ar]\frac{J_0[ar_w]}{a^2 + q^2}\frac{\cosh[\eta z]}{c\eta\sinh[\eta b] + \cosh[\eta b]}da. \quad (5.133)
$$

The integrand is an odd function of a, so the technique used for Example 3H in Chapter 3 is applicable. Let $J_0[ar] = 0.5(H_0^{(1)}[ar] + H_0^{(2)}[ar])$. Noting that $H_0^{(2)}[-ar] = H_0^{(1)}[ar]$, the integral becomes

$$A = \frac{1}{2} \int_{-\infty}^{\infty} \frac{a J_0[ar_w] H_0^{(1)}[ar]}{a^2 + q^2} \frac{\cosh[\eta z]}{c\eta \sinh[\eta b] + \cosh[\eta b]} da. \tag{5.134}$$

Let $a = i\xi$, then use the relations

$$H_0^{(1)}[ix] = -\frac{i2}{\pi} K_0[x], \quad J_0[ix] = I_0[x] \tag{5.135}$$

to get

$$
\begin{aligned}
A &= \frac{1}{2} \int_{-i\infty}^{i\infty} \frac{i\xi J_0[i\xi r_w] H_0^{(1)}[i\xi r]}{\xi^2 - q^2} \frac{\cos[\zeta z]}{-c\zeta \sin[\zeta b] + \cos[\zeta b]} di\xi \\
&= \frac{-i}{\pi} \int_{-i\infty}^{i\infty} \frac{\xi I_0[\xi r_w] K_0[\xi r]}{\xi^2 - q^2} \frac{\cos[\zeta z]}{c\zeta \sin[\zeta b] - \cos[\zeta b]} d\xi, \tag{5.136} \\
\zeta^2 &= \left(\xi^2 - q^2\right) / k_{\mathrm{D}}.
\end{aligned}
$$

(Also, see Examples 3H and 5H in Chapter 3.)

The above integrand is an even function of ζ, so there is no branch cut. Take a clockwise integration in the right half of the ξ-plane. The contribution at singular point $\xi = q$ is $I_0[qr_w] K_0[qr]$. The contribution at $\xi = \xi_n$ can be obtained from the case of a well with infinitesimal diameter. The net result is

$$\boxed{A = I_0[qr_w] K_0[qr] - 2 \sum_{n=0}^{\infty} \lambda_n I_0[\xi_n r_w] K_0[\xi_n r] \cos[\zeta_n z]}. \tag{5.137}$$

Key 3

Recognizing the relations that

$$\frac{\partial J_0[x]}{\partial x} = -J_1[x], \quad \frac{\partial I_0[x]}{\partial x} = I_1[x] \tag{5.138}$$

we obtain the desired transform by differentiating the A in Problem 2 with respect to r_w:

$$r_w \frac{\partial A}{\partial r_w} = r_w \frac{\partial}{\partial r_w} \int_0^{\infty} a J_0[ar] \frac{J_0[ar_w]}{a^2 + q^2} \frac{\cosh[\eta z]}{(\alpha_y \eta / p) \sinh[\eta b] + \cosh[\eta b]} da$$

$$= r_w \frac{\partial}{\partial r_w} \left\{ I_0[qr_w] K_0[qr] \right.$$

$$\left. - 2 \sum_{n=0}^{\infty} \lambda_n I_0[\xi_n r_w] K_0[\xi_n r] \cos[\zeta_n z] \right\}. \tag{5.139}$$

Simplification leads to

$$\int_0^\infty \frac{a J_0[ar]}{a^2 + q^2} \frac{a r_w J_1[ar_w] \cosh[\eta z]}{(\alpha_y \eta / p) \sinh[\eta b] + \cosh[\eta b]} da$$

$$= -q r_w I_1[qr_w] K_0[qr]$$

$$+ 2 \sum_{n=0}^{\infty} \lambda_n \xi_n r_w I_1[\xi_n r_w] K_0[\xi_n r] \cos[\zeta_n z]. \tag{5.140}$$

Key 4

Substituting

$$\overline{\Delta h}[r, z, p] = \sum_{n=0}^{\infty} A_n K_0[\xi_n r] \cos[\zeta_n z] \tag{5.141}$$

into Condition 3k yields

$$\sum_{n=0}^{\infty} A_n \left\{ \xi_n r_w K_1[\xi_n r_w] \cos[\zeta_n z] \right\} = \Omega \delta. \tag{5.142}$$

Here we encounter one obstacle in determining A_n. The factor $\overline{\Delta h}[r_w, z, p]\delta$ in Condition 3k prevents the orthogonal property of $\cos[\zeta_n z]$ from being applied over the screen interval. To proceed further, assume that drawdown at the well face is independent of z, for example,

$$\overline{\Delta h}[r_w, z, p] = \overline{\Delta h}[r_w, b, p], \tag{5.143}$$

implying no vertical flow along the well face.

Under the assumption of equation (5.143), multiplying equation (5.142) by $\cos[\zeta_m z]$, integrating the result with respect to z from 0 to b, and using the orthogonal property of $\cos[\zeta_n z]$ yields

$$A_n = \frac{2\Omega^{part} \left(\sin[\zeta_n Z_{top}] - \sin[\zeta_n Z_{bot}] \right)}{\xi_n r_w K_1[\xi_n r_w] \left(\zeta_n b + 0.5 \sin[2\zeta_n b] \right)}. \tag{5.144}$$

This A_n is identical to equation (5.69) in form. However, Ω^{part} bears implicitly a factor of drawdown $\overline{\Delta h}[r_{\text{w}}, b, p]$ at the well face. Therefore, the problem cannot be solved this way. An alternative solution can be pursued, however, by following Moench's (1997) procedures for a slightly different problem.

• **Suggested Readings**

See the suggested readings in Chapter 4.

5.6   Notations

Symbol	Definition	SI Unit	Dimension
a	Hankel transform parameter	m^{-1}	L^{-1}
b	Aquifer thickness	m	L
c	α_y/p	m	L
C_w	Cross-sectional area of well bore	m^2	L^2
D	Eq. (5.56),	-	-
D_s, D_s^{part}	Eq. (5.88), (5.118)	$m^2 s^{-1}$	$L^2 T^{-1}$
E_n	Eq. (5.52)	-	-
F_m	Eq. (5.100)	ms	LT
h	Hydraulic head	m	L
h_{w0}	Slug height	m	L
G, G^{part}	Eq. (5.55), (5.119)	-	-
$H_0^{(1)}$, $H_0^{(2)}$	Hankel functions	-	-
I_0, I_1	Modified Bessel function, 1st kind	-	-
J_0, J_1	Bessel function	-	-
k, k_z, k_r	Hydraulic conductivity	ms^{-1}	LT^{-1}
k_D	k_z/k_r	-	-
K_0, K_1	Modified Bessel function, 2nd kind	-	-
O_{top}, O_{bot}	Top or bottom of observation screen	m	L
p, p_j	Laplace parameter, $p_j = (j \ln 2)/t$	s^{-1}	T^{-1}
P_n	Eq. (5.115)		
Q	Discharge rate	$m^3 s^{-1}$	$L^3 T^{-1}$
q, q_m	$q^2 = p/\alpha_s, \quad q_m^2 = q^2 + k_D(m\pi/b)^2$	m^{-1}	L^{-1}
r	Radial distance	m	L
r_s	Skin thickness (radius)	m	L
r_w	Well radius	m	L
S_s	Specific storage	m^{-1}	L^{-1}
S_y	Specific yield	-	-
t	Time	s	T
t_s	Time, $t_s = Tt/r^2 S$	-	-
T	Transmissivity, kb or $k_r b$	$m^2 s^{-1}$	$L^2 T^{-1}$
u	$r^2 S/4Tt = 1/4t_s$	-	-
x, z	Distance	m	L
Z_{top}, Z_{bot}	Top or bottom of pumping screen	m	L

Symbol	Definition	SI Unit	Dimension
$\overline{Q}, \overline{Q}^{\text{part}}$	Eq. (5.87), (5.117)	ms	LT
α_s	Hydraulic diffusivity, $T/S = k_r/S_s$,	$m^2 s^{-1}$	$L^2 T^{-1}$
α_y	k_z/S_y	ms^{-1}	LT^{-1}
β	$(r/b)^2 k_D$	-	-
$\gamma, \gamma^{\text{part}}$	Eq. (5.58), (5.121)	-	-
Δh	Drawdown	m	L
Δh_w	Change of water level in well	m	L
$\overline{\Delta h}$	Laplace transform of Δh	ms	LT
$\overline{\Delta h}^*$	Hankel transform of $\overline{\Delta h}$	$m^{-1}s$	$L^{-1}T$
ϵ	Well efficiency indicator	s	T
$\zeta_n b$	Root of $\tan[\zeta_n b] = 1/c\zeta_n$	-	-
η	$\sqrt{(a^2 + q^2)/k_D}$	m^{-1}	L^{-1}
θ_n	$\theta_n = \zeta_n b - n\pi, \quad n = 0, 1, 2...$	-	-
$\lambda_n, \lambda_n^{\text{ave}}, \lambda_n^{\text{part}}$	Eq. (5.16), (5.21), (5.38)	-	-
ξ_n	$\sqrt{\zeta_n^2 k_D + q^2}$	m^{-1}	L^{-1}
σ	S/S_y	-	-
$\omega, \omega^{\text{part}}$	Eq. (5.81), (5.98)	$m^2 s^{-1}$	$L^2 T^{-1}$
$\Omega, \Omega^{\text{part}}$	Eq. (5.81), (5.98)	ms	LT
Ω_n^{F}	Eq. (5.114)	ms	LT

Chapter 6

HEAT TRANSFER AND GROUNDWATER FLOW

The Earth's temperature at shallow depth is influenced by diurnal and seasonal temperature variations at the ground surface. Below the depth influenced by seasonal variations, the temperature generally increases with depth. To first approximation, the temperature-depth curve is usually linear in the depth zones that are devoid of fluid flow. Perturbations to the linear temperature-depth relation occur where thermal conductivity changes or heat sources are present. Topographic relief or climatic change may also disturb the linear relation at shallow depths. In places where significant fluid flow exists, the isotherms can be so severely altered that the geothermal gradient is locally reversed. We will evaluate the effect of fluid flow on temperature distribution and ultimately infer fluid flow pattern from the temperature distribution. Note that *geotherm* means an isotherm in the subsurface, but is frequently referred to as a temperature-depth profile. The latter usage is adopted here for convenience.

Fluid can also be driven to flow by a strong temperature gradient or a density gradient that causes gravitational instability. Such free convection is not the subject of this chapter. Nevertheless the last section will be devoted to discussion of fluid pressurization induced by heating. Here, we are mainly interested in how groundwater flow driven by a hydraulic gradient can affect the thermal regime that is otherwise dominated by conductive heat transfer. We will call the process of heat transfer by groundwater flow *advective heat transfer* – it is a type of convective heat transfer induced by the force of hydraulic gradient.

6.1   Advective Heat Transfer

• **Equations**

In an advective system, heat flux $\mathbf{q}$ consists of conductive $\mathbf{q}^c$ and advective $\mathbf{q}^a$ fluxes:

$$
\begin{aligned}
\mathbf{q} &= \mathbf{q}^c + \mathbf{q}^a, \\
\mathbf{q}^c &= -k_T \nabla T, \\
\mathbf{q}^a &= \rho c \mathbf{v} \left(T - T_o \right),
\end{aligned}
\tag{6.1}
$$

where k_T is the thermal conductivity of a fluid-saturated solid $[Wm^{-1}K^{-1}]$, ρ is the mass density $[kg\,m^{-3}]$, c is the specific heat $[J\,kg^{-1}K^{-1}]$ at constant volume, ρc is the volumetric heat capacity of fluid $[Jm^{-3}K^{-1}]$, T is the temperature $[K]$, T_o is a reference temperature, $\mathbf{v}$ is Darcy's velocity $[ms^{-1}]$, and $\mathbf{q}$ is in units of Wm^{-2}.

According to the principle of energy conservation, the net energy outflow through the surface of an elementary control volume is equal to the decline of energy within the control volume

$$
\iint \mathbf{q} \cdot \hat{\mathbf{n}} ds = - \iiint (\rho c)_s \frac{\partial T}{\partial t} dx dy dz,
\tag{6.2}
$$

where $\hat{\mathbf{n}}$ is the outward unit normal, $(\rho c)_s$ is the volumetric heat capacity of the fluid-rock mixture, and t is the time $[s]$. Following the divergence theorem

$$
\iint \mathbf{q} \cdot \hat{\mathbf{n}} ds = \iiint \nabla \cdot \mathbf{q} dx dy dz,
\tag{6.3}
$$

we have

$$
\boxed{ \nabla \cdot \mathbf{q} = -(\rho c)_s \frac{\partial T}{\partial t} }.
\tag{6.4}
$$

Substituting for $\mathbf{q}$ and simplifying the above relation yields the advective heat transfer equation

$$
\boxed{ \nabla \cdot (k_T \nabla T) - \rho c \mathbf{v} \cdot \nabla T = (\rho c)_s \frac{\partial T}{\partial t} }
\tag{6.5}
$$

in a steady field of incompressible groundwater flow in which

$$
\nabla \cdot \mathbf{v} = 0.
\tag{6.6}
$$

If thermal conductivity k_T is independent of position, then

$$\nabla^2 T - \frac{\rho c}{k_T}\mathbf{v}\cdot\nabla T = \frac{(\rho c)_s}{k_T}\frac{\partial T}{\partial t}. \tag{6.7}$$

• **Dimensionless Relations**

Now if the temperature, distance, and time are scaled by some characteristic temperature T_c, length L, and time t_c, respectively, to define a set of dimensionless parameters

$$T' = T/T_c, \qquad x' = x/L, \qquad t' = t/t_c, \tag{6.8}$$

the advective heat transfer equation becomes

$$\nabla'^2 T' - \frac{\rho c}{k_T}L\mathbf{v}\cdot\nabla' T' = \frac{(\rho c)_s}{k_T t_c}L^2\frac{\partial T'}{\partial t'}, \tag{6.9}$$

where ∇'^2 and ∇' are, respectively, the Laplacian and gradient operators in the primed coordinates. From this relation, we can define a dimensionless *thermal Peclet number*

$$N_{Pe} = \frac{\rho c}{k_T}Lv = \frac{\rho c v(T'-T'_o)}{k_T\,(T'-T'_o)/L} \tag{6.10}$$

as a measure of the relative strength of the advective and conductive heat transfer, where T'_o is a dimensionless reference temperature.

Darcy's velocity for fluid with a variable density is

$$\mathbf{v} = -\frac{K}{\mu}\nabla\,(\rho g h) = -\frac{K}{\mu}\nabla\,(P + \rho g z)\,, \tag{6.11}$$

where K is the permeability of the solid medium $[m^2]$, μ is the dynamic viscosity of the fluid $[Nm^{-2}s]$, P is the pressure $[Pa]$, and g is the gravitational acceleration $[ms^{-2}]$.

Let us assume that density ρ is a function of temperature only,

$$\rho = \rho_o[1 - \alpha(T - T_o)] \tag{6.12}$$

and that the pressure is hydrostatic,

$$P = \rho_o g(z_s - z), \tag{6.13}$$

where α is the thermal expansivity $[K^{-1}]$, z is upward positive, and z_s is the surface elevation). Then, Darcy's velocity becomes

$$\begin{aligned}
\mathbf{v} &= \frac{K}{\mu} \rho_0 g \nabla \left[\alpha \left(T - T_0\right) z\right] \\
&= \frac{K}{\mu} \rho_0 g \alpha \left[\left(T - T_0\right) + z \nabla T\right].
\end{aligned} \tag{6.14}$$

The first term in $\mathbf{v}$ is due to buoyancy, $g\rho_0 \alpha \left(T - T_0\right)/\mu$, induced by density change. Thus if the v in the Peclet number is replaced by the buoyancy term, the resulting number

$$N_{\mathrm{Ra}} = \frac{\rho c L}{k_{\mathrm{T}}} \frac{K g \rho_0 \alpha \left(T - T_0\right)}{\mu} \tag{6.15}$$

is the *Rayleigh number* for free thermal convection. Note that in an active convective system, geothermal gradient $\partial T/\partial z$ is very close to an adiabatic gradient and the second term is typically very small compared to the first term.

6.1.1   Steady Vertical Flow

• **Formulation**

Consider a steady vertical flow between $z = Z_1$ and $z = Z_2$. Equation (6.7) becomes

$$\frac{\partial^2 T}{\partial z^2} - \frac{\rho c v}{k_{\mathrm{T}}} \frac{\partial T}{\partial z} = 0. \tag{6.16}$$

This is to be solved under the boundary conditions (Figure 6.1) that

$$\begin{aligned}
T &= T_1 \quad \text{at} \quad z = Z_1 \\
T_2 &= T_2 \quad \text{at} \quad z = Z_2.
\end{aligned} \tag{6.17}$$

• **Solution**

The ordinary differential equation (6.16) has a general solution of the form

$$T[z] = A + B e^{\gamma z}, \quad \gamma = \frac{\rho c v}{k_{\mathrm{T}}}. \tag{6.18}$$

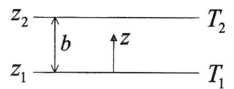

Figure 6.1: Geometry for the effect of a steady vertical flow of groundwater on temperature.

Imposing the boundary conditions yields

$$A = \frac{T_1 e^{\gamma Z_2} - T_2 e^{\gamma Z_1}}{e^{\gamma Z_2} - e^{\gamma Z_1}},$$

$$B = \frac{T_2 - T_1}{e^{\gamma Z_2} - e^{\gamma Z_1}}. \tag{6.19}$$

Using T_1 as the reference temperature, the normalized temperature is

$$\frac{T - T_1}{T_2 - T_1} = \frac{\exp\left[N_{\mathrm{Pe}}\left(z - Z_1\right)/b\right] - 1}{\exp\left(N_{\mathrm{Pe}}\right) - 1}, \tag{6.20}$$

$$N_{\mathrm{Pe}} = \frac{\rho c v b}{k_{\mathrm{T}}}, \quad b = Z_2 - Z_1.$$

Figure 6.2 depicts the temperature-depth relations at different thermal Peclet numbers.

The geothermal gradient is

$$\frac{\partial T}{\partial z} = \gamma B e^{\gamma z} = \gamma (T - A) = \frac{N_{\mathrm{Pe}}}{b}(T - A), \tag{6.21}$$

which can be normalized

$$\begin{aligned}
\frac{b}{T_2 - T_1}\frac{\partial T}{\partial z} &= N_{\mathrm{Pe}}\frac{T - T_1}{T_2 - T_1} + N_{\mathrm{Pe}}\frac{T_1 - T + B e^{\gamma z}}{T_2 - T_1} \\
&= N_{\mathrm{Pe}}\frac{T - T_1}{T_2 - T_1} - \frac{\rho c v\left(T - T_1\right) - k_{\mathrm{T}}\gamma B e^{\gamma z}}{k_{\mathrm{T}}\left(T_2 - T_1\right)/b} \\
&= N_{\mathrm{Pe}}\frac{T - T_1}{T_2 - T_1} - \frac{q}{k_{\mathrm{T}}(T_2 - T_1)/b}. \tag{6.22}
\end{aligned}$$

The total heat flux in the depth zone considered is

$$\boxed{q = \rho c v\left(T - T_1\right) - k_{\mathrm{T}}\frac{\partial T}{\partial z}.} \tag{6.23}$$

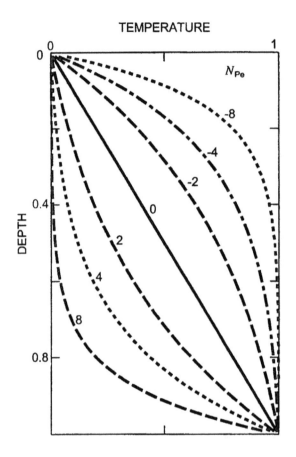

Figure 6.2: Profiles of temperature versus depth at different thermal Peclet numbers.

In the absence of a heat source or lateral groundwater flow, this equation implies that q remains constant within the depth range between Z_1 and Z_2. Problem 1 shows that this expression of energy conservation can also be used to derive equation (6.22).

- **Application**

  □ According to equation (6.22), a linear regression of normalized geothermal gradient versus normalized temperature yields the Peclet number N_{Pe} which can then be used to estimate Darcy's velocity v if thermal conductivity k_T of the saturated rock is measured.

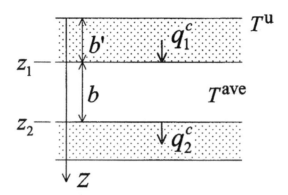

Figure 6.3: Geometry for the effect of horizontal groundwater flow on temperature distribution.

☐ From the intercept $-qb/k_T(T_2 - T_1)$, the total heat flux q (with respect to reference temperature T_1) can be estimated (Mansure and Reiter, 1979).

6.1.2 Steady Horizontal Flow

• Formulation

Consider the effect of horizontal groundwater flow on temperature distribution in a confined aquifer bounded between depths $z = Z_1$ and $z = Z_2$ (where z is downward positive, Figure 6.3). The governing differential equation is

$$k_T \frac{\partial^2 T}{\partial x^2} + k_T \frac{\partial^2 T}{\partial z^2} - \rho c v \frac{\partial T}{\partial x} = 0. \qquad (6.24)$$

Its solution depends on the chosen boundary conditions.

Let us deal with the mean temperature between depths Z_1 and Z_2,

$$\frac{1}{b} \int_{Z_1}^{Z_2} \left\{ k_T \frac{\partial^2 T}{\partial x^2} + k_T \frac{\partial^2 T}{\partial z^2} - \rho c v \frac{\partial T}{\partial x} \right\} dz = 0, \qquad (6.25)$$

where $b = Z_2 - Z_1$ is the aquifer thickness. Integrating the above relation results in

$$\frac{\partial^2 T^{\text{ave}}}{\partial x^2} - \frac{\rho c v}{k_T} \frac{\partial T^{\text{ave}}}{\partial x} = \frac{q_2^c - q_1^c}{b k_T}, \qquad (6.26)$$

where T^{ave} is the mean temperature

$$T^{\text{ave}} = \frac{1}{b} \int_{z_1}^{z_2} T \, dz \tag{6.27}$$

and q^c is the conductive heat flux in the overlying or underlying aquitard $q^c = -k_T \frac{\partial T}{\partial z}$.

• Case 1: Constant Difference in Flux

If $\Delta q^c = q_2^c - q_1^c$ is constant, equation (6.26) has a solution of the form

$$T^{\text{ave}}[x] = A + B \exp\left[\frac{\rho c v x}{k_T}\right] - \frac{\Delta q^c}{\rho c v b} x, \tag{6.28}$$

where A and B are undetermined coefficients. Because the horizontal temperature gradient is

$$\frac{\partial T^{\text{ave}}}{\partial x} = B \frac{\rho c v}{k_T} \exp\left[\frac{\rho c v x}{k_T}\right] - \frac{\Delta q^c}{\rho c v b}, \tag{6.29}$$

one obtains

$$
\begin{aligned}
T^{\text{ave}}[x] &= A + \frac{k_T}{\rho c v} \frac{\partial T^{\text{ave}}}{\partial x} + \frac{\Delta q^c}{\rho c v b}\left(\frac{k_T}{\rho c v} - x\right) \\
&= A + \frac{b}{N_{\text{Pe}}} \frac{\partial T^{\text{ave}}}{\partial x} + \frac{\Delta q^c}{k_T} N_{\text{Pe}} \left(\frac{b}{N_{\text{Pe}}} - x\right).
\end{aligned}
\tag{6.30}
$$

• Application

☐ A linear regression of $T^{\text{ave}}[x]$ versus $\partial T^{\text{ave}}/\partial x$ and x yields coefficient b/N_{Pe} or $k_T/\rho c v$ from which Darcy's velocity v can be estimated if $k_T/\rho c$ is given.

☐ Δq^c can be estimated from the regression coefficient for the x term. It can also be estimated if heat flow measurements are made in the overlying and underlying aquitards.

• Solution 2: Constant Basal Flux

If q_2^c is constant but q_1^c varies laterally, let the temperature T^u at the top of the upper aquitard be steady (Figure 6.3). Then equation (6.26) becomes

$$\frac{\partial^2 T^{\text{ave}}}{\partial x^2} - \frac{\rho c v}{k} \frac{\partial T^{\text{ave}}}{\partial x} - \frac{k'}{bk} \frac{(T^{\text{ave}} - T^u)}{b'} = \frac{q_2^c}{bk}, \tag{6.31}$$

where k is the thermal conductivity of saturated rock (k_T), ρc is the heat capacity of water, k' is the thermal conductivity of aquitard, b' is upper aquitard thickness, and b is the thickness of the confined aquifer.

The general solution is

$$T^{ave}[x] \;\; = \;\; Ae^{\eta^- x} + Be^{\eta^+ x} + T_r, \tag{6.32}$$

$$\eta^{\pm} = \frac{\rho cv}{2k}\left(1 \pm \sqrt{1 + \frac{4k'k}{bb'(\rho cv)^2}}\right),$$

$$T_r = T^u - q_2^c b'/k',$$

where $q_2^c b'/k'$ would represent the temperature difference across the upper aquitard if $v = 0$ (and in this case $q_1^c = q_2^c$).

Consider the case that $T^{ave} = T_0$ at $x = 0$ and that T^{ave}_∞ is finite at $x = \infty$. Since $\eta^+ > 0$, then $B = 0$ and

$$A = T_0 - T_r. \tag{6.33}$$

Simplification yields

$$\frac{T^{ave} - T_r}{T_0 - T_r} = e^{\eta^- x}. \tag{6.34}$$

In general, if $T^{ave} = T_0$ at $x = 0$ and $T^{ave} = T_1$ at $x = x_1$, the coefficients are

$$A \;\; = \;\; \frac{(T_1 - T_r) - (T_0 - T_r)\exp[\eta^+ x]}{\exp[\eta^- x_1] - \exp[\eta^+ x_1]},$$

$$B \;\; = \;\; \frac{-(T_1 - T_r) + (T_0 - T_r)\exp[\eta^- x]}{\exp[\eta^- x_1] - \exp[\eta^+ x_1]}. \tag{6.35}$$

6.1.3 Lateral Gradient $\partial T/\partial x = $ Constant

• Formulation

If the lateral gradient $\partial T/\partial x = \Gamma$ is a constant, then equation (6.24) becomes

$$\frac{\partial^2 T}{\partial z^2} = \frac{\rho cv\Gamma}{k}. \tag{6.36}$$

In the case that

$$T = T_1 \;\; \text{at} \;\; z = Z_1 \quad \text{and} \quad T = T_2 \;\; \text{at} \;\; z = Z_2, \tag{6.37}$$

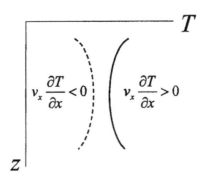

Figure 6.4: The temperature-depth curve is concave to the right in the direction of the postive T-axis if temperature increases in the groundwater flow direction; otherwise it is concave to the left.

the solution is

$$T = Az^2 + Bz + C, \qquad (6.38)$$

$$A = \frac{\rho c v \Gamma}{2k}, \quad B = \frac{(T_2 - AZ_2^2) - (T_1 - AZ_1^2)}{Z_2 - Z_1}$$

$$C = \frac{Z_2(T_1 - AZ_1^2) - Z_1(T_2 - AZ_2^2)}{Z_2 - Z_1}.$$

- **Application**

☐ If $v\Gamma > 0$, then $\partial^2 T/\partial z^2 = 2A > 0$ and the T versus z curve is concave toward the positive $T-$axis (Figure 6.4), i.e., temperature increases in the direction of groundwater flow.

☐ If $v\Gamma < 0$, then $\partial^2 T/\partial z^2 < 0$ and the T versus z curve is concave toward the negative $T-$axis.

☐ From a set of geotherms at different locations, we know the sign of Γ ($= \partial T/\partial x$). So the sign of v can be determined to tell the direction of groundwater flow. Actually, one can qualitatively infer the flow direction by examining equation (6.36) without going through the mathematical exercise.

It is cautioned that vertical groundwater flow can also cause the geotherm to be concave toward or away from the temperature axis as depicted in Figure 6.2. If several profiles are available, the influences by horizontal or vertical flow should be distinguishable.

• **Field Example**

Figure 6.5 depicts a set of geotherms from a superfund site in California. Their relative locations of observation are shown in Figure 6.6. All measurements were made in alluvium at depths below 45 m so that the effect of seasonal temperature variations at the ground surface is negligible. Observation wells 1 and 16 demonstrate two extreme examples. Well 1 reveals the incursion of cold water whereas Well 16 appears to be free of groundwater influence and its temperature gradient is likely regional in nature (a value of 5°C has been subtracted from the temperature readings at Well 16 to allow plotting). All profiles but Well 16 show the influence of groundwater flow. The clustering of the geotherms shown in Figure 6.5 indicates different extent of influence. The lines of equal influence are depicted approximately as dashed lines in Figure 6.6. As inferred from Figures 6.4 and 6.5, the mass of the cold groundwater flows northwest and gains heat along its travel path. Apparently Well 16 is isolated from the groundwater influence by a to-be-located groundwater barrier. Well 14 is slightly affected and Well 15 appears to have been affected by a reversed flow.

6.2 Heat Sources in Regional Groundwater Flow

We will assume that the aquifer is homogeneous and isotropic and that the groundwater flow is steady. The solutions obtained here are also applicable to some solute transport problems.

6.2.1 Instantaneous Point Source

Assume that heat energy M is released instantaneously at point $(0, 0, 0)$ in a homogeneous, isotropic, infinite medium with a uniform regional groundwater flow velocity **v**. The temperature is

$$T[x, y, z, t] = \frac{M}{(4\pi t)^{3/2}(D_x D_y D_z)^{1/2}(\rho c)_s} E, \qquad (6.39)$$

where

$$E = \exp\left[-\frac{X^2}{4D_x t} - \frac{Y^2}{4D_y t} - \frac{Z^2}{4D_z t}\right], \qquad (6.40)$$

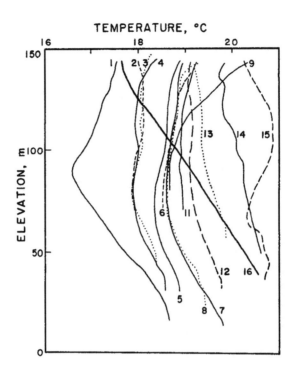

Figure 6.5: Temperature-depth profiles, showing the influence of lateral ground-water flow. (5°C has been subtracted from Profile 16 for plotting.)

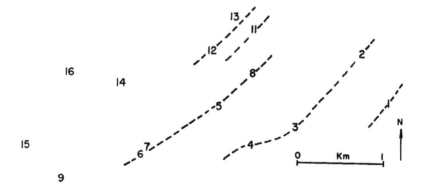

Figure 6.6: Locations of temperature observation wells. Dashed lines show the approximate locations of equal influence on the temperature-depth profiles by the northwest flowing groundwater.

$$X \;=\; x - \bar{v}_x t, \quad \bar{v}_x = \frac{v_x}{\phi},$$

$(\rho c)_s$ is the volumetric heat capacity of the saturated medium $[Jm^{-3}K^{-1}]$, D_x is the thermal dispersion coefficient in the x direction $[m^2 s^{-1}]$, v_x is Darcy's velocity in the x direction $[ms^{-1}]$, and $\bar{v}_x$ is the linearized velocity to account for the medium with a porosity ϕ. Other parameters can be defined by the cyclic permutation of subscripts. This solution is modified from the thermal point source solution (Section 2.1.1).

Equation (6.39) can be viewed as the result for a moving point source. Alternatively it can be regarded as the medium (groundwater) moving over a source point in the space.

6.2.2 Continuous Point Source

By convolution, the instantaneous point source solution can be extended to the solution for a continuous point source,

$$\begin{aligned} T[x,y,z,t] \;=\;& \frac{1}{(4\pi)^{3/2}(D_x D_y D_z)^{1/2}(\rho c)_s} \\ & \cdot \int_0^t \frac{M[t-\tau]}{\tau^{3/2}} \exp\left[-\frac{(x - \bar{v}_x \tau)^2}{4 D_x \tau} - \ldots \right] d\tau, \quad (6.41) \end{aligned}$$

where $M[t]$ is now the heating rate, $[W]$, and ... means the rest of the y and z terms.

The integrand can be simplified by expanding and rearranging the exponent of the exponential term,

$$E \;=\; \exp[\ldots] = \exp\left[\frac{x \bar{v}_x}{2 D_x} + \ldots \right] \exp\left[-\frac{\alpha^2}{\tau} - \beta^2 \tau \right], \quad (6.42)$$

$$\alpha^2 = \frac{x^2}{4 D_x} + \ldots, \qquad \beta^2 = \frac{\bar{v}_x^2}{4 D_x} + \ldots.$$

For a constant heating rate, we obtain

$$\begin{aligned} T[x,y,z,t] \;=\;& \frac{M}{8\pi^{3/2}\sqrt{D_x D_y D_z}(\rho c)_s} \exp\left[\left(\frac{\bar{v}_x x}{2 D_x} + \ldots \right) \right] \\ & \cdot \int_0^t \frac{\exp\left[-\alpha^2/\tau - \beta^2 \tau \right]}{\tau^{3/2}} d\tau. \quad (6.43) \end{aligned}$$

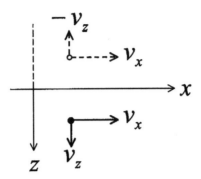

Figure 6.7: Image source (open circle) of a moving point source in a semi-infinite medium.

Let $\xi = \alpha/\sqrt{\tau}$. Then the integral can be simplified further to yield

$$T[x, y, z, t] = \frac{M}{8\pi^{3/2}\sqrt{D_x D_y D_z}(\rho c)_s} \exp\left[\left(\frac{\bar{v}_x x}{2D_x} + \ldots\right)\right]$$
$$\cdot \frac{2}{\alpha} \int_{\alpha/\sqrt{t}}^{\infty} \exp\left[-\xi^2 - \frac{\alpha^2 \beta^2}{\xi^2}\right] d\xi. \tag{6.44}$$

According to the result in Problem 4, the final integration is

$$\boxed{T[x, y, z, t] = \frac{M}{8\pi^{3/2}\sqrt{D_x D_y D_z}(\rho c)_s} \exp\left[\left(\frac{\bar{v}_x x}{2D_x} + \ldots\right)\right] F}, \tag{6.45}$$

where

$$F = \frac{\sqrt{\pi}}{2\alpha}\left\{e^{2\alpha\beta}\text{erfc}\left[\frac{\alpha + \beta t}{\sqrt{t}}\right] + e^{-2\alpha\beta}\text{erfc}\left[\frac{\alpha - \beta t}{\sqrt{t}}\right]\right\}. \tag{6.46}$$

(see also Sections 3.2.7 and 7.2.1.)

6.2.3   Influence of Ground Surface

Temperature rise originating from a point source at $(0, 0, \zeta)$ in a semi-infinite medium can be obtained by superposition of an image source at $(0, 0, -\zeta)$ to make the temperature rise vanish at the ground surface

$$\boxed{T^{\text{semi}}[x, y, z, t; \zeta] = T[x, y, z - \zeta - \bar{v}t, t] - T[x, y, z + \zeta + \bar{v}t, t]} \tag{6.47}$$

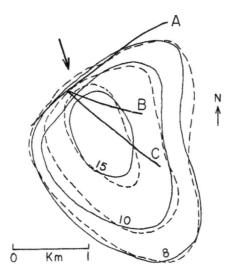

Figure 6.8: Isotherms for observed (solid curves) and simulated (dash curves) temperature at the depth of 150 m in the overburden above a fracture-controlled geothermal reservoir. Temperature was normalized by an arbitrary value. Heavy lines marked by A, B, and C are contact traces of simulated fractures and cap rocks as represented by a series of point sources. The southeast-pointing arrow is the direction of simulated groundwater flow.

(Figure 6.7). Solutions for line, plane, or other sources can be easily obtained from the point source solution. See Problem 6 for an example of a cylindrical canister that releases radiogenic heat.

• Application

In general, a series of point source solutions can be used to simulate the disturbance of temperature distribution by groundwater flow. Figure 6.8 depicts the simulation result for a shallow geothermal anomaly caused by flow of hot fluid along fractures in crystalline rocks. The fractures are sealed by cap rocks from extending to the ground surface. The cap rocks are, in turn, covered by overburden. Shown are isotherms (solid curves in normalized temperature units) at the depth of 150 m as contoured from shallow borehole temperature data. The purpose of the simulation was to recommend candidate sites for exploratory slant drilling to intercept the fractures. A series of point sources with different heating rates were placed along fractures at the base of the cap rocks to simulate the pattern of the observed isotherms. Source strengths, fracture locations (traces marked by A, B, and C), and the regional groundwater flow vector (direction depicted by an arrow) were estimated by trial-and-error simulations. The influence on the temperature distribution in the overburden by the southeast-flowing regional groundwater flow is clearly demonstrated by the simulation results (dashed isotherms). However, it is cautioned that this forward-modeling result is by no means unique; other model interpretations are possible.

6.3   Topography-Controlled Flow

Groundwater is commonly recharged at high elevation and discharged at low topography in a watershed. Along the descending flow path, the groundwater is heated by the ambient rocks but is cooled by the surroundings when it ascends to the discharge area. In this section we will determine analytically how the topography-controlled groundwater flow may disturb a thermal regime that otherwise would be dominated by a linear geotherm. The model to be described is based on the work of Domenico and Palciauskas (1973). Besides showing the regional influence of groundwater flow on temperature distribution, this section demonstrates by example how a linear *perturbation theory* works.

6.3.1   Conservation Equations

We are to solve three conservation equations simultaneously. The energy equation is

$$\nabla \cdot [-k\nabla T + \rho c \mathbf{v} (T - T_o)] = -(\rho c)_s \frac{\partial T}{\partial t}. \qquad (6.48)$$

The momentum equation is approximated by Darcy's law,

$$\mathbf{v} = -\frac{K}{\mu}\nabla (P + \rho g z) \approx -\frac{K}{\mu}\rho g \nabla h, \qquad (6.49)$$

where P is the pressure $[Pa]$, g is the gravitational acceleration $[m\,s^{-2}]$, K is the permeability of the medium $[m^2]$, μ is the dynamic viscosity $[Pa \cdot s]$ of water, h is the hydraulic head $[m]$, and z is the elevation (upward positive). It is noted that we have assumed that the density of water is independent

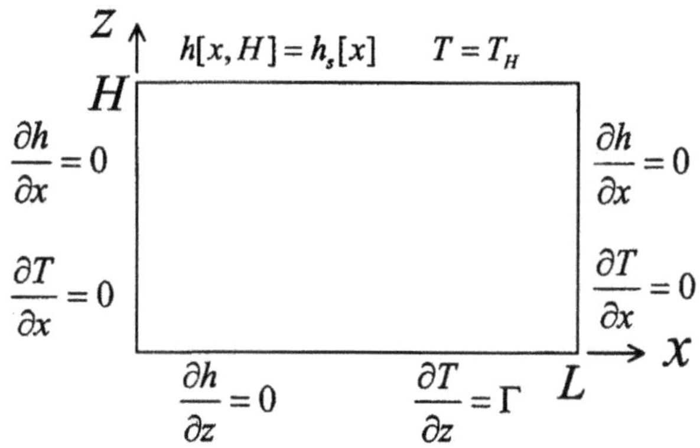

Figure 6.9: Boundary conditions for steady groundwater flow and temperature distribution.

of temperature so that $\nabla(P + \rho g z)$ can be approximated by the hydraulic gradient and the groundwater flow field determined by the hydraulic gradient only.

The mass conservation equation is represented by

$$\nabla \cdot (\rho \mathbf{v}) = -\frac{\partial(\rho \phi)}{\partial t}, \tag{6.50}$$

where ϕ is the porosity.

For an incompressible fluid-flow or a time-independent flow problem,

$$\nabla \cdot \mathbf{v} = 0, \tag{6.51}$$

and the three conservation equations are simplified to

$$
\begin{aligned}
\nabla^2 h &= 0, \\
\nabla^2 T - \frac{\rho c}{k} \mathbf{v} \cdot \nabla T &= 0.
\end{aligned}
\tag{6.52}
$$

The former is Laplace's equation and the latter is the steady advective heat flow equation.

We are to solve the partially coupled set of equations by the following steps: first, find h from the Laplace equation; second, determine $\mathbf{v}$; and third, obtain T from the advective equation under some appropriate boundary conditions. This is a one-way coupled equation, meaning that the $\mathbf{v}$ field affects the T field, but not vice versa.

6.3.2   Steady Groundwater Flow

Consider a two-dimensional rectangular block with width L and height H (Figure 6.9). Coordinate z is upward positive. The steady groundwater flow is governed by the Laplace equation, subject to the boundary conditions

$$\text{Condition 1h:} \qquad \left.\frac{\partial h}{\partial x}\right|_{x=0} = 0,$$

$$\text{Condition 2h:} \qquad \left.\frac{\partial h}{\partial x}\right|_{x=L} = 0,$$

$$\text{Condition 3h:} \qquad \left.\frac{\partial h}{\partial z}\right|_{z=0} = 0,$$

$$\text{Condition 4h:} \qquad h[x, H] = h_{\rm s}[x]. \qquad (6.53)$$

Conditions 1h and 2h imply that the 2D block is bounded by groundwater divides at $x = 0$ and $x = L$. Condition 3h implies an impermeable lower boundary. The upper boundary at $z = H$ is set for a condition of prescribed head (Dirichlet condition); the boundary is not the water table, but the head at $z = H$ is influenced by the water table, that to first approximation mimics the topography.

• **Solution**

Let us use the technique of separation of variables to solve $\nabla^2 h = 0$. Assume that

$$h[x, z] = X[x]Z[z]. \qquad (6.54)$$

Then put it into the Laplace equation and divide the resultant by XZ to yield

$$\frac{1}{X}\frac{\partial^2 X}{\partial x^2} + \frac{1}{Z}\frac{\partial^2 Z}{\partial z^2} = 0. \qquad (6.55)$$

The first term does not depend on z and the second term does not depend on x; so the two can be related by a constant only. Let this separation constant be a^2, say

$$\frac{1}{X}\frac{\partial^2 X}{\partial x^2} = -a^2, \qquad \frac{1}{Z}\frac{\partial^2 Z}{\partial z^2} = a^2. \qquad (6.56)$$

The above two ordinary differential equations have the general solutions

$$X[x] = A_{\rm a}\sin[ax] + B_{\rm a}\cos[ax],$$
$$Z[z] = C_{\rm a}\sinh[az] + D_{\rm a}\cosh[az], \qquad (6.57)$$

where coefficients A_a through D_a are to be determined from the boundary conditions.

• Imposition of Boundary Conditions

To make the no-flow condition at $x = 0$ and $z = 0$ (Conditions 1h and 3h), A_a and C_a must be set to zero. The no-flow condition at $x = L$ (Condition 2h) yields

$$Z[z] \, B_a a \sin[aL] = 0. \tag{6.58}$$

This is satisfied if

$$a = \frac{n\pi}{L}, \quad n = 0, 1, 2, ... \tag{6.59}$$

After combining B_a and D_a into coefficient E_n, we obtain the general solution,

$$h[x, z] = E_0 + \sum_{n=1}^{\infty} E_n \cos \frac{n\pi x}{L} \cosh \frac{n\pi z}{L}. \tag{6.60}$$

Now, impose Condition 4h at $z = H$ to yield

$$h_s[x] = E_0 + \sum_{n=1}^{\infty} E_n \cos \frac{n\pi x}{L} \cosh \frac{n\pi H}{L}. \tag{6.61}$$

Then multiply the above relation by $\cos[n\pi x/L]$ and integrate x from 0 to L, and finally use the orthogonality property of cosine functions to determine

$$
\begin{aligned}
E_0 &= \frac{1}{L} \int_0^L h_s[x] dx, \\
E_n &= \frac{2}{L \cosh[n\pi H/L]} \int_0^L h_s[x] \cos \frac{n\pi x}{L} dx.
\end{aligned}
\tag{6.62}
$$

Thus our general solution (6.60) is fully determined if the functional form of $h_s[x]$ is available.

• Special Case

Consider a simple functional form for hydraulic head at $z = H$,

$$h_s[x] = A - B \cos \frac{\pi x}{L}. \tag{6.63}$$

Substituting this relation into E_n yields

$$E_0 = A, \quad E_1 = -\frac{B}{\cosh[\pi H/L]},$$
$$E_n = 0 \quad \text{for} \quad n \geq 2. \tag{6.64}$$

The hydraulic head for this special case can be summarized as

$$h[x, z] = A - B\frac{\cosh[\pi z/L]}{\cosh[\pi H/L]} \cos \frac{\pi x}{L}. \tag{6.65}$$

The flow path starting from point (x_i, z_i) to (x_{i+1}, z_{i+1}) is

$$x_{i+1} = x_i + v_x\Delta t, \quad z_{i+1} = z_i + v_z\Delta t, \tag{6.66}$$

where v_x and v_z are components of Darcy's velocity in the x and z directions, respectively, and Δt is a small time increment.

6.3.3   Temperature Distribution

We will now solve the steady advective heat transport equation (6.52), using perturbation theory. Let the temperature be split into two parts,

$$\boxed{T[x, z] = T^c[x, z] + T^\Delta[x, z]}, \tag{6.67}$$

where T^c is the temperature distribution due to heat conduction only and T^Δ represents the part due to advective groundwater flow, a perturbation from the conductive field. Perturbation theory works as follows: we know how to get the normal part T^c and wish to obtain the perturbed part T^Δ. We need to choose the two parts in such a way that together both satisfy the differential equation and boundary conditions.

• **Solution: Steady Field**

Temperature T^c satisfies a steady heat conduction equation (another Laplace's equation)

$$\nabla^2 T^c[x, z] = 0. \tag{6.68}$$

The boundary conditions are lateral heat flow across the groundwater divides vanishes, the top surface of the rectangular block is maintained at constant temperature T_H, and the temperature gradient at the base is fixed at Γ. In other expressions, the conditions are

$$\left.\frac{\partial T^c}{\partial x}\right|_{x=0} = 0, \quad \left.\frac{\partial T^c}{\partial x}\right|_{x=L} = 0,$$

$$T^c[x, H] = T_\mathrm{H}, \quad \left.\frac{\partial T^c}{\partial z}\right|_{z=0} = \Gamma. \tag{6.69}$$

These conditions lead to the solution,

$$T^c[x, z] = T_H + (z - H)\Gamma. \tag{6.70}$$

• **Solution: Perturbed Field**

To find a differential equation for T^Δ, substitute the newly derived T^c into equation (6.67), then put the result into the advective equation, yielding

$$\nabla^2 T^\Delta - \frac{\rho c}{k} \mathbf{v} \cdot \nabla \left(T^c + T^\Delta \right) = 0. \tag{6.71}$$

Noting that $\mathbf{v} \cdot \nabla T^c = v_z \Gamma$ and neglecting the term $\mathbf{v} \cdot \nabla T^\Delta$ for small perturbation (see remarks later) yields

$$\nabla^2 T^\Delta - \frac{\rho c}{k} v_z \Gamma = 0 \tag{6.72}$$

or

$$\nabla^2 T^\Delta + \beta \sum_{n=1}^{\infty} n E_n \sinh \frac{n\pi z}{L} \cos \frac{n\pi x}{L} = 0, \tag{6.73}$$

$$\beta = \frac{\rho c}{k} \frac{K \rho g \pi}{\mu L} \Gamma.$$

The boundary conditions for T^Δ are

$$\left.\frac{\partial T^\Delta}{\partial x}\right|_{x=0} = 0, \quad \left.\frac{\partial T^\Delta}{\partial x}\right|_{x=L} = 0,$$

$$\left.\frac{\partial T^\Delta}{\partial z}\right|_{z=0} = 0, \quad T^\Delta\big|_{z=H} = 0. \tag{6.74}$$

To solve equation (6.73), let us try a solution of the form,

$$T^\Delta[x, z] = \sum_{n=1}^{\infty} F_n[z] \cos \frac{n\pi x}{L}. \tag{6.75}$$

Then, equation (6.73) becomes

$$\sum_{n=1}^{\infty} \left[\frac{\partial^2}{\partial z^2} F_n - \left(\frac{n\pi}{L} \right)^2 F_n \right] \cos \frac{n\pi x}{L} = -\beta \sum_{n=1}^{\infty} n E_n \sinh \frac{n\pi z}{L} \cos \frac{n\pi x}{L}. \quad (6.76)$$

Equating the coefficients of $\cos [n\pi x/L]$ (by orthogonal property) yields

$$\frac{\partial^2}{\partial z^2} F_n - \left(\frac{n\pi}{L} \right)^2 F_n = -\beta n E_n \sinh \frac{n\pi z}{L}. \quad (6.77)$$

The inhomogeneous part can be determined from a trial solution,

$$(Mz + N)e^{n\pi z/L} + (Pz + Q)e^{-n\pi z/L},$$

which upon substitution into equation (6.77) yields the particular solution

$$-\frac{\beta E_n}{2\pi/L} z \cosh \frac{n\pi z}{L}.$$

The homogeneous part of equation (6.77) has a complementary solution of the form

$$G^+ \exp[\frac{n\pi z}{L}] + G^- \exp[-\frac{n\pi z}{L}],$$

where coefficients G^+ and G^- are to be determined by the boundary conditions.

Adding the complementary and particular solutions together gives

$$F_n[z] = G^+ \exp[\frac{n\pi z}{L}] + G^- \exp[-\frac{n\pi z}{L}] - \frac{\beta E_n}{2\pi/L} z \cosh \frac{n\pi z}{L}. \quad (6.78)$$

Imposing the boundary conditions at $z = 0$ and H yields

$$G^+ = \frac{(n\pi H/L) \cosh [n\pi H/L] + \exp[-n\pi H/L]}{(2n\pi/L) \cosh[n\pi H/L]} \frac{\beta E_n}{2\pi/L},$$

$$G^- = \frac{(n\pi H/L) \cosh [n\pi H/L] - \exp[n\pi H/L]}{(2n\pi/L) \cosh[n\pi H/L]} \frac{\beta E_n}{2\pi/L}. \quad (6.79)$$

Simplify the result to obtain

$$F_n[z] = \frac{\beta L E_n}{2\pi} \left[(H - z) \cosh \frac{n\pi z}{L} + \frac{L}{n\pi} \frac{\sinh [n\pi (z - H)/L]}{\cosh [n\pi H/L]} \right]. \quad (6.80)$$

- **Complete Solution**

The complete solution is

$$
\begin{aligned}
T[x, z] &= T^c[x, z] + T^\Delta[x, z] \\
&= T_H - (H - z)\Gamma + \sum_{n=1}^{\infty} F_n[z] \cos \frac{n\pi x}{L}
\end{aligned}
\tag{6.81}
$$

and the geothermal gradient is

$$
\begin{aligned}
\frac{\partial T[x, z]}{\partial z} &= \Gamma + \frac{\beta L}{2\pi} \sum_{n=1}^{\infty} E_n \left\{ -\cosh \frac{n\pi z}{L} + \frac{(H - z)\, n\pi}{L} \sinh \frac{n\pi z}{L} \right. \\
&\quad \left. + \frac{\cosh \left[n\pi (z - H)/L \right]}{\cosh \left[n\pi H/L \right]} \right\} \cos \frac{n\pi x}{L}.
\end{aligned}
\tag{6.82}
$$

- **Special Case**

For the special case that

$$
h_s[x] = A - B \cos[\pi x/L],
\tag{6.83}
$$

we obtain

$$
\begin{aligned}
E_1 &= -B/\cosh[\pi H/L], \\
E_n &= 0 \quad \text{for} \quad n \geq 2.
\end{aligned}
\tag{6.84}
$$

The temperature gradient at $z = H$ is

$$
\begin{aligned}
\left. \frac{\partial T[x, z]}{\partial z} \right|_{z=H} &= \Gamma + \frac{\beta L B}{2\pi} \tanh^2 \frac{\pi H}{L} \cos \frac{\pi x}{L} \\
&= \Gamma \left\{ 1 + B \frac{\rho c}{2} \frac{k_{\text{hydra}}}{k_{\text{therm}}} \tanh^2 \frac{\pi H}{L} \cos \frac{\pi x}{L} \right\},
\end{aligned}
\tag{6.85}
$$

where hydraulic conductivity

$$
k_{\text{hydra}} = K\rho g/\mu.
\tag{6.86}
$$

The second term in equation (6.85) represents perturbation due to groundwater flow.

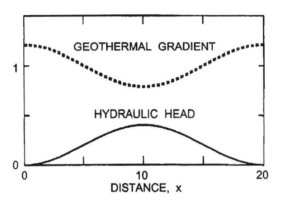

Figure 6.10: Sketch showing negative correlation between geothermal gradient $\partial T/\partial z$ (dotted curve) and hydraulic head $h_s[x]$ at $z = H$.

Though the temperature at $z = H$ is maintained at T_H, geothermal gradient varies laterally. In the discharge area $x = 0$, the gradient is the highest while in the recharge area $x = L$, the gradient is the least. At the hinge line (midline, $x = L/2$), the disturbance vanishes. Figure (6.10) depicts the hydraulic head $h_s(x)$ and the normalized geothermal gradient at $z = H$. As expected, the variations of the two are 180° out of phase.

- **Remark**

After computing T^Δ, if $\mathbf{v} \cdot \nabla T^\Delta$ for equation (6.73) turns out to be significant, this T^Δ can be considered as the first guess value. So, add $\mathbf{v} \cdot \nabla T^\Delta$ to equation (6.71), then solve the updated equation for a new T^Δ. Presumably the process can be iterated.

6.4   Heating and Pressurization

In the previous sections, we have discussed how groundwater flow may influence temperature distribution. Now we address the reverse process: how temperature rise may affect fluid flow. The effects of thermal convection cannot be adequately addressed in one section, so we will narrow our discussion to a small class of temperature-induced flow problems. The flow we will deal with is essentially horizontal and occurs before the onset of thermal convection. Heating events are short in duration compared to the thermal

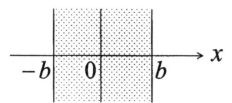

Figure 6.11: One-dimensional model for heating-pressurization problems. Heating zone is shaded.

diffusion time for the physical system under consideration, for example, frictional heating during fault slip.

6.4.1 Equation for Fluid Pressurization

To investigate the fluid pressurization due to a temperature rise, we assume that there is no feedback effect of fluid flow on temperature distribution, and that the problems are one-dimensional (Figure 6.11). Here we follow the work by Lee (1996) who addresses the geophysical implications of frictional heating in fault zones.

The principle of conservation of mass, as represented by the equation of continuity, gives

$$\frac{\partial}{\partial t}(\rho\phi) = -\frac{\partial}{\partial x}(\rho v), \tag{6.87}$$

where ρ is the fluid density and ϕ is the porosity. The velocity is given by Darcy's law

$$\boxed{v = -\frac{K}{\mu}\frac{\partial P}{\partial x}} \tag{6.88}$$

as an approximation to the principle of conservation of momentum, where P is the pressure, $[Pa]$, K is the permeability, $[m^2]$, and μ is the dynamic viscosity, $[Pa \cdot s]$.

Combining the above two equations yields a relation between fluid flow and rates of change in fluid density and porosity

$$\frac{K}{\mu\phi}\frac{\partial^2 P}{\partial x^2} = \frac{1}{\rho}\frac{\partial\rho}{\partial t} + \frac{1}{\phi}\frac{\partial\phi}{\partial t} \tag{6.89}$$

in which the second-order term $v\partial\rho/\partial x \propto (\partial P/\partial x)(\partial T/\partial x)$ has been neglected.

By definition of compressibility β_f and thermal expansivity α_f of fluid, the rate of fractional density change is

$$\frac{1}{\rho}\frac{\partial\rho}{\partial t} = \frac{1}{\rho}\left(\frac{\partial\rho}{\partial P}\frac{\partial P}{\partial t} + \frac{\partial\rho}{\partial T}\frac{\partial T}{\partial t}\right) = \beta_f\frac{\partial P}{\partial t} - \alpha_f\frac{\partial T}{\partial t}. \tag{6.90}$$

The rate of fractional porosity change is

$$\frac{1}{\phi}\frac{\partial\phi}{\partial t} = \frac{1}{\phi}\left(\frac{\partial\phi}{\partial P}\frac{\partial P}{\partial t} + \frac{\partial\phi}{\partial T}\frac{\partial T}{\partial t}\right), \tag{6.91}$$

where

$$\frac{\partial\phi}{\partial P} = (1-\phi)\beta_b \tag{6.92}$$

is given in Section 1.2.3 and β_b is the bulk compressibility. The porosity change due to temperature variation is given by

$$\begin{aligned}\frac{\partial\phi}{\partial T} &= \frac{\partial}{\partial T}\left(1 - \frac{V_g}{V_b}\right) = -\frac{1}{V_b^2}\left(V_b\frac{\partial V_g}{\partial T} - V_g\frac{\partial V_b}{\partial T}\right) \\ &= \frac{V_g}{V_b}(\alpha_b - \alpha_g) = (1-\phi)(\alpha_b - \alpha_g)\end{aligned} \tag{6.93}$$

where α_b and α_g are, respectively, the thermal expansivity of bulk rock and grain.

Substituting the rates of porosity and density changes into equation (6.89) yields the fluid-pressurization equation,

$$\boxed{\frac{\partial P}{\partial t} = \frac{\alpha}{\beta}\frac{\partial T}{\partial t} + \omega\frac{\partial^2 P}{\partial x^2}}, \tag{6.94}$$

where

$$\begin{aligned}\alpha &= \alpha_f\phi - (1-\phi)(\alpha_b - \alpha_g) \\ \beta &= \beta_f\phi + (1-\phi)\beta_b \\ \omega &= K/\mu\beta.\end{aligned} \tag{6.95}$$

Treating the temperature term $(\alpha/\beta\ \partial T/\partial t)$ as the pressure source, equation (6.94) is the diffusion equation for the pressure rise. Temperature rise induces pressurization through the temperature-pressure transformation factor

α/β, which is the ratio of effective thermal expansivity α, $[K^{-1}]$, to effective compressibility β, $[Pa^{-1}]$. The intensity of pressurization is reduced by the net fluid flow $\omega \partial^2 P/\partial x^2$, of which ω is the hydraulic diffusivity, $[m^2 s^{-1}]$.

The pressurization equation together with the heat conduction equation,

$$\frac{\partial T}{\partial t} = \kappa \frac{\partial^2 T}{\partial x^2}, \qquad (6.96)$$

governs the temperature-pressure relation, where κ is the thermal diffusivity, $[m^2 s^{-1}]$. The feedback of fluid flow on temperature distribution (i.e., the advective term) is hereby omitted.

6.4.2 Pressurization

Lee (1996) has solved equations (6.94) and (6.96) for the cases that the physical properties within the zone of heat sources are different from those outside the zone. Here, for mathematical simplicity, the medium is treated as being homogeneous.

The initial and boundary conditions are

$$\begin{aligned} T[x,0] &= 0, \quad P[x,0] = 0, \\ T[\infty,t] &= 0, \quad P[\infty,t] = 0. \end{aligned} \qquad (6.97)$$

The conditions for T and P at $x = 0$ will depend on how the heating function is implemented.

In the Laplace-transform domain, the two differential equations are

$$\begin{aligned} \frac{\partial^2 \overline{T}}{\partial x^2} - \frac{s}{\kappa}\overline{T} &= 0, \\ \frac{\partial^2 \overline{P}}{\partial x^2} - \frac{s}{\omega}\overline{P} &= -\frac{\alpha s}{\beta \omega}\overline{T}, \end{aligned} \qquad (6.98)$$

where s is the Laplace transform parameter.

• Instantaneous Heating: Plane Source

In this case the temperature rise is generated by an instantaneous plane source. The mathematical expression can be easily obtained from the line source solution in Section 2.1.1

$$T[x,t] = \frac{W}{(\rho c)_s \sqrt{4\pi\kappa t}} \exp\left[-\frac{x^2}{4\kappa t}\right], \qquad (6.99)$$

where W is the amount of heat released per unit area, $[J\,m^{-2}]$ and $(\rho c)_s$ is the heat capacity of the fluid-saturated rock, $[J\,m^3 K^{-1}]$. By symmetry of pressurization due to a symmetric temperature distribution, there is no fluid flow across the plane at $x = 0$. Hence,

$$\frac{\partial P}{\partial x} = 0 \quad \text{at} \quad x = 0. \tag{6.100}$$

According to the result in Section 3.2.5, the Laplace transform of equation (6.99) is

$$\overline{T}[x, s] = \frac{W}{2\kappa(\rho c)_s} \frac{1}{\sqrt{s/\kappa}} \exp\left[-\sqrt{\frac{s}{\kappa}}x\right]. \tag{6.101}$$

Substituting $\overline{T}[x, s]$ into equation (6.98) yields

$$\frac{\partial^2 \overline{P}}{\partial x^2} - \frac{s}{\omega}\overline{P} = -\frac{\alpha}{\beta\omega}\frac{W}{2k}\frac{s}{\sqrt{s/\kappa}} \exp\left[-\sqrt{\frac{s}{\kappa}}x\right], \tag{6.102}$$

where $k = \kappa(\rho c)_s$ is the fluid-saturated thermal conductivity. The general solution is

$$\overline{P} = Ae^{-\sqrt{s/\omega}x} - \frac{\alpha}{\beta}\frac{W}{2k}\frac{\exp\left[-\sqrt{s/\kappa}x\right]}{\omega\,(1/\kappa - 1/\omega)\sqrt{s/\kappa}}. \tag{6.103}$$

Imposing the no-flow boundary condition at $x = 0$ yields A and

$$\overline{P} = \frac{\alpha}{\beta}\frac{W}{2(\rho c)_s\,(\omega - \kappa)}\left\{\frac{\exp\left[-\sqrt{s/\omega}x\right]}{\sqrt{s/\omega}} - \frac{\exp\left[-\sqrt{s/\kappa}x\right]}{\sqrt{s/\kappa}}\right\}. \tag{6.104}$$

Its inverse Laplace transform is

$$\begin{aligned}
P[x, t] &= \frac{\alpha}{\beta}\frac{W}{2(\rho c)_s\,(\omega - \kappa)} \\
&\quad \cdot\left\{\sqrt{\frac{\omega}{\pi t}}\exp\left[-\frac{x^2}{4\omega t}\right] - \sqrt{\frac{\kappa}{\pi t}}\exp\left[-\frac{x^2}{4\kappa t}\right]\right\}. \tag{6.105}
\end{aligned}$$

This is the pore pressure rise due to instantaneous release of heat from a plane source, i.e., it is the impulse response.

Redesignating $T[x, t]$ in equation (6.99) as $T[x, t; \kappa]$ and letting $\eta = \kappa/\omega$ yields the final pressure-temperature relation

$$\boxed{P[x, t] = \frac{\alpha}{\beta}\frac{1}{1 - \eta}\left\{T[x, t; \omega] - \eta T[x, t; \kappa]\right\}.} \tag{6.106}$$

• **Instantaneous Strip Source**

If the heating occurs in a strip $-b \leq x \leq b$, the temperature distribution is

$$T[x, t; \kappa] = \frac{\Delta T}{2} \left\{ \mathrm{erf}\left[\frac{x+b}{\sqrt{4\kappa t}}\right] - \mathrm{erf}\left[\frac{x-b}{\sqrt{4\kappa t}}\right] \right\}, \qquad (6.107)$$

where $\Delta T = W/2b(\rho c)_s$ represents the instantaneous temperature rise within the heating zone, with W being the heat generation per unit volume, $[J\,m^{-3}]$.

The pressure is

$$
\begin{aligned}
P[x, t] &= \frac{\alpha}{\beta} \frac{1}{1-\eta} \frac{W}{2b(\rho c)_s} \int_{-b}^{b} \left\{ \frac{1}{\sqrt{4\pi\omega t}} \exp\left[-\frac{(x-x')^2}{4\omega t}\right] \right. \\
&\qquad \left. - \frac{\eta}{\sqrt{4\pi\kappa t}} \exp\left[-\frac{(x-x')^2}{4\kappa t}\right] \right\} dx' \\
&= \frac{\alpha}{\beta} \frac{1}{1-\eta} \frac{W}{4b(\rho c)_s} \left\{ \mathrm{erf}\left[\frac{x+b}{\sqrt{4\omega t}}\right] - \mathrm{erf}\left[\frac{x-b}{\sqrt{4\omega t}}\right] \right. \\
&\qquad \left. -\eta\,\mathrm{erf}\left[\frac{x+b}{\sqrt{4\kappa t}}\right] + \eta\,\mathrm{erf}\left[\frac{x-b}{\sqrt{4\kappa t}}\right] \right\}. \qquad (6.108)
\end{aligned}
$$

As expected, the pressure-temperature relation is of the form in equation (6.106).

• **Remark**

Since $\eta \ll 1$ for most porous media, we get approximately

$$P[x, t] \approx \frac{\Delta P}{2} \left\{ \mathrm{erf}\left[\frac{x+b}{\sqrt{4\kappa t/\eta}}\right] - \mathrm{erf}\left[\frac{x-b}{\sqrt{4\kappa t/\eta}}\right] \right\}, \qquad (6.109)$$

$$\Delta P = (\alpha/\beta)\Delta T/(1-\eta)$$

for the strip-heating. The pressure response behaves like a temperature response but the pressure front migrates ahead of the temperature front by a factor of $1/\eta$ or ω/κ. Such behavior is demonstrated by the normalized temperature and pressure in Figure 6.12 as drawn according to equations (6.107) and (6.108). This conclusion is also applicable to other heating functions. It is noted that both $\partial T/\partial x$ and $\partial P/\partial x$ vanish at $x = 0$ for the strip heating model.

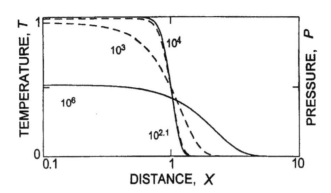

Figure 6.12: Rises in normalized temperature (solid curves) and pressure at different time t, shown for $\kappa = 10^{-6}$ and $\eta = 10^{-2}$ (all in SI units). Note that the pressure curve at $t = 10^{2.1}$ follows essentially the temperature curve at $t = 10^4$, indicating that the pressure front advances ahead of the temperature front by a factor of $1/\eta$.

Figure 6.13 depicts the normalized pressure at time $t = 10^{2.01}$ and 10^3 for a heating duration of 10^2. The solution can be easily obtained by convolution (see Problem 7). Also shown for comparison is the pressure due to instantaneous heating. As illustrated in the figure, the two sets are hardly distinguishable.

6.5 Problems, Keys, and Suggested Readings

• Problems

1. Derive equation (6.22) from the principle of constant heat flux

$$q = \rho c v (T - T_1) - k_T \frac{\partial T}{\partial z} \qquad (6.110)$$

assuming that there is no heat source, lateral heat transfer, or lateral mass transport.

2. Peclet number varies with the characteristic thickness, which is somewhat arbitrarily defined between $z = Z_1$ and $z = Z_2$ within an aquifer (Section 6.1.1). If Z_2 is changed, how will the plotting of $T[z]$ be modified? Using Figure 6.2 as a reference, plot the evolving path of one point on a given geotherm as Z_2 is changed.

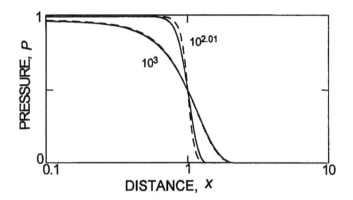

Figure 6.13: Normalized pressures at two different times for a heating duration of 10^2 (dashed curves). For comparision, the pressures due to instantaneous heating are also shown (solid curves). $\eta = 10^{-2}$.

3. In Section 6.1.2 for horizontal steady flow, if heat fluxes in the upper and lower aquitards vary laterally, what is the temperature distribution? This is Solution 3 (Figure 6.14).

4. Show that

$$\int_0^y \exp[-a^2 x^2 - b^2/x^2]dx = \frac{\sqrt{\pi}}{4a}$$
$$\cdot \left\{ e^{2ab}\mathrm{erf}\left[ay + \frac{b}{y}\right] + e^{-2ab}\mathrm{erf}\left[ay - \frac{b}{y}\right] - e^{2ab} + e^{-2ab}\right\}. \quad (6.111)$$

5. Solve $T[x,t]$ for

$$\frac{\partial T}{\partial t} = \kappa \frac{\partial^2 T}{\partial x^2}, \quad 0 \leq x \leq l \qquad (6.112)$$

under the conditions that

$$\begin{aligned} T[x,0] &= 0, \\ T[0,t] &= 0, \\ -k\frac{\partial T}{\partial x}\bigg|_{x=l} &= \sigma T[l,t] - F, \end{aligned} \qquad (6.113)$$

where κ is the thermal diffusivity, k is the thermal conductivity, σ is the radiative heat transfer constant, and F is a constant influx.

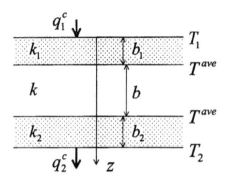

Figure 6.14: Sketch diagram for heat fluxes q_1^c and q_2^c that vary laterally.

6. A rectangular canister contains radioactive material with decay constant λ. If the canister is buried in a region with groundwater flow velocity $\mathbf{v}$, determine the temperature distribution.

7. If the duration of heating is τ for the strip heating model (assuming a constant heating rate), find the pressure in terms of the repeated complementary error function. Discuss whether the fluid flow can reverse its direction after the heating terminates.

- **Keys**

Key 1

Rearrange terms and multiply both sides by $L/k_T(T_2 - T_1)$ to obtain

$$
\begin{aligned}
\frac{L}{T_2 - T_1}\frac{\partial T}{\partial z} &= -\frac{qL}{k_T(T_2 - T_1)} + \frac{\rho c v(T - T_1)L}{k_T(T_2 - T_1)} \\
&= -\frac{q}{k_T(T_2 - T_1)/L} + N_{\text{Pe}}\frac{T - T_1}{T_2 - T_1}.
\end{aligned} \tag{6.114}
$$

Key 3

If q_1^c and q_2^c vary laterally (Figure 6.14),

$$
q_2^c = -k_2\frac{T_2 - T^{\text{ave}}}{b_2} \quad \text{and} \quad q_1^c = -k_1\frac{T^{\text{ave}} - T_1}{b_1}, \tag{6.115}
$$

then the differential equation (6.26) becomes

$$\frac{\partial^2 T^{\mathrm{ave}}}{\partial x^2} - \frac{\rho c v}{k}\frac{\partial T^{\mathrm{ave}}}{\partial x} - \left(\frac{k_2}{b_2} + \frac{k_1}{b_1}\right)\frac{T^{\mathrm{ave}}}{bk} = -\frac{1}{bk}\left(\frac{k_2 T_2}{b_2} + \frac{k_1 T_1}{b_1}\right). \qquad (6.116)$$

Its general solution is

$$T^{\mathrm{ave}}[x] = Ae^{\eta^- x} + Be^{\eta^+ x} + T_{\mathrm{r}}, \qquad (6.117)$$

where

$$\eta^{\pm} = \frac{\rho c v}{k}\left\{1 \pm \sqrt{1 + \frac{4k}{b(\rho c v)^2}\left(\frac{k_2}{b_2} + \frac{k_1}{b_1}\right)}\right\}$$

$$T_r = \left(\frac{k_2 T_2}{b_2} + \frac{k_1 T_1}{b_1}\right)\Big/\left(\frac{k_2}{b_2} + \frac{k_1}{b_1}\right). \qquad (6.118)$$

The coefficients A and B are to be determined from the boundary conditions. For example, if

$$T^{\mathrm{ave}} = T_0 \text{ at } x = 0 \quad \text{and} \quad T^{\mathrm{ave}} \neq \infty \text{ at } x = \infty, \qquad (6.119)$$

then $B = 0$ because $\eta^+ > 0$ (assuming $v > 0$). In this case, $A = T_0 - T_{\mathrm{r}}$ such that

$$T^{\mathrm{ave}} = (T_0 - T_{\mathrm{r}})\, e^{\eta^- x} + T_{\mathrm{r}}, \qquad (6.120)$$

which has the same functional form as Case 2 in Section 6.1.2.

Key 4

Let the integral be represented by I. Then, construct two quadratic relations to yield

$$I = \frac{e^{2ab}}{2}\int_0^y e^{-(ax+b/x)^2}\,dx + \frac{e^{-2ab}}{2}\int_0^y e^{-(ax-b/x)^2}\,dx. \qquad (6.121)$$

Letting

$$u = ax + b/x \quad \text{and} \quad v = ax - b/x \qquad (6.122)$$

results in

$$I = \frac{e^{2ab}}{2}\left\{\int_{\infty}^{u_y} e^{-u^2}\frac{du}{a} + \int_0^y \frac{be^{-(ax+b/x)^2}}{ax^2}\,dx\right\}$$

$$+\frac{e^{-2ab}}{2}\left\{\int_{-\infty}^{v_y} e^{-v^2}\frac{dv}{a} - \int_0^y \frac{be^{-(ax-b/x)^2}}{ax^2}\,dx\right\}. \qquad (6.123)$$

The two x-integrations cancel each other. Hence,

$$
\begin{aligned}
I &= -\frac{e^{2ab}}{2}\int_{u_y}^{\infty}e^{-u^2}\frac{du}{a}+\frac{e^{-2ab}}{2}\int_{-v_y}^{\infty}e^{-v^2}\frac{dv}{a}\\
&= -\frac{e^{2ab}\sqrt{\pi}}{4a}\operatorname{erfc}[u_y]+\frac{e^{-2ab}\sqrt{\pi}}{4a}\left(1+\int_{-v_y}^{0}e^{-v^2}\frac{dv}{a}\right)\\
&= \frac{\sqrt{\pi}}{4a}\left\{e^{2ab}\left(\operatorname{erf}\left[ay+b/y\right]-1\right)\right.\\
&\quad\left.+\,e^{-2ab}\left(\operatorname{erf}\left[ay-b/y\right]+1\right)\right\}.
\end{aligned}
\tag{6.124}
$$

Key 5

Applying the Laplace transform to the differential equations and boundary conditions yields

$$
\frac{d^2\overline{T}}{dx^2} = q^2\overline{T}, \quad q^2 = \frac{s}{\kappa},
\tag{6.125}
$$

$$
\overline{T}[0,s] = 0,
\tag{6.126}
$$

$$
-k\frac{\partial\overline{T}}{\partial x}\bigg|_{x=l} = \sigma\overline{T}[l,s]-\frac{F}{s},
\tag{6.127}
$$

where s is the Laplace transform parameter.

The general solution is

$$
\overline{T}[x,s] = A\cosh[qx]+B\sinh[qx].
\tag{6.128}
$$

The condition at $x=0$ requires that $A=0$. Substituting $\overline{T}[x,s]$ into the advective condition at $x=l$ yields B and

$$
\overline{T}[x,s] = \frac{F\sinh[qx]}{s\left(kq\cosh[ql]+\sigma\sinh[ql]\right)}.
\tag{6.129}
$$

Inverse Laplace Transform: Take the inverse Laplace transform of $\overline{T}[x,s]$

$$
T[x,t] = \frac{F}{i2\pi}\int_{a-i\infty}^{a+i\infty}\frac{\sinh[qx]\exp[st]}{s\left(kq\cosh[ql]+\sigma\sinh[ql]\right)}ds.
\tag{6.130}
$$

Since the integrand is an even function of q, there is no branch cut.

Find the residue at $s = 0$,

$$\text{Res}\big|_{s=0} = \lim_{s \to 0} s \frac{\sinh[qx]\exp[st]}{s\left(kq\cosh[ql] + \sigma\sinh[ql]\right)}$$

$$= \frac{\frac{\partial}{\partial s}\left(\sinh[qx]\exp[st]\right)}{\frac{\partial}{\partial s}\left(kq\cosh[ql] + \sigma\sinh[ql]\right)}\Bigg|_{s=0}$$

$$= \frac{x}{k + \sigma l}. \tag{6.131}$$

The residues at the roots of

$$kq\cosh[ql] + \sigma\sinh[ql] = 0 \tag{6.132}$$

are found as follows. Let $i\beta = q = \sqrt{s/\kappa}$, then root equation becomes

$$k\beta\cos[\beta l] = -\sigma\sin[\beta l] \tag{6.133}$$

or

$$\tan[\beta_n l] = -\frac{k\beta_n}{\sigma}. \tag{6.134}$$

Because k/σ is positive, the root $\beta_n l$ is constrained in the second and fourth quadrants of a circular angle from 0 to 2π (or their integer multiples)

$$\left(n + \frac{1}{2}\right)\pi \leq \beta_n l \leq (n+1)\pi, \quad n = 0, 1, 2, \ldots \tag{6.135}$$

The residues at $s = -\kappa\beta_n^2$

$$\text{Res}\big|_{s=-\kappa\beta_n^2} = \lim_{s \to -\kappa\beta_n^2} \frac{\partial\left(s + \kappa\beta_n^2\right)/\partial s \ \sinh[qx]\exp[st]}{s\partial\left(kq\cosh[ql] + \sigma\sinh[ql]\right)/\partial s}$$

$$= \frac{i\sin[\beta_n x]\exp[-\kappa\beta_n^2 t]}{s\left(k\cos[\beta_n l] - k\beta_n l\sin[\beta_n l] + \sigma l\cos[\beta_n l]\right)\frac{\partial q}{\partial s}}$$

$$= \frac{2\sigma l}{\beta_n}\frac{\sin[\beta_n x]}{\cos[\beta_n l]}\frac{\exp[-\kappa\beta_n^2 t]}{\sigma l\left(k + \sigma l\right) + (k\beta_n l)^2}. \tag{6.136}$$

Finally,

$$\boxed{T[x,t] = \frac{Fx}{k + \sigma l} + 2\sigma l F\sum_{n=0}^{\infty} \frac{\sin[\beta_n x]\exp[-\kappa\beta_n^2 t]}{\beta_n\left[\sigma l\left(k + \sigma l\right) + (k\beta_n l)^2\right]\cos[\beta_n l]}.}$$

$$\tag{6.137}$$

Special Case: $\sigma = 0$.   As $\sigma \to 0$, then

$$\beta_n l \to (n + \frac{1}{2})\pi. \tag{6.138}$$

Use l'Hopital's rule to evaluate

$$T[x, t] = \frac{Fx}{k} + 2lF \sum_{n=0}^{\infty} \frac{\sin[\beta_n x] \exp[-\kappa \beta_n^2 t]}{\beta_n \left[\sigma l (k + \sigma l) + (k\beta_n l)^2\right] \frac{\partial}{\partial \sigma} \cos[\beta_n l]}, \tag{6.139}$$

where $\frac{\partial}{\partial \sigma} \cos[\beta_n l]$ can be obtained from equation (6.133)

$$k\beta_n \frac{\partial}{\partial \sigma} \cos[\beta_n l] = -\sin[\beta_n l] = -(-1)^n \quad \text{as} \quad \sigma \to 0. \tag{6.140}$$

Therefore,

$$
\begin{aligned}
T[x, t] &= \frac{Fx}{k} - 2lF \sum_{n=0}^{\infty} (-1)^n \frac{\sin[\beta_n x] \exp[-\kappa \beta_n^2 t]}{k(\beta_n l)^2} \\
&= \frac{Fx}{k} - \frac{8Fl}{k\pi^2} \sum_{n=0}^{\infty} \frac{(-1)^n}{(2n + 1)^2} \\
&\quad \cdot \sin \frac{(2n + 1)\pi x}{2l} \exp\left[-\frac{\kappa(2n + 1)^2 \pi^2 t}{4l^2}\right].
\end{aligned}
\tag{6.141}
$$

This is the temperature rise in a slab ($0 \le x \le l$) with zero initial temperature, constant flux F into the slab at $x = l$, and zero temperature at $x = 0$ (Carslaw and Jaeger, 1959, p. 113).

Key 6

Rectangular Prism: Let the z-axis of the rectangular canister be vertical, the canister be bounded by (x_1, x_2), (y_1, y_2) and (z_1, z_2), and $\overline{\mathbf{V}} = \mathbf{v}/\phi$ (ϕ is the effective porosity). We will start from the instantaneous point source solution in a homogeneous, isotropic, infinite medium, equation (6.39), with the following modifications

$$X = x - \overline{v_x}t - x', \quad Y = y - \overline{v_y}t - y', \quad Z = z - \overline{v_z}t - z'. \tag{6.142}$$

Integrating equation (6.39) with respect to z' from z_1 to z_2 yields the impulse response for a source lying between z_1 and z_2,

$$
T^I = \frac{M \exp\left[-\frac{X^2}{4D_x t} - \frac{Y^2}{4D_y t}\right]}{(4\pi t)^{3/2}(D_x D_y D_z)^{1/2}(\rho c)_s} \int_{z_1}^{z_2} \exp\left[-\frac{(z - \overline{v_z}t - z')^2}{4D_z t}\right] dz
$$

$$
= \frac{M f_z}{8\pi t (D_x D_y)^{1/2}(\rho c)_s} \exp\left[-\frac{X^2}{4D_x t} - \frac{Y^2}{4D_y t}\right],
$$

where

$$
f_z[z, t] = \mathrm{erfc}\left[\frac{z - \overline{v_z}t - z_2}{\sqrt{4D_z t}}\right] - \mathrm{erfc}\left[\frac{z - \overline{v_z}t - z_1}{\sqrt{4D_z t}}\right]. \tag{6.143}
$$

Extending the integration to x' and y' yields the impulse response

$$
T^I[x, y, z, t] = \frac{M}{8} f_x f_y f_z, \tag{6.144}
$$

where f_x and f_y are defined by replacing z in f_z with x and y, respectively. By the convolution theorem we obtain

$$
T[x, y, z, t] = G \int_0^t e^{-\lambda(t-\tau)} T^I[x, y, z, \tau] d\tau, \tag{6.145}
$$

where $G \exp[-\lambda t]$ is the radioactive heat generation rate per unit volume.
If the canister is buried near the ground surface, we use

$$
\begin{aligned}
f_z = \; & \mathrm{erfc}\left[\frac{z - \overline{v_z}t - z_2}{\sqrt{4D_z t}}\right] - \mathrm{erfc}\left[\frac{z - \overline{v_z}t - z_1}{\sqrt{4D_z t}}\right] \\
& + \mathrm{erfc}\left[\frac{z + \overline{v_z}t + z_2}{\sqrt{4D_z t}}\right] - \mathrm{erfc}\left[\frac{z + \overline{v_z}t + z_1}{\sqrt{4D_z t}}\right]
\end{aligned} \tag{6.146}
$$

such that the temperature at $z = 0$ is constant (f_x and f_y remain unchanged).
A cylindrical canister would be more practical but its solution is more complicated. One can convert the Cartesian coordinates into polar coordinates, then integrate over the circular surface.

Radial Flow v_r and v_z: If the flow field is axi-symmetric and if D does not depend on v_r, we can use polar coordinates to get

$$
X^2 + Y^2 = (r - \overline{v_r})^2 + r'^2 - 2(r - \overline{v_r})r' \cos[\theta - \theta'] \tag{6.147}
$$

and the impulse response

$$
\begin{aligned}
T^I[r,z,t] &= \frac{M f_z}{8\pi t (D_x D_y)^{1/2} (\rho c)_s} \\
&\quad \cdot \int_0^a r' dr' \int_0^{2\pi} \exp\left[-\frac{(r - \overline{v_r})^2 + r'^2 - 2(r - \overline{v_r})r' \cos[\theta']}{4 D_r t}\right] d\theta' \\
&= \frac{M f_z}{8\pi t (D_x D_y)^{1/2} (\rho c)_s} \\
&\quad \cdot \int_0^a r' \exp\left[-\frac{(r - \overline{v_r})^2 + r'^2}{4 D_r t}\right] I_0\left[\frac{(r - \overline{v_r})r'}{2 D_r t}\right] dr',
\end{aligned}
\tag{6.148}
$$

where I_0 is the modified Bessel function of the first kind. As a reminder, v_r is radial flow toward or away from the cylinder; it is not the radial component of the flow around the cylinder. This simple solution is not the solution for a cylindrical canister buried in a region with groundwater flow velocity **v**, which does not give a constant radial component v_r.

• Suggested Readings

Most problems in advective heat transfer are handled by numerical solutions. *Hydrogeological Regimes and Their Subsurface Thermal Effects* edited by Beck, Garven, and Stegena (1989) is a good starting reference on this topic. Domenico and Schwartz's (1998) *Physical and Chemical Hydrogeology* provides additional coverage on the subject. Furbish's (1997) *Fluid Physics in Geology* is recommended for those who want to know the geological implications.

6.6 Notations

Symbol	Definition	SI Unit	Dimension
$D, D_x, \dots$	Thermal dispersion coefficient	$m^2 s$	$L^2 T$
F	Heat flux	$W\,m^{-2}$	$ML^{-1}T^{-2}$
g	Gravitational acceleration	$m\,s^{-2}$	LT^{-2}
h	Hydraulic head	m	L
h_s	Hydraulic head at upper model surface	m	L
H	Height	m	L
k, k_T	Thermal conductivity of saturated rock	$W\,m^{-1}K^{-1}$	$MLT^{-3}K^{-1}$
k_{hydra}	Hydraulic conductivity	$m\,s^{-1}$	LT^{-1}
K	Permeability	m^2	L^2
L	Length	m	L
M	Heating rate		
N_{Pe}	Peclet number	-	-
N_{Ra}	Rayleigh number	-	-
P	Pressure	Pa	$ML^{-1}T^{-2}$
q	Total heat flux	$W\,m^{-2}$	MT^{-3}
q	$\sqrt{s/\kappa}$ in Problem 5 only	m^{-1}	L^{-1}
q^{a}	Advective heat flux	$W\,m^{-2}$	MT^{-3}
q^{c}	Conductive heat flux	$W\,m^{-2}$	MT^{-3}
s	Laplace transform parameter	s^{-1}	T^{-1}
t	Time	s	T
$T,$	Temperature	K	K
T^{ave}	Average temperature	K	K
T^{c}	Unperturbed temperature	K	K
T^{Δ}	Perturbed temperature		
$\overline{T}$	Laplace-transformed temperature	$K\,s$	TK
$v, \overline{v}, \mathbf{v}$	Velocity	$m\,s^{-1}$	LT^{-1}
$V_{\text{b}}, V_{\text{g}}$	Volume of bulk rock, or grain	m^3	L^3
W	Heat generation per unit area	$J m^{-2}$	MT^{-2}
x	Horizontal coordinate	m	L
z, Z	Vertical coordinate, distance	m	L

Symbol	Definition	SI Unit	Dimension
α	Thermal expansivity	K^{-1}	K^{-1}
$\alpha_b, \alpha_f, \alpha_g$	α for bulk rock, fluid, grain	K^{-1}	K^{-1}
β, β_b, β_f	Compressibility	Pa^{-1}	$M^{-1}LT^2$
∇	Del, gradient operator	m^{-1}	L^{-1}
∇^2	Laplacian operator	m^{-2}	L^{-2}
Γ	Temperature gradient	$K\,m^{-1}$	$L^{-1}K$
η	κ/ω	-	-
κ	Thermal diffusivity	m^2s^{-1}	L^2T^{-1}
μ	Dynamic viscosity	$Nm^{-2}s$	$ML^{-1}T^{-1}$
ρ	Density of water	$kg\,m^{-3}$	ML^{-3}
ρc	Heat capacity of water	$J\,m^{-3}K^{-1}$	$ML^{-1}T^{-2}K^{-1}$
$(\rho c)_s$	Heat capacity of saturated rock	$J\,m^{-3}K^{-1}$	$ML^{-1}T^{-2}K^{-1}$
ϕ	Porosity	-	-
σ	Radiative heat transfer constant	$Wm^{-2}K^{-1}$	$ML^{-1}T^{-3}K^{-1}$
ω	Hydraulic diffusivity	m^2s^{-1}	L^2T^{-1}

Chapter 7

SOLUTE TRANSPORT

This chapter deals with the mass balance equation for the transport of dissolved constituents. Suspended particles are not covered. It is assumed that the groundwater flow field is known and the solute concentration is so low that the mass movement of solute does not affect the flow field. Factors affecting solute transport are advection, dispersion, adsorption, and sources. A general review on solute transport can be found in Mercer and Waddell (1993).

We will emphasize analytic solutions for one-dimensional problems, use the z-transform to simulate system response, and demonstrate how one solution can be obtained from another solution. Two-dimensional problems will be described briefly and references to three-dimensional problems will be cited.

7.1 Formulation of Equations

Advection is the primary means of solute transport in groundwater. Solutes are carried by fluid movement, and thus are transported on average at the same rate as the fluid flowing through the porous media, provided that the solutes do not react with the media and the dispersion is not severe. The average rate $\bar{v}$ is Darcy's velocity v divided by effective porosity ϕ, i.e., $\bar{v} = v/\phi$. $\bar{v}$ is also called average pore fluid velocity, average linear velocity, or seepage velocity.

Dispersion modifies the pattern of advective transport. It causes an initially sharply defined pulse of solute input to become fuzzy at its leading and trailing edges of a solute plume. Dispersion arises from both molecular diffusion and mechanical mixing. Diffusion follows the solute concentration gradient and a solute may thus diffuse against the flow direction. Mixing can result from variation of flow velocity. A flow conduit can branch to several conduits and several flow conduits can converge to one. Flow velocity also varies with conduit size. Because of internal friction and friction along the fluid-solid interface, flow velocity varies across a physical flow tube. Flow paths can be tortuous, hence fluid particles starting at the same time at a given point may arrive at other points with the same linearized distance at different times. All of these factors contribute to nonuniformity in velocity and solute dispersion but are difficult to quantify individually.

Because of dispersion, solute transport lacks a well-defined migration front for the determination of transport velocity. A velocity can be defined, for example, as the migrating rate of an *isopleth* (line or surface of given concentration) for a given solute. That isopleth can be a fraction of the source concentration or other threshold value set by the concerned users. A "breakthrough" for a given solute occurs when the chosen isopleth reaches a given location. The apparent velocity for the breakthrough varies with distances even for a one-dimensional, one-component solute transport model (Figure 7.1).

Empirically, the hydrodynamic dispersion coefficient D_{ij} is related to flow velocity $\bar{v}_i$ and molecular diffusion coefficient D by

$$\boxed{D_{ij} = \alpha_{ij}\bar{v}_i + D}, \tag{7.1}$$

where α_{ij} is the dispersivity. Both D_{ij} and α_{ij} are tensors. The dispersivity is a scale-dependent parameter (Domenico and Robin, 1984; Mercer and Waddell, 1993), hence a laboratory-determined α_{ij} should be applied with care to field studies. Because of the D_{ij}-dependency on flow velocity $\bar{v}_i$, the dispersion coefficient D_{ij} is always anisotropic (except for one-dimensional problems) even for a steady flow in a homogeneous, isotropic matrix medium. Usually D_{ij} is largest in the longitudinal (flow) direction, intermediate in the area (horizontal) transverse direction, and least in the vertical transverse direction.

The sources include negative sources – sinks. A source can be injection, extraction, adsorption, desorption, chemical reaction, radioactive decay, or biodegradation.

A mass conservation equation in an elementary or control volume can be readily formulated to account for all factors symbolically,

$$\nabla \cdot [-D\nabla(\phi C)] + \mathbf{v} \cdot \nabla C - v_{\text{in}} C_{\text{in}} + v_{\text{out}} C - G = -\frac{\partial(\phi C)}{\partial t}, \qquad (7.2)$$

where C is the solute concentration. The first term in equation (7.2) represents the net dispersive efflux of solute (see Section 1.4); the second term represents advection of solute; the third term represents the influxing fluid of concentration C_{in} at a rate of v_{in}; the fourth term represents effluxing of fluid at a rate of v_{out}; and G is the solute production rate (+ for source, - for sink). Depending on physical and chemical conditions, the source term G (in unit of mass per volume) can have a variety of forms.

For steady, incompressible groundwater flow,

$$\nabla \cdot \mathbf{v} = 0. \qquad (7.3)$$

7.1.1 Adsorption

• Linear Equilibrium

Sorbate concentration S_c is usually expressed as the mass ratio of solute absorbed X to the mass of sorbent M. Assuming that S_c is linearly proportional to the solute concentration,

$$S_c = \frac{X}{M} = K_d C, \qquad (7.4)$$

the rate of mass absorption per unit control volume is

$$G = -\frac{\partial (S_c \rho_{\text{dry}})}{\partial t} = -K_d \rho_{\text{dry}} \frac{\partial C}{\partial t}, \qquad (7.5)$$

where K_d is the distribution coefficient and ρ_{dry} is the dry bulk density. G is the rate of solute removal from the solution.

For a one-dimensional adsorption problem with no source, constant D, and steady flow, equation (7.2) becomes

$$D\frac{\partial^2 C}{\partial x^2} - \bar{v}\frac{\partial C}{\partial x} = R\frac{\partial C}{\partial t}, \qquad (7.6)$$

where R is the retardation factor (see Problem 4 in Chapter 1)

$$R = 1 + K_d \rho_{\text{dry}}/\phi. \qquad (7.7)$$

• **Decay of Solute**

If the decay rate of solute is proportional to the mass of solute in the control volume, the source term is

$$G = \frac{\partial(\phi C)}{\partial t} = -\lambda \phi C,$$ (7.8)

where λ is the decay constant. This leads to

$$\boxed{D\frac{\partial^2 C}{\partial x^2} - \bar{v}\frac{\partial C}{\partial x} - \lambda C = \frac{\partial C}{\partial t}}.$$ (7.9)

Note that this $\partial(\phi C)/\partial t$ term of G is the rate of solute change due to processes such as radioactive decay, and it should not be confused with the storage term on the right-hand side of equation (7.2).

• **Radioactivity and Retardation**

If the solute is radioactive and the system is retardative, the decay rate is proportional to the solute and the sorbate mass in the control volume

$$\begin{aligned}
G &= \frac{\partial(C\phi)}{\partial t} - \frac{\partial(S_c\rho_{\text{dry}})}{\partial t} \\
&= -\lambda\left(C\phi + \rho_{\text{dry}}S_c\right) - \frac{\partial(S_c\rho_{\text{dry}})}{\partial t} \\
&= -\lambda\left(1 + \rho_{\text{dry}}K_d/\phi\right)C\phi - K_d\rho_{\text{dry}}\frac{\partial C}{\partial t} \\
&= -\lambda R C\phi - K_d\rho_{\text{dry}}\frac{\partial C}{\partial t}.
\end{aligned}$$ (7.10)

The transport equation becomes

$$\boxed{D\frac{\partial^2 C}{\partial x^2} - \bar{v}\frac{\partial C}{\partial x} - \lambda R C = R\frac{\partial C}{\partial t}}.$$ (7.11)

• **Nonlinear Equilibrium Adsorption**

Freundlich equilibrium adsorption assumes a power law

$$S_c = KC^n,$$ (7.12)

such that

$$G = -\frac{\partial \left(S_{\mathrm{c}}\rho_{\mathrm{dry}}\right)}{\partial t} = -K\rho_{\mathrm{dry}}nC^{n-1}\frac{\partial C}{\partial t}. \tag{7.13}$$

This leads to a nonlinear differential equation, which requires numerical solution and will not be pursued hereafter.

Langmuir equilibrium adsorption is another frequently assumed nonlinear relation

$$S_{\mathrm{c}} = \frac{aC}{1 + bC}, \tag{7.14}$$

where a and b are empirical constants.

7.2 One-Dimensional Problems

Here we use the results in Chapter 3 to find analytic solutions for one-dimensional problems. All problems are restricted to linear equilibrium adsorption only.

7.2.1 Example 1: A Step Input

• **Equation and Conditions**

Solve

$$D\frac{\partial^2 C}{\partial x^2} - \overline{v}\frac{\partial C}{\partial x} = \frac{\partial C}{\partial t} \tag{7.15}$$

subject to the conditions that

$$\begin{aligned}
&\text{Condition 1:} && C[x,0] = 0, \\
&\text{Condition 2:} && C[0,t] = C_0, \\
&\text{Condition 3:} && C[\infty,t] = 0.
\end{aligned} \tag{7.16}$$

• **Solution**

Take the Laplace transform of the governing differential equation to obtain

$$D\frac{d^2\overline{C}}{dx^2} - \overline{v}\frac{d\overline{C}}{dx} = p\overline{C} - C[x,0], \tag{7.17}$$

where p is the Laplace transform parameter. The general solution is

$$\overline{C}_1[x,p] = A\exp\left[\left(\overline{v} - \sqrt{\overline{v}^2 + 4Dp}\right)x/2D\right], \tag{7.18}$$

which satisfies the first and third conditions. Here we use subscript i of $\overline{C}_i[x,p]$ to denote the ith example, e.g., $\overline{C}_1$ is the solute concentration in Example 1.

Imposing the second condition yields coefficient A and

$$\frac{\overline{C}_1[x,p]}{C_0} = \frac{\exp[\overline{v}x/2D]}{p} \exp\left[-\sqrt{p + \overline{v}^2/4D}x/\sqrt{D}\right]. \qquad (7.19)$$

Now, use the parameter-shifting procedure to obtain the transform pair

$$\frac{C_1[x,t]}{C_0} \leftrightarrow \exp\left[\frac{2\overline{v}x - \overline{v}^2 t}{4D}\right] \mathcal{L}^{-1}\left\{\frac{1}{p - \overline{v}^2/4D}\exp\left[-\sqrt{p/D}x\right]\right\} \qquad (7.20)$$

so that Example 4 in Chapter 3 can be invoked to yield the desired inversion

$$\begin{aligned}
\frac{C_1[x,t]}{C_0} &= \frac{\exp[\overline{v}x/2D]}{2}\left\{\exp[-\frac{\overline{v}x}{2D}]\text{erfc}\left[\frac{x - \overline{v}t}{\sqrt{4Dt}}\right]\right. \\
&\left. + \exp[\frac{\overline{v}x}{2D}]\text{erfc}\left[\frac{x + \overline{v}t}{\sqrt{4Dt}}\right]\right\}.
\end{aligned} \qquad (7.21)$$

This is the solution for a step input (Ogata, 1970). The solution is a special case of equation 6.22 when $y = z = 0$ and $v_y = v_z = 0$. An alternative but more difficult way to obtain the time-domain solution is by means of convolution from equation (7.19).

• **Dimensionless Variables**

Introduce the following dimensionless variables

$$\mathcal{C}_1[\xi,\tau] = C_1[x,t]/C_0, \quad \xi = \overline{v}x/D, \quad \tau = \overline{v}^2 t/D. \qquad (7.22)$$

Then equation (7.21) becomes

$$\boxed{\mathcal{C}_1[\xi,\tau] = e^{\xi/2}\mathcal{G}_0[\xi,\tau]}, \qquad (7.23)$$

where

$$\mathcal{G}_0[\xi,\tau] = \frac{1}{2}\left\{e^{-\xi/2}\text{erfc}\left[\frac{\xi - \tau}{2\sqrt{\tau}}\right] + e^{\xi/2}\text{erfc}\left[\frac{\xi + \tau}{2\sqrt{\tau}}\right]\right\}. \qquad (7.24)$$

Function $\mathcal{G}_0[\xi,\tau]$ serves as a generating function from which solutions for other cases of solute transport can be obtained. Step response $\mathcal{C}_1[\xi,\tau]$

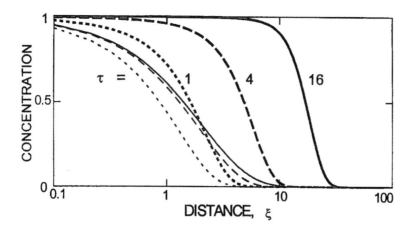

Figure 7.1: Comparison of $C_1[\xi, \tau]$ (heavy curves) and $\mathcal{G}_0[\xi, \tau]$ (light curves) at $\tau = 1$, 4, and 16.

modifies $\mathcal{G}_0[\xi, \tau]$ with a multiplication factor of $\exp[\xi/2]$. Both are shown in Figure (7.1) for comparison.

If the 0.5-isopleth is chosen for estimating the breakthrough time of solute transport in this model, Figure (7.1) indicates the time is not a linear function of distance. In other words, the breakthrough velocity changes with distance.

7.2.2 Example 2: Exponentially Decaying Input

• Equation and Conditions

If the input at $x = 0$ decays exponentially with time,

$$\text{Condition 2a:} \quad C[0, t] = C_0 \exp[-\alpha t], \tag{7.25}$$

the second boundary condition in the Laplace domain is

$$\overline{C}_2[0, p] = \frac{C_0}{p + \alpha}. \tag{7.26}$$

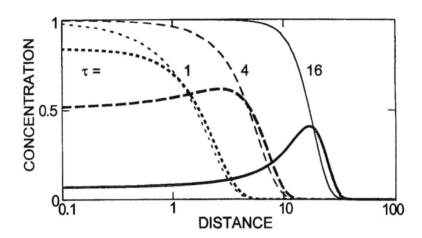

Figure 7.2: Comparison between $\mathcal{C}_2[\xi, \tau, \alpha D/\overline{v}^2]$ (heavy curves) and $\mathcal{C}_1[\xi, \tau]$ (light curves) at $\tau = 1, 4, 16$. Note the migration of the peak concentration location. Also, note that ξ and τ are defined differently for the two models.

- **Solution**

Replace $1/p$ in equation (7.19) for the step-input by $1/(p + \alpha)$ to get

$$\frac{\overline{C}_2[x, p]}{C_0} = \frac{\exp[\overline{v}x/2D]}{p + \alpha} \exp\left[-\sqrt{p + \overline{v}^2/4D}x/\sqrt{D}\right]. \qquad (7.27)$$

Following similar procedures for Example 1, the inverse is

$$\begin{aligned}
\frac{C_2[x, t]}{C_0} &= \frac{\exp[-\alpha t + \overline{v}x/2D]}{2} \\
&\cdot \left\{ \exp\left[-\frac{\sqrt{\overline{v}^2 - 4D\alpha}x}{2D}\right] \operatorname{erfc}\left[\frac{x - \sqrt{\overline{v}^2 - 4D\alpha}t}{\sqrt{4Dt}}\right] \right. \\
&\left. + \exp\left[\frac{\sqrt{\overline{v}^2 - 4D\alpha}x}{2D}\right] \operatorname{erfc}\left[\frac{x + \sqrt{\overline{v}^2 - 4D\alpha}t}{\sqrt{4Dt}}\right] \right\}. \quad (7.28)
\end{aligned}$$

- **Dimensionless Variables**

Introducing the following dimensionless variables

$$\mathcal{C}_2\left[\xi, \tau, \frac{\alpha D}{\overline{v}^2}\right] = \frac{C_2[x, t]}{C_0}, \quad \xi = \frac{\sqrt{\overline{v}^2 - 4\alpha D}}{D}x, \quad \tau = \frac{\overline{v}^2 - 4\alpha D}{D}t, \qquad (7.29)$$

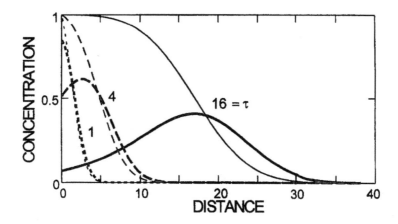

Figure 7.3: Comparison of $C_2[\xi, \tau, \alpha D/\overline{v}^2]$ (heavy curves) and $C_1[\xi, \tau]$ (light curves). Here the distance ξ is plotted at a linear scale instead of the logarithmic scale of Figure 7.2. Note the changing shapes of the concentration profiles at different times. Also, note that ξ and τ are defined differently for the two models.

we obtain

$$
C_2\left[\xi, \tau, \frac{\alpha D}{\overline{v}^2}\right] = \exp\left\{-\frac{\tau}{4\left(\overline{v}^2/4\alpha D - 1\right)} + \frac{\xi}{2\sqrt{1 - 4\alpha D/\overline{v}^2}}\right\} G_0[\xi, \tau].
$$

(7.30)

In Figure (7.2), $C_2[\xi, \tau, \alpha D/\overline{v}^2]$ is compared with $C_1[\xi, \tau]$. This shows that the concentration peak may migrate from the source location if the source concentration is decreasing with time. The extent of such migration depends on the value of $\alpha D/\overline{v}^2$. This view of the changing concentration is distorted by the semilog plot. An undistorted view is seen in the linear plot shown in Figure 7.3. Note that ξ and τ are defined differently for C_1 and C_2. A true comparison should be made with the proper physical dimensions.

7.2.3 Example 3: A System with Retardation

• Equation and Conditions

The solute transport equation for a system with adsorptive retardation is

$$
D\frac{\partial^2 C}{\partial x^2} - \overline{v}\frac{\partial C}{\partial x} = R\frac{\partial C}{\partial t},
$$

(7.31)

where R is the retardation factor (assumed to be constant for a linear equilibrium model). Because

$$R\frac{\partial C}{\partial t} = \frac{\partial C}{\partial t/R} = \frac{\partial C}{\partial t'}, \quad t' = \frac{t}{R}, \tag{7.32}$$

a retardative system can be viewed in a time frame scaled by the retardation factor.

• **Solution**

To obtain the solution in a retarded system from the equivalent solution in the nonretardative system, rewrite the differential equation as

$$\frac{D}{R}\frac{\partial^2 C}{\partial x^2} - \frac{\bar{v}}{R}\frac{\partial C}{\partial x} = \frac{\partial C}{\partial t}. \tag{7.33}$$

The solutions can be obtained by replacing D with D/R and $\bar{v}$ with $\bar{v}/R$. Note that replacing t by t/R as suggested by equation (7.32) is not recommended for getting the new solution because a term with an explicit time dependence (e.g., input function) may not be retardative and its t should not be replaced by t/R (see Example 10).

7.2.4   Example 4: A Reactive System

• **Equation and Solution**

If the solute decays with a decay constant λ, the solute transport equation is

$$D\frac{\partial^2 C}{\partial x^2} - \bar{v}\frac{\partial C}{\partial x} - \lambda C = \frac{\partial C}{\partial t}. \tag{7.34}$$

Its general solution in the Laplace domain is

$$\overline{C}_4[x,p] = A\exp\left[\left(\bar{v} - \sqrt{\bar{v}^2 + 4D(p+\lambda)}\right)x/2D\right]. \tag{7.35}$$

The time-domain solution is obtained by replacing $\bar{v}$ with $\sqrt{\bar{v}^2 + 4D\lambda}$ in equations (7.21) and (7.28) except that the $\bar{v}$ in the first factor outside the braces $\{..\}$ should not be replaced.

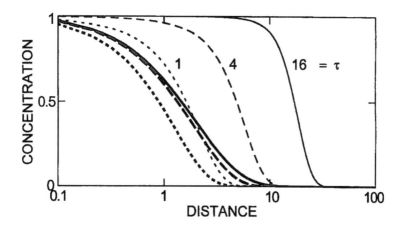

Figure 7.4: Comparison of $\mathcal{C}_4[\xi, \tau, \lambda D/\bar{v}^2]$ (heavy curves) and $\mathcal{C}_1[\xi, \tau]$ at different time τ for $\lambda D/\bar{v}^2 = 1$.

For example, the case with a step input $C[0, t] = C_0$ at $x = 0$ has the step response

$$
\begin{aligned}
\frac{C_4[x, t]}{C_0} &= \frac{\exp[\bar{v}x/2D]}{2} \left\{ \exp\left[-\frac{\sqrt{\bar{v}^2 + 4D\lambda}\,x}{2D} \right] \operatorname{erfc}\left[\frac{x - \sqrt{\bar{v}^2 + 4D\lambda}\,t}{\sqrt{4Dt}} \right] \right. \\
&\left. + \exp\left[\frac{\sqrt{\bar{v}^2 + 4D\lambda}\,x}{2D} \right] \operatorname{erfc}\left[\frac{x + \sqrt{\bar{v}^2 + 4D\lambda}\,t}{\sqrt{4Dt}} \right] \right\}.
\end{aligned}
\tag{7.36}
$$

• **Dimensionless Variables**

Introducing the following dimensionless variables,

$$
\mathcal{C}_4\left[\xi, \tau, \frac{\lambda D}{\bar{v}^2}\right] = \frac{C_4[x, t]}{C_0}, \quad \xi = \frac{\sqrt{\bar{v}^2 + 4D\lambda}}{D}, \quad \tau = \frac{\bar{v}^2 + 4D\lambda}{D}t
\tag{7.37}
$$

yields

$$
\boxed{\mathcal{C}_4\left[\xi, \tau, \frac{\lambda D}{\bar{v}^2}\right] = \exp\left[\frac{\xi}{2\sqrt{1 + 4\lambda D/\bar{v}^2}} \right] \mathcal{G}_0\left[\xi, \tau\right],}
\tag{7.38}
$$

which is illustrated in Figure (7.4) for $\lambda D/\bar{v}^2 = 0.1$ at different times τ.

- **Impulse Response**

Differentiate the right-hand side of equation (7.36) with respect to time t, yielding the impulse response (Section 2.2.1)

$$C^I[x,t] = \frac{x}{2\sqrt{\pi D}t^{3/2}} \exp\left[-\lambda t - \frac{(x-vt)^2}{4Dt}\right], \qquad (7.39)$$

which can be convolved with any input functions to obtain the system responses.

7.2.5   Example 5: A Reactive and Retardative System

See Problem 1.

7.2.6   Example 6: Advection at Boundary, $\lambda \neq \alpha$

- **Equation and Conditions**

Consider a reactive and retardative system,

$$D\frac{\partial^2 C}{\partial x^2} - \bar{v}\frac{\partial C}{\partial x} - \lambda RC = R\frac{\partial C}{\partial t}, \qquad (7.40)$$

with the boundary conditions:

$$
\begin{aligned}
&\text{Condition 1:} \quad C[x,0] = 0,\\
&\text{Condition 2:} \quad -D\frac{\partial C}{\partial x} + \bar{v}C\bigg|_{x=0} = \bar{v}C_0\exp[-\alpha t],\\
&\text{Condition 3:} \quad \frac{\partial C[x,t]}{\partial x} = 0 \quad \text{as } x \to \infty \qquad (7.41)
\end{aligned}
$$

(van Genuchten and Alves, 1982). Condition 2 equates a dispersive and advective flow at the boundary with influx for which the concentration decays with time, $C_0\exp[-\alpha t]$. This advective condition is an example of a mixed-type boundary condition (type 3 or Cauchy condition).

• Solution in the Laplace Domain

In the Laplace domain, the general solution is

$$\overline{C}_6[x,p] = A \exp\left[\left(\overline{v} - \sqrt{\overline{v}^2 + 4DR(p+\lambda)}\right)x/2D\right], \tag{7.42}$$

which satisfies conditions 1 and 3. According to the advective boundary condition 2, coefficient A is

$$A = \frac{\overline{v}C_0}{p+\alpha} \cdot \frac{2}{\overline{v} + \sqrt{\overline{v}^2 + 4DR(p+\lambda)}}. \tag{7.43}$$

Hence

$$\frac{\overline{C}_6[x,p]}{C_0} = \frac{2\overline{v}\exp[\frac{\overline{v}x}{2D}]}{p+\alpha} \cdot \frac{\exp\left[-\sqrt{\overline{v}^2 + 4DR(p+\lambda)}x/2D\right]}{\overline{v} + \sqrt{\overline{v}^2 + 4DR(p+\lambda)}}. \tag{7.44}$$

• Time-Domain Solution

The time-domain solution is obtained as follows. Rationalize the above relation to yield

$$
\begin{aligned}
\frac{\overline{C}_6[x,p]}{C_0} &= -\overline{v}\exp\left[\frac{\overline{v}x}{2D}\right]\left(\overline{v} - \sqrt{\overline{v}^2 + 4DR(p+\lambda)}\right) \\
&\quad \cdot \frac{\exp\left[-\sqrt{\overline{v}^2 + 4DR(p+\lambda)}x/2D\right]}{2DR(p+\alpha)(p+\lambda)}.
\end{aligned} \tag{7.45}
$$

We see that if

$$\overline{B}[x,p] = \frac{\exp\left[-\sqrt{\overline{v}^2 + 4DR(p+\lambda)}x/2D\right]}{(p+\alpha)(p+\lambda)} \tag{7.46}$$

is introduced, then

$$\frac{\overline{C}_6[x,p]}{C_0} = -\frac{\overline{v}^2\exp[\frac{\overline{v}x}{2D}]}{2DR}\left\{\overline{B}[x,p] + \frac{2D}{\overline{v}}\frac{\partial\overline{B}[x,p]}{\partial x}\right\}. \tag{7.47}$$

By the partial fraction method, $\overline{B}[x,p]$ is split into

$$\overline{B}[x,p] = \overline{B}_\alpha[x,p] - \overline{B}_\lambda[x,p], \tag{7.48}$$

where

$$\overline{B}_\alpha[x,p] = \frac{1}{(\lambda - \alpha)(p + \alpha)} \exp\left[-x\sqrt{\left(\frac{\overline{v}^2}{4DR} + \lambda + p\right)\frac{R}{D}}\right],$$

$$\overline{B}_\lambda[x,p] = \frac{1}{(\lambda - \alpha)(p + \lambda)} \exp\left[-x\sqrt{\left(\frac{\overline{v}^2}{4DR} + \lambda + p\right)\frac{R}{D}}\right]. \quad (7.49)$$

Use the parameter shifting procedure (see Basic Rule 4 in Chapter 3) to obtain

$$B_\alpha[x,t] = \frac{\exp\left[-(\lambda + \overline{v}^2/4DR)t\right]}{\lambda - \alpha} \mathcal{L}^{-1}\left\{\frac{\exp\left[-\sqrt{pR/D}x\right]}{p + \alpha - \lambda - \overline{v}^2/4DR}\right\},$$

$$B_\lambda[x,t] = \frac{\exp\left[-(\lambda + \overline{v}^2/4DR)t\right]}{\lambda - \alpha} \mathcal{L}^{-1}\left\{\frac{\exp\left[-\sqrt{pR/D}x\right]}{p - \overline{v}^2/4DR}\right\}. \quad (7.50)$$

According to Example 4 in Chapter 3, the inverse Laplace transforms are

$$B_\alpha[x,t] = \frac{\exp\left[-\alpha t\right]}{2(\lambda - \alpha)}\left\{e^{x\sqrt{\beta R/D}}\mathrm{erfc}\left[\frac{x}{\sqrt{4Dt/R}} + \sqrt{\beta t}\right]\right.$$
$$\left. + e^{-x\sqrt{\beta R/D}}\mathrm{erfc}\left[\frac{x}{\sqrt{4Dt/R}} - \sqrt{\beta t}\right]\right\}, \quad (7.51)$$

and

$$B_\lambda[x,t] = \frac{\exp\left[-\lambda t\right]}{2(\lambda - \alpha)}\left\{e^{x\overline{v}/2D}\mathrm{erfc}\left[\frac{x}{\sqrt{4Dt/R}} + \sqrt{\frac{\overline{v}^2 t}{4DR}}\right]\right.$$
$$\left. + e^{-x\overline{v}/2D}\mathrm{erfc}\left[\frac{x}{\sqrt{4Dt/R}} - \sqrt{\frac{\overline{v}^2 t}{4DR}}\right]\right\}, \quad (7.52)$$

where $\beta = \lambda - \alpha + \overline{v}^2/4DR$.

The derivatives of B_α and B_λ are, respectively,

$$\frac{\partial B_\alpha}{\partial x} = \frac{e^{-\alpha t}}{2(\lambda - \alpha)}\left\{\sqrt{\frac{\beta R}{D}}e^{x\sqrt{\beta R/D}}\mathrm{erfc}\left[\frac{x}{\sqrt{4Dt/R}} + \sqrt{\beta t}\right] - \sqrt{\frac{\beta R}{D}}\right.$$
$$\left. \cdot e^{-x\sqrt{\beta R/D}}\mathrm{erfc}\left[\frac{x}{\sqrt{4Dt/R}} - \sqrt{\beta t}\right] - \frac{2e^{x^2 R/4Dt - \beta t}}{\sqrt{\pi Dt/R}}\right\} \quad (7.53)$$

and

$$\frac{\partial B_\lambda}{\partial x} = \frac{e^{-\lambda t}}{2(\lambda - \alpha)} \left\{ \frac{\bar{v}}{2D} e^{x\bar{v}/2D} \text{erfc} \left[\frac{x}{\sqrt{4Dt/R}} + \sqrt{\frac{\bar{v}^2 t}{4DR}} \right] - \frac{\bar{v}}{2D} e^{-x\bar{v}/2D} \right.$$

$$\left. \cdot \text{erfc} \left[\frac{x}{\sqrt{4Dt/R}} - \sqrt{\frac{\bar{v}^2 t}{4DR}} \right] - \frac{2 \exp \left[-\frac{x^2 R}{4Dt} - \frac{\bar{v}^2 t}{4DR} \right]}{\sqrt{\pi Dt/R}} \right\}. \quad (7.54)$$

Adding all components together in accordance with equation (7.47) yields

$$\frac{C_6}{C_0} = -\frac{\bar{v}^2 \exp \left[-\alpha t + \bar{v} x/2D \right]}{4DR (\lambda - \alpha)}$$

$$\cdot \left\{ \left(1 + \sqrt{4DR\beta}/\bar{v} \right) e^{x\sqrt{\beta R/D}} \text{erfc} \left[\frac{x}{\sqrt{4Dt/R}} + \sqrt{\beta t} \right] \right.$$

$$\left. + \left(1 - \sqrt{4DR\beta}/\bar{v} \right) e^{-x\sqrt{\beta R/D}} \text{erfc} \left[\frac{x}{\sqrt{4Dt/R}} - \sqrt{\beta t} \right] \right\}$$

$$+ \frac{\bar{v}^2 e^{-\lambda t + \bar{v} x/2D}}{2DR (\lambda - \alpha)} \left\{ e^{x\bar{v}/2D} \text{erfc} \left[\frac{x}{\sqrt{4Dt/R}} + \sqrt{\frac{\bar{v}^2 t}{4DR}} \right] \right\}. \quad (7.55)$$

The relation can be simplified further by defining

$$U = \sqrt{4DR\beta} = \sqrt{4DR (\lambda - \alpha + \bar{v}^2/4DR)}, \quad (7.56)$$

$$\beta = U^2/4DR, \quad x\sqrt{\beta R/D} = xU/2D, \quad \alpha \leq \lambda + \bar{v}^2/4DR, \quad (7.57)$$

to yield

$$\boxed{\frac{C_6[x, t]}{C_0} = \exp \left[-\alpha t \right] A_6[x, t]}, \quad \lambda \neq \alpha, \quad (7.58)$$

where

$$A_6[x, t] = \frac{\bar{v}^2}{2DR (\alpha - \lambda)} \left\{ \frac{\bar{v} - U}{2\bar{v}} \exp \left[\frac{(\bar{v} - U) x}{2D} \right] \text{erfc} \left[\frac{x - Ut/R}{\sqrt{4Dt/R}} \right] \right.$$

$$+ \frac{\bar{v} + U}{2\bar{v}} \exp \left[\frac{(\bar{v} + U) x}{2D} \right] \text{erfc} \left[\frac{x + Ut/R}{\sqrt{4Dt/R}} \right]$$

$$\left. - \exp \left[-(\lambda - \alpha) t + \frac{x\bar{v}}{D} \right] \text{erfc} \left[\frac{x + \bar{v}t/R}{\sqrt{4Dt/R}} \right] \right\}. \quad (7.59)$$

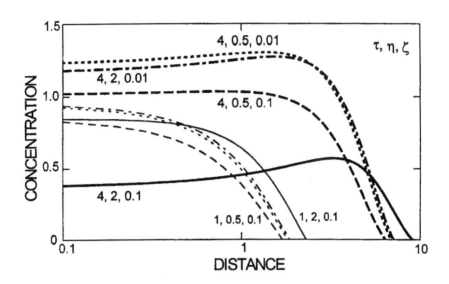

Figure 7.5: Concentration $\mathcal{C}_6[\xi, \tau]$ for different values of (τ, η, ζ). Unlabeled light curves are at $\tau = 1$.

This result can be shown to agree with that of van Genuchten and Alves (1982) (also quoted by Javandel et al., 1984, p. 15). See Example 8 for an input of finite duration.

• **Dimensionless Variables**

Introducing the following dimensionless variables

$$\mathcal{C}_6\left[\xi, \tau,\right] = \frac{C_6[x,t]}{C_0}, \quad \xi = \frac{Ux}{D}, \quad \tau = \frac{U^2 t}{DR} \tag{7.60}$$

yields

$$
\begin{aligned}
\mathcal{C}_6\left[\xi, \tau\right] &= \frac{\bar{v}^2 \exp\left[-\alpha D R \tau / U^2\right]}{2 D R \left(\alpha - \lambda\right)} \\
&\quad \cdot \left\{ e^{\bar{v}\xi/2U} \left(\frac{\bar{v} - U}{2\bar{v}} e^{-\xi/2} \text{erfc} \left[\frac{\xi - \tau}{2\sqrt{\tau}} \right] + \frac{\bar{v} + U}{2\bar{v}} e^{\xi/2} \text{erfc} \left[\frac{\xi + \tau}{2\sqrt{\tau}} \right] \right) \right. \\
&\quad \left. - e^{\xi} \text{erfc} \left[\frac{\xi + \bar{v}^2 \tau / U^2}{2\bar{v}\sqrt{\tau}/U} \right] \right\}, \quad \alpha \neq \lambda. \tag{7.61}
\end{aligned}
$$

Note that $\mathcal{C}_6\left[\xi, \tau\right]$ has reduced the number of independent variables in $C_6[x,t]$ from 7 (i.e., x, t, D, R, α, λ, and $\bar{v}$) to 4 (i.e., ξ, τ, α/λ, and $\lambda D R/\bar{v}^2$)

The usage of

$$\eta = \frac{\alpha}{\lambda}, \quad \zeta = \frac{\lambda DR}{\overline{v}^2} \tag{7.62}$$

leads to

$$C_6[\xi, \tau] = \frac{\exp\left[-\eta\zeta\tau\overline{v}^2/U^2\right]}{2\zeta(\eta - 1)} B_6, \tag{7.63}$$

where

$$B_6 = \left\{ e^{\overline{v}\xi/2U} \left(\frac{\overline{v} - U}{2\overline{v}} e^{-\xi/2} \text{erfc} \left[\frac{\xi - \tau}{2\sqrt{\tau}} \right] + \frac{\overline{v} + U}{2\overline{v}} e^{\xi/2} \text{erfc} \left[\frac{\xi + \tau}{2\sqrt{\tau}} \right] \right) \right.$$
$$\left. - e^{\xi} \text{erfc} \left[\frac{\xi + \overline{v}^2\tau/U^2}{2\overline{v}\sqrt{\tau}/U} \right] \right\} \tag{7.64}$$

and

$$\frac{U}{\overline{v}} = \sqrt{4\zeta - 4\eta\zeta + 1}, \quad \eta < 1 + 0.25\zeta. \tag{7.65}$$

Figure 7.5 depicts the result for a combination of τ, η, and ζ values.

7.2.7 Example 7: Advection at Boundary, $\lambda = \alpha$

• **Solution**

As $\lambda \to \alpha$ in Example 6, $U \to \overline{v}$. At this limit, $C/C_0 \to 0/0$. Applying l'Hopital's rule to the first term in equation (7.58) and using the relation

$$\lim_{\lambda \to \alpha} \frac{\partial U}{\partial \lambda} = \frac{2DR}{\overline{v}} \tag{7.66}$$

yields

$$\frac{\overline{v}^2}{4DR} \text{erfc} \left[\frac{x - \overline{v}t/R}{\sqrt{4Dt/R}} \right] \frac{2DR}{\overline{v}} = \frac{1}{2} \text{erfc} \left[\frac{x - \overline{v}t/R}{\sqrt{4Dt/R}} \right].$$

Similarly, the second term of equation (7.58) becomes

$$\sqrt{\frac{\overline{v}^2 t}{\pi DR}} \exp\left[-\left(\frac{x - \overline{v}t/R}{\sqrt{4Dt/R}} \right)^2 \right] - \frac{D + \overline{v}x}{2D} \exp\left[\frac{\overline{v}x}{D} \right] \text{erfc} \left[\frac{x + \overline{v}t/R}{\sqrt{4Dt/R}} \right]$$

and the third term becomes

$$\frac{-\overline{v}^2 t \exp\left[-(\lambda - \alpha)t + x\overline{v}/D \right]}{2DR} \text{erfc} \left[\frac{x + \overline{v}t/R}{\sqrt{4Dt/R}} \right].$$

Add the three components together to give

$$\boxed{\frac{C_7[x,t]}{C_0} = \exp[-\alpha t] A_7[x,t]}, \quad \alpha = \lambda, \tag{7.67}$$

where

$$\begin{aligned}
A_7[x,t] &= \frac{1}{2}\text{erfc}\left[\frac{x - \overline{v}t/R}{\sqrt{4Dt/R}}\right] + \sqrt{\frac{\overline{v}^2 t}{\pi DR}}\exp\left[-\left(\frac{x - \overline{v}t/R}{\sqrt{4Dt/R}}\right)^2\right] \\
&\quad - \frac{1}{2}\left(1 + \frac{\overline{v}x}{D} + \frac{\overline{v}^2 t}{DR}\right)\exp\left[\frac{\overline{v}x}{D}\right]\text{erfc}\left[\frac{x + \overline{v}t/R}{\sqrt{4Dt/R}}\right].
\end{aligned} \tag{7.68}$$

This agrees with the result of van Genuchten and Alves (1982).

• Dimensionless Variables

In dimensionless form for $\alpha = \lambda$,

$$\begin{aligned}
\mathcal{A}_7[\xi,\tau] &= \frac{1}{2}\text{erfc}\left[\frac{\xi - \tau}{2\sqrt{\tau}}\right] + \sqrt{\frac{\tau}{\pi}}\exp\left[-(\xi - \tau)^2/4\tau\right] \\
&\quad - \frac{1}{2}\left(1 + \xi + \tau\right)e^{\xi}\text{erfc}\left[\frac{\xi + \tau}{2\sqrt{\tau}}\right],
\end{aligned} \tag{7.69}$$

where ξ and τ are defined earlier in Example 6.

7.2.8   Example 8: A Pulse Input

In Examples 1 through 6, all step responses C/C_0 (including special cases when $\alpha = 0$) can be used to obtain the response to a pulse input of duration t_p,

$$\begin{aligned}
C^p[x,t] &= C[x,t] \quad \text{for } t \le t_p, \\
C^p[x,t] &= C[x,t] - C[x,t - t_p] \quad \text{for } t \ge t_p
\end{aligned} \tag{7.70}$$

because the equations are linear.

For a step-pulse input as an example, the concentration is

$$\mathcal{C}_8[\xi,\tau,\tau_p] = \mathcal{C}_1[\xi,\tau] - \mathcal{C}_1[\xi,\tau - \tau_p]H[\tau - \tau_p] \tag{7.71}$$

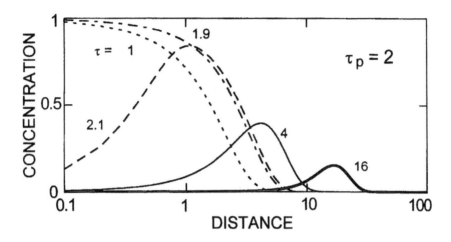

Figure 7.6: Concentration $C_8[\xi, \tau]$ due to a step-pulse input of duration $\tau_p = 2$.

where $H[\tau - \tau_p]$ is the Heaviside unit step function. Figure 7.6 depicts the result for a pulse duration of 2 (i.e., $\tau_p = 2$). Note the migration and decay of the peak concentration.

For an arbitrary input function $f[t]$, we need the impulse function

$$C^I[x, t] = \frac{\partial C[x, t]}{C_0 \partial t} \tag{7.72}$$

to obtain the desired distribution from the convolution formula

$$C^f[x, t] = C_0 \int_0^t f[\tau] C^I[x, t - \tau] d\tau. \tag{7.73}$$

7.2.9 Example 9: Advection, Transfer Function

Set $\alpha = 0$ in equation (7.44) in Example 6 to get the transform of the step response. Then, remove the factor of $1/p$ from the result to obtain the transfer function

$$
\begin{aligned}
\overline{C}^I[x, p] &= 2\overline{v} \exp\left[\frac{\overline{v}x}{2D}\right] \frac{\exp\left[-\sqrt{\overline{v}^2 + 4DR(p + \lambda)}x/2D\right]}{\overline{v} + \sqrt{\overline{v}^2 + 4DR(p + \lambda)}} \\
&= \frac{-\overline{v}\exp[\overline{v}x/2D]}{2DR}\left(\overline{v} - \sqrt{\overline{v}^2 + 4DR(p + \lambda)}\right)\overline{B}
\end{aligned}
$$

$$= \frac{-\overline{v}^2 \exp[\overline{v}x/2D]}{2DR} \left(\overline{B} + \frac{2D}{\overline{v}} \frac{\partial \overline{B}}{\partial x} \right), \tag{7.74}$$

where

$$\overline{B}[x, p] = \frac{\exp\left[-\sqrt{\overline{v}^2 + 4DR(p + \lambda)}x/2D\right]}{p + \lambda}. \tag{7.75}$$

The inverse of $\overline{B}[x, p]$ is

$$B[x, t] = \exp\left[-\left(\lambda + \overline{v}^2/4DR\right)t\right] \mathcal{L}^{-1} \left\{ \frac{\exp\left[-\sqrt{pR/D}x\right]}{p - \overline{v}^2/4DR} \right\} = \frac{e^{-\lambda t}}{2}$$

$$\cdot \left\{ e^{x\overline{v}/2D} \operatorname{erfc} \left[\frac{x + \overline{v}t/R}{\sqrt{4Dt/R}} \right] + e^{-x\overline{v}/2D} \operatorname{erfc} \left[\frac{x - \overline{v}t/R}{\sqrt{4Dt/R}} \right] \right\}. \tag{7.76}$$

Its derivative is

$$\frac{\partial B[x, t]}{\partial x} = \frac{\overline{v} \exp[-\lambda t]}{4D}$$

$$\cdot \left\{ e^{x\overline{v}/2D} \operatorname{erfc} \left[\frac{x + \overline{v}t/R}{\sqrt{4Dt/R}} \right] - e^{-x\overline{v}/2D} \operatorname{erfc} \left[\frac{x - \overline{v}t/R}{\sqrt{4Dt/R}} \right] \right\}$$

$$- \frac{e^{-\lambda t}}{2\sqrt{\pi Dt/R}} \left\{ \exp\left[-\frac{(x + \overline{v}t/R)^2}{4Dt/R} \right] e^{x\overline{v}/2D} \right.$$

$$\left. + \exp\left[-\frac{(x - \overline{v}t/R)^2}{4Dt/R} \right] e^{-x\overline{v}/2D} \right\}. \tag{7.77}$$

Therefore, the impulse response is

$$C^{\mathrm{I}}[x, t] = \frac{-\overline{v}^2 \exp[\frac{\overline{v}x}{2D} - \lambda t]}{2DR}$$

$$\cdot \left\{ e^{x\overline{v}/2D} \operatorname{erfc} \left[\frac{x + \overline{v}t/R}{\sqrt{4Dt/R}} \right] - 2\sqrt{\frac{DR}{\pi \overline{v}^2 t}} \exp\left[-\frac{x^2 + \overline{v}^2 t^2/R^2}{4Dt/R} \right] \right\}$$

$$= \frac{\overline{v}^2 e^{-\lambda t}}{DR} \left\{ \sqrt{\frac{DR}{\pi \overline{v}^2 t}} \exp\left[-\frac{(x - vt/R)^2}{4Dt/R} \right] \right.$$

$$\left. - \frac{e^{x\overline{v}/D}}{2} \operatorname{erfc} \left[\frac{x + \overline{v}t/R}{\sqrt{4Dt/R}} \right] \right\}. \tag{7.78}$$

7.3 Two-Dimensional Problems

Some analytical solutions to solute transport in two-dimensional problems have been listed by Cleary and Ungs (1978) but the derivations were not provided. What follow are derivations based on the superposition of one-dimensional solutions. Some solution formats are also improved for better stability in numerical integration and implementation of boundary conditions.

7.3.1 Example 10: A Plane Dispersion Model

• **Equation and Conditions**

Consider the governing differential equation,

$$D_x \frac{\partial^2 C}{\partial x^2} + D_y \frac{\partial^2 C}{\partial y^2} - \bar{v}_x \frac{\partial C}{\partial x} - \lambda C = \frac{\partial C}{\partial t}, \tag{7.79}$$

and the initial and boundary conditions (**Figure 7.7**),

$$
\begin{aligned}
C[x, y, 0] &= 0, \\
C[0, y, t] &= \begin{cases} C_0 \exp[-\alpha t] & \text{if} \quad a \le y \le b \\ 0 & \text{otherwise} \end{cases}, \\
\lim_{y \to \pm\infty} \frac{\partial C}{\partial y} &= 0, \\
\lim_{x \to \infty} \frac{\partial C}{\partial x} &= 0.
\end{aligned}
\tag{7.80}
$$

The conditions restrict the solute source in a strip of width $b - a$. In this example, with reference to steady fluid flow in the x direction, D_x and D_y are, respectively, the longitudinal and transverse dispersion coefficients.

• **Solution**

We will use the technique of separation of variables to obtain impulse response, then use the convolution theorem to obtain the desired solution for an exponentially decaying input source function. Let

$$C[x, y, t] = C^x[x, t] C^y[y, t]. \tag{7.81}$$

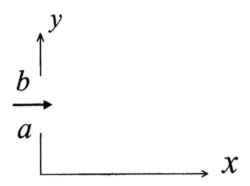

Figure 7.7: A plane dispersion model. Input flux $C[0, y, t] = C_0 \exp[-\alpha t]$ between $y = a$ and $y = b$. For Example 10, $\partial C/\partial y \to 0$ as $y \to \pm\infty$. For Example 11, $\partial C/\partial y \to 0$ at $y = 0$.

Substituting this relation into the governing equation yields

$$D_x \frac{\partial^2 C^x}{\partial x^2} - \bar{v}_x \frac{\partial C^x}{\partial x} - \lambda C^x = \frac{\partial C^x}{\partial t} \qquad (7.82)$$

and

$$D_y \frac{\partial^2 C^y}{\partial y^2} = \frac{\partial C^y}{\partial t}. \qquad (7.83)$$

Note that C^x has the same dimension as C but C^y is dimensionless.

The impulse response for C^x is

$$C_I^x[x, t] = \frac{x}{2\sqrt{\pi D_x} t^{3/2}} \exp\left[-\lambda t - \frac{(x - \bar{v}_x t)^2}{4 D_x t}\right] \qquad (7.84)$$

according to Example 4.

The impulse response for C^y can be obtained by integrating the instantaneous point source solution with respect to x and z, each from $-\infty$ to ∞ to obtain the plane-source solution

$$C_I^y[y, t] = \frac{\exp[-(y - y')^2/4 D_y t]}{2\sqrt{\pi D_y t}}, \qquad (7.85)$$

(See Section 2.1) where y' is the plane source location and C_I^y has a dimension of L^{-1}. Integrating y' over the strip between a and b yields

$$C_{IS}^y[y, t] = \frac{1}{2}\left(\text{erf}\left[\frac{b - y}{\sqrt{4 D_y t}}\right] - \text{erf}\left[\frac{a - y}{\sqrt{4 D_y t}}\right]\right). \qquad (7.86)$$

Both $C_I^x[x,t]$ and $C_{IS}^y[y,t]$ are impulse responses. Their product is an impulse response too. Therefore, the convolution of the two with the input function $C_0 \exp[-\alpha t]$ gives

$$C_{10}[x,y,t] = C_0 \int_0^t \exp[-\alpha(t-\tau)]C_I^x[x,\tau]C_{IS}^y[y,\tau]d\tau, \qquad (7.87)$$

where subscript 10 is now attached for problem identification. Simplification leads to

$$\begin{aligned}
\frac{C_{10}[x,y,t]}{C_0} &= \frac{x}{4\sqrt{\pi D_x}}\exp\left[\frac{\overline{v}_x x}{2D_x} - \alpha t\right] \\
&\quad \cdot \int_0^t \exp\left[-\left(\lambda - \alpha + \frac{\overline{v}_x^2}{4D_x}\right)\tau - \frac{x^2}{4D_x\tau}\right]\tau^{-3/2} \\
&\quad \cdot \left(\mathrm{erf}\left[\frac{b-y}{\sqrt{4D_y\tau}}\right] - \mathrm{erf}\left[\frac{a-y}{\sqrt{4D_y\tau}}\right]\right)d\tau.
\end{aligned} \qquad (7.88)$$

● **Case For Transverse Flow**

If the **v** has y component, $C_I^y[y,t]$ is modified as

$$C_I^y[y,t] = \frac{\exp[-(y - \overline{v}_y t - y')^2/4D_y t]}{2\sqrt{\pi D_y t}}. \qquad (7.89)$$

The solution can be obtained from equation (7.88) by replacing y with $y - \overline{v}_y t$:

$$\begin{aligned}
\frac{C_{10}[x,y,t]}{C_0} &= \frac{x}{4\sqrt{\pi D_x}}\exp\left[\frac{\overline{v}_x x}{2D_x} - \alpha t\right] \\
&\quad \cdot \int_0^t \exp\left[-\left(\lambda - \alpha + \frac{\overline{v}_x^2}{4D_x}\right)\tau - \frac{x^2}{4D_x\tau}\right]\tau^{-3/2} \\
&\quad \cdot \left(\mathrm{erf}\left[\frac{b-y+\overline{v}_y\tau}{\sqrt{4D_y\tau}}\right] - \mathrm{erf}\left[\frac{a-y+\overline{v}_y\tau}{\sqrt{4D_y\tau}}\right]\right)d\tau,
\end{aligned} \qquad (7.90)$$

which was listed by Cleary and Ungs (1978) in a slightly different form.

● **Case with Retardation**

For a retardative system

$$D_x\frac{\partial^2 C}{\partial x^2} + D_y\frac{\partial^2 C}{\partial y^2} - \overline{v}_x\frac{\partial C}{\partial x} - \lambda RC = R\frac{\partial C}{\partial t}, \qquad (7.91)$$

the impulse responses can be obtained by replacing D and $\bar{v}_x$ with D/R and $\bar{v}_x/R$, respectively

$$C_{\mathrm{I}}^{\mathrm{x}}[x,t] = \frac{xR^{1/2}}{2\sqrt{\pi D_x}t^{3/2}} \exp\left[-\lambda t - \frac{(x - \bar{v}_x t/R)^2}{4D_x t/R}\right], \tag{7.92}$$

$$C_{\mathrm{IS}}^{\mathrm{y}}[y,t] = \frac{1}{2}\left(\mathrm{erf}\left[\frac{b-y}{\sqrt{4D_y t/R}}\right] - \mathrm{erf}\left[\frac{a-y}{\sqrt{4D_y t/R}}\right]\right). \tag{7.93}$$

The convolution integral becomes

$$\frac{C_{10}[x,y,t]}{C_0} = \frac{x}{4\sqrt{\pi D_x}} \exp\left[\frac{\bar{v}_x x}{2D_x} - \alpha t\right]$$

$$\cdot \int_0^t \exp\left[-\left(\lambda - \alpha + \frac{\bar{v}_x^2}{4D_x R}\right)\tau - \frac{x^2}{4D_x \tau/R}\right]\frac{R^{1/2}}{\tau^{3/2}}$$

$$\cdot \left(\mathrm{erf}\left[\frac{b-y}{\sqrt{4D_y \tau/R}}\right] - \mathrm{erf}\left[\frac{a-y}{\sqrt{4D_y \tau/R}}\right]\right) d\tau. \tag{7.94}$$

To conform with the formula listed by Javandel et al. (1984), substitute τ/R by τ' in the above formula, then omit the prime symbol to have

$$\frac{C_{10}[x,y,t]}{C_0} = \frac{x}{4\sqrt{\pi D_x}} \exp\left[\frac{\bar{v}_x x}{2D_x} - \alpha t\right]$$

$$\cdot \int_0^{t/R} \exp\left[-\left((\lambda - \alpha)R + \frac{\bar{v}_x^2}{4D_x}\right)\tau - \frac{x^2}{4D_x \tau}\right]$$

$$\cdot \left(\mathrm{erf}\left[\frac{b-y}{\sqrt{4D_y \tau}}\right] - \mathrm{erf}\left[\frac{a-y}{\sqrt{4D_y \tau}}\right]\right)\frac{d\tau}{\tau^{3/2}}. \tag{7.95}$$

• Similarity Transform

All the solutions listed or cited in this section yield zero concentration at $x = 0$. However, as demanded by the boundary condition at $x = 0$ for $a \le y \le b$, the concentration should be $C_0 \exp[-\alpha t]$, not zero. Also an apparent singularity at $\tau = 0$ for the factor of $\tau^{-3/2}$ needs to be removed for numerical integration.

To remedy these defects, let us introduce a similarity transform

$$u = \frac{x}{\sqrt{4D_x \tau}} \tag{7.96}$$

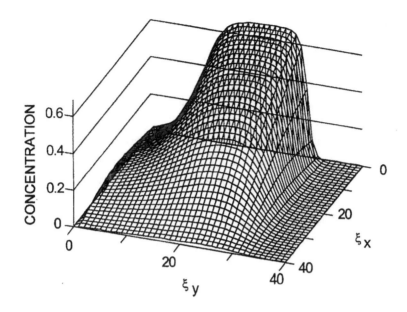

Figure 7.8: A three-dimensional view of solute concentration $C_{10}[\xi, \tau]$. ($\xi_x = \xi$ in the text.) Distances are plotted as $10^{0.03n}$ where n is the unit shown. $\eta = 0.1$, $\zeta = 0.2$, $\zeta_b = 8$, $\zeta_a = 0.2$, and $D_x/D_y = 4$.

into equation (7.95) to obtain

$$
\begin{aligned}
\frac{C_{10}[x, y, t]}{C_0} &= \frac{1}{\sqrt{\pi}} \exp\left[\frac{\bar{v}_x x}{2D_x} - \alpha t\right] \\
&\quad \cdot \int_{x/\sqrt{4D_x t/R}}^{\infty} \exp\left[-\left((\lambda - \alpha) R + \frac{\bar{v}_x^2}{4D_x}\right)\frac{x^2}{4D_x u^2} - u^2\right] \\
&\quad \cdot \left\{ \mathrm{erf}\left[\frac{u(b-y)}{x D_x^y}\right] - \mathrm{erf}\left[\frac{u(a-y)}{x D_x^y}\right] \right\} du,
\end{aligned}
\tag{7.97}
$$

where $D_x^y = \sqrt{D_y/D_x}$.

Now the boundary condition at $x = 0$ is satisfied,

$$
\begin{aligned}
\frac{C[0, y, t]}{C_0} &= \frac{\exp[-\alpha t]}{\sqrt{\pi}} \int_0^{\infty} e^{-u^2} \delta du \\
&= \frac{\delta}{2} \exp[-\alpha t],
\end{aligned}
\tag{7.98}
$$

where

$$
\begin{aligned}
\delta &= \text{erf}[-\infty] - \text{erf}[-\infty] = 0 \quad \text{if} \quad y > b, \\
\delta &= \text{erf}[\infty] - \text{erf}[-\infty] = 2 \quad \text{if } a \le y \le b, \\
\delta &= \text{erf}[\infty] - \text{erf}[\infty] = 0 \quad \text{if} \quad y < a.
\end{aligned}
\tag{7.99}
$$

Also, there is no apparent singularity at $t = 0$ and the integration can converge rapidly. Therefore equation (7.97) is the preferred solution, which satisfies the boundary conditions explicitly and is good for any $[x, y, t]$.

• **Dimensionless Variables**

Introducing the following dimensionless variables (similar to those for Example 6)

$$
\begin{aligned}
\xi &= \frac{\bar{v}_x x}{D_x}, \quad \tau = \frac{\bar{v}_x^2 t}{D_x R}, \quad \eta = \frac{\alpha}{\lambda}, \quad \zeta = \frac{\lambda D_x R}{\bar{v}_x^2}, \\
\xi_a &= \frac{\bar{v}_x a}{D_x}, \quad \xi_b = \frac{\bar{v}_x b}{D_x}, \quad \xi_y = \frac{\bar{v}_y y}{D_x},
\end{aligned}
\tag{7.100}
$$

equation (7.97) becomes

$$
\boxed{\; \mathcal{C}_{10}[\xi, \tau] = \frac{1}{\sqrt{\pi}} \exp[\xi/2 - \eta \zeta \tau] F \;}, \tag{7.101}
$$

where

$$
F = \int_{\xi/2\sqrt{\tau}}^{\infty} \exp\left[-\frac{(\zeta - \zeta \eta + 1/4)\xi^2}{4u^2} - u^2 \right]
$$
$$
\cdot \left\{ \text{erf}\left[\frac{u}{\xi D_x^y} (\xi_b - \xi_y) \right] - \text{erf}\left[\frac{u}{\xi D_x^y} (\xi_a - \xi_y) \right] \right\} du. \tag{7.102}
$$

Figure 7.8 depicts a three-dimensional view of $\mathcal{C}_{10}[\xi, \tau]$ for the parameters given in the figure caption.

7.3.2   Example 11: One Impermeable Boundary

Consider the case that an impermeable boundary exists at $y = 0$; otherwise the conditions are the same as described in Example 10. The $C_I^x[x, t]$ remains the same. The $C_{\mathrm{I}}^{y}[y,t]$ is modified by adding an image plane source (Figure 7.7) to ensure that

$$\frac{\partial C_{\mathrm{I}}^{y}[y,t]}{\partial y} = 0 \quad \text{at} \quad y = 0. \tag{7.103}$$

The instantaneous plane source solution is

$$C_{\mathrm{I}}^{y}[y,t] = \frac{\exp[-(y-y')^2/4D_y t]}{2\sqrt{\pi D_y t}} + \frac{\exp[-(y+y')^2/4D_y t]}{2\sqrt{\pi D_y t}}. \tag{7.104}$$

It is apparent that equation (7.97) can be modified by adding the image-source term

$$
\begin{aligned}
\frac{C_{11}[x,y,t]}{C_0} &= \frac{1}{\sqrt{\pi}} \exp\left[\frac{\bar{v}_x x}{2D_x} - \alpha t\right] \int_{x/\sqrt{4D_x t/R}}^{\infty} H[u] \\
&\quad \cdot \exp\left[-\frac{(4(\lambda-\alpha)RD_x + \bar{v}_x^2)x^2}{16D_x^2 u^2} - u^2\right] du, \tag{7.105}
\end{aligned}
$$

where

$$
\begin{aligned}
H[u] &= \operatorname{erf}\left[\frac{u(b-y)}{x D_{\mathrm{x}}^{y}}\right] - \operatorname{erf}\left[\frac{u(a-y)}{x D_{\mathrm{x}}^{y}}\right] \\
&\quad + \operatorname{erf}\left[\frac{u(b+y)}{x D_{\mathrm{x}}^{y}}\right] - \operatorname{erf}\left[\frac{u(a+y)}{x D_{\mathrm{x}}^{y}}\right]. \tag{7.106}
\end{aligned}
$$

7.3.3 Example 12: Bounded Flow

The boundary of no flow is described by Figure 7.9

$$\frac{\partial C_{\mathrm{I}}^{y}}{\partial y} = 0 \quad \text{at} \quad y = 0 \quad \text{and} \quad y = w. \tag{7.107}$$

The influx of solute is still

$$C^{x}[0,t] = C_0 \exp[-\alpha t], \quad a \leq y \leq b. \tag{7.108}$$

We will use the method of images again. If the plane source is located at y', its image across the mirror at $y = 0$ is located at $-y'$. This image reflected at $y = w$ is produced as if a second image is located at $2w + y'$ (see Problem 2 in Chapter 2). The third image is at $-(2w + y')$ and the fourth at $4w + y'$. Now we have a set of two series representing the image locations

$$(2nw + y') \quad \text{and} \quad -(2nw + y'), \quad n = 0, 1, 2, \ldots \tag{7.109}$$

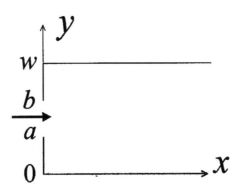

Figure 7.9: Flow bounded between $y = 0$ and $y = w$ for Example 12. Solute inputs between $y = a$ and $y = b$. For Problem 11, the boundary $w \to \infty$.

of which the first series includes the real source location at y'.

Another set of images starts from reflection at $y = w$. The images are located at

$$(2mw - y') \text{ and } -(2mw - y'), \quad m = 1, 2, \ldots \qquad (7.110)$$

By analogy to Example 11, the solution can be written by modifying $H[u]$ in equation (7.105) as follows

$$
\begin{aligned}
H[u] &= \sum_{n=0}^{\infty} \left\{ \operatorname{erf} \left[\frac{u\,(b + 2nw - y)}{x D_{\mathrm{x}}^{\mathrm{y}}} \right] - \operatorname{erf} \left[\frac{u\,(a + 2nw) - y}{x D_{\mathrm{x}}^{\mathrm{y}}} \right] \right. \\
&\quad \left. + \operatorname{erf} \left[\frac{u\,(b + 2nw + y)}{x D_{\mathrm{x}}^{\mathrm{y}}} \right] - \operatorname{erf} \left[\frac{u\,(a + 2nw + y)}{x D_{\mathrm{x}}^{\mathrm{y}}} \right] \right\} \\
&\quad + \sum_{m=1}^{\infty} \left\{ \operatorname{erf} \left[\frac{u\,(b - 2mw + y)}{x D_{\mathrm{x}}^{\mathrm{y}}} \right] - \operatorname{erf} \left[\frac{u\,(a - 2mw + y)}{x D_{\mathrm{x}}^{\mathrm{y}}} \right] \right. \\
&\quad \left. + \operatorname{erf} \left[\frac{u\,(b - 2mw - y)}{x D_{\mathrm{x}}^{\mathrm{y}}} \right] - \operatorname{erf} \left[\frac{u\,(a - 2mw - y)}{x D_{\mathrm{x}}^{\mathrm{y}}} \right] \right\} . \qquad (7.111)
\end{aligned}
$$

Note that this formula is different in form from that of Cleary and Ungs (1978).

7.4 Three-Dimensional Problems

Generally, analytic solutions for three-dimensional problems can be derived by two different techniques. The first starts from a Green's function or an impulse response and the second uses integral transforms.

7.4.1 Green's Function

Analytic solutions are obtained by convolving source functions with impulse response in time and space. Such solutions are usually in integral forms that require numerical integration. Solutions for various conditions are available in Carslaw and Jaeger (1959) by converting heat sources into solute sources. For example, the solution for instantaneous point-heat source (Section 6.2.1) can be converted into a solution for solute

$$C[x, y, z, t] = \frac{G}{(4\pi t)^{3/2}(D_x D_y D_z)^{1/2}} E, \qquad (7.112)$$

by replacing $M/(\rho c)_s$ with G where

$$E = \exp\left[-\frac{X^2}{4D_x t} - \frac{Y^2}{4D_y t} - \frac{Z^2}{4D_z t}\right] \qquad (7.113)$$
$$X = x - \bar{v}_x t - x', \; ...$$

and G, $[kg]$, is the total amount of solute released at source point $[x', y', z']$.

By integrating x', y', and z' over the appropriate source region, one can obtain the impulse response from which the system response to any input functions can be formulated through convolution. In this case, G is the solute mass production per unit volume.

The technique can handle problems with vanishing concentration at infinity, but it is difficult to adopt it for problems with the mixed-type boundary condition (Cauchy condition).

7.4.2 Integral Transforms

The technique of integral transforms is more versatile for different types of boundary conditions. For example, Leij et al. (1991) present a comprehensive list of analytic solutions for problems in a semi-infinite medium. They apply the Laplace transform to time and the x-dimension (the x-axis points into the medium from a plane surface) and the Fourier transform to the remaining y- and z-dimensions. For radial symmetric problems, the Hankel transform instead of the Fourier transform is used.

By means of these integral transforms, a differential equation becomes an algebraic equation and the solute concentration in the transformed domains is obtained. Inverse transforms are then applied one by one to the transformed concentration to return to the space and time domains. As usual for integral-transform methods, the resulting solutions are given in closed forms that may bear triple or quadruple integrals except for a few simple cases. The final step is to perform numerical integration.

• An Example

Following is the result for case 2 of Leij et al. (1991). Initially the region of interest is free of solute. At time zero for a type 1 boundary condition (Dirichlet condition), a constant solute concentration C_0 is imposed within a rectangular area bounded by $|y| \leq a$ and $|z| \leq b$ on the ground surface ($x = 0$). Subsequently the concentration is described by

$$\frac{C[x,y,z,t]}{C_0} = \frac{1}{4} \int_0^t \frac{x}{\tau} \Lambda_1[\tau] \Gamma_2[\tau] d\tau + \frac{G}{2R} \int_0^t \Lambda_2[\tau] d\tau, \qquad (7.114)$$

where G is the mass production rate per unit volume (λ of Leij et al.), and

$$
\begin{aligned}
\Gamma_2[\tau] &= \left(\text{erfc} \left[\frac{y-a}{\sqrt{4D_y\tau/R}} \right] - \text{erfc} \left[\frac{y+a}{\sqrt{4D_y\tau/R}} \right] \right) \\
&\quad \cdot \left(\text{erfc} \left[\frac{z-b}{\sqrt{4D_z\tau/R}} \right] - \text{erfc} \left[\frac{z+b}{\sqrt{4D_z\tau/R}} \right] \right), \qquad (7.115)
\end{aligned}
$$

$$\Lambda_1[\tau] = \sqrt{\frac{R}{4\pi D_x \tau}} \exp \left[-\frac{\lambda \tau}{R} - \frac{(Rx - \bar{v}\tau)^2}{4RD_x\tau} \right], \qquad (7.116)$$

and

$$
\begin{aligned}
\Lambda_2[\tau] &= \exp \left[-\frac{\lambda \tau}{R} + \frac{vx}{2D_x} \right] \left\{ \exp \left[-\frac{vx}{2D_x} \right] \text{erfc} \left[\frac{v\tau - Rx}{\sqrt{4RD_x\tau}} \right] \right. \\
&\quad \left. - \exp \left[\frac{vx}{2D_x} \right] \text{erfc} \left[\frac{v\tau + Rx}{\sqrt{4RD_x\tau}} \right] \right\} \qquad (7.117)
\end{aligned}
$$

with λ being the decay constant (μ of Leij et al.)

It is noted that $C[x,y,z,t] = 0$ for all y and z at $x = 0$. This result violates the boundary condition that the concentration is C_0 in the source area bounded by $-a \leq y \leq a$ and $-b \leq z \leq b$. We can remedy this apparent defect at $x = 0$ by performing the similarity transform, $u = x/\sqrt{4D_x\tau/R}$, as was done in Section 7.3.1. This similarity transform, however, is not needed for their solutions with a type 3 condition (or advective condition).

7.5 Radial Dispersion

All the analytical solutions we have presented so far deal with steady ground-water flow v and constant dispersion coefficient D. Now we consider the case that D varies in a steady radial flow.

7.5.1 Formulation

Let steady radial flow from an injection well be established such that the radial flow velocity is

$$\bar{v} = \frac{Q}{2\pi r b \phi}, \tag{7.118}$$

where Q is the injection rate, ϕ is the effective porosity, and b is the aquifer thickness. Neglecting molecular diffusion, equation (7.1) becomes

$$D = \alpha\bar{v} = \frac{\alpha B}{r}, \tag{7.119}$$

where $B = Q/2\pi b\phi$, α is the radial dispersivity (assumed to be independent of distance) and D is now the dispersion coefficient that varies inversely with radial distance.

For an axi-symmetric radial flow, the transport equation (7.2) becomes

$$\frac{1}{r}\frac{\partial}{\partial r}\left(rD\frac{\partial(\phi C)}{\partial r}\right) - v\frac{\partial C}{\partial r} = \frac{\partial(\phi C)}{\partial t}. \tag{7.120}$$

Simplification leads to

$$\frac{\alpha B}{r}\frac{\partial^2 C}{\partial r^2} - \frac{B}{r}\frac{\partial C}{\partial r} = \frac{\partial C}{\partial t}. \tag{7.121}$$

Next, we set the following initial and boundary conditions

$$\begin{aligned}
C[r,0] &= 0, \\
\bar{v}C - D\frac{\partial C}{\partial r} &= \bar{v}C_0 \quad \text{at} \quad r = r_{\text{w}}, \\
C[\infty, t] &\to 0,
\end{aligned} \tag{7.122}$$

where r_w is the well radius.

Substituting the following dimensionless variables

$$\mathcal{C} = C/C_0, \quad \rho = r/\alpha, \quad \tau = tB/\alpha^2 \tag{7.123}$$

into the differential equation and boundary conditions yields

$$\frac{\partial^2 \mathcal{C}}{\partial \rho^2} - \frac{\partial \mathcal{C}}{\partial \rho} = \rho \frac{\partial \mathcal{C}}{\partial \tau} \tag{7.124}$$

and

$$
\begin{aligned}
\mathcal{C}[\rho, 0] &= 0, \\
\mathcal{C} - \frac{\partial \mathcal{C}}{\partial \rho} &= 1 \quad \text{at} \quad \rho = \rho_w = r_w/\alpha, \\
\mathcal{C}[\infty, \tau] &\to 0.
\end{aligned}
\tag{7.125}
$$

7.5.2 Solution

Applying the Laplace transform to equations (7.124) and (7.125) yields

$$\frac{\partial^2 \overline{\mathcal{C}}}{\partial \rho^2} - \frac{\partial \overline{\mathcal{C}}}{\partial \rho} - \rho p \overline{\mathcal{C}} = 0 \tag{7.126}$$

and

$$
\begin{aligned}
\overline{\mathcal{C}} - \frac{\partial \overline{\mathcal{C}}}{\partial \rho} &= \frac{1}{p} \quad \text{at} \quad \rho = \rho_w, \\
\overline{\mathcal{C}}[\infty, p] &\to 0.
\end{aligned}
\tag{7.127}
$$

Note that the solution for the case of type 1 condition (i.e., $\mathcal{C} = 1$ or $\overline{\mathcal{C}} = 1/p$ at $\rho = \rho_w$) has been derived by Moench and Ogata (1981) in terms of Airy functions Ai$[x]$

$$\boxed{\overline{\mathcal{C}}[\rho, p] = \frac{1}{p} \exp\left[\frac{\rho - \rho_w}{2}\right] \frac{\text{Ai}[Y]}{\text{Ai}[Y_w]},} \tag{7.128}$$

where

$$
\begin{aligned}
Y &= (p\rho + 1/4)\, p^{-2/3}, \\
Y_w &= (p\rho_w + 1/4)\, p^{-2/3}.
\end{aligned}
\tag{7.129}
$$

They used Stehfest's method of numerical inverse Laplace transform to obtain the time-domain solution. Hsieh (1986) obtained a closed-form time-domain solution for type 1 condition. What follows is the solution given by Chen (1987) for type 3 condition.

• Elimination of the First-Derivative Term

A second-order differential equation can often be solved by eliminating the first-derivative term (Matthew and Walker, 1970). Let

$$\overline{C} = \Phi e^{\rho/2} \tag{7.130}$$

(see Problem 4), then equation (7.126) becomes

$$\frac{\partial^2 \Phi}{\partial \rho^2} - \left(\rho p + \frac{1}{4} \right) \Phi = 0. \tag{7.131}$$

Let $\zeta = \rho p + 1/4$ to simplify the equation for Φ

$$p^2 \frac{\partial^2 \Phi}{\partial \zeta^2} - \zeta \Phi = 0, \tag{7.132}$$

which looks like an Airy equation but not exactly.

So, take another transformation (by trial-and-error) to obtain

$$\frac{\partial^2 \left(\Phi p^{2/3} \right)}{\partial \left(\zeta / p^{2/3} \right)^2} - \frac{\zeta \Phi p^{2/3}}{p^{2/3}} = 0. \tag{7.133}$$

Now, let

$$\Psi = \Phi p^{2/3}, \quad \eta = \zeta p^{-2/3} \tag{7.134}$$

to get the Airy equation

$$\frac{\partial^2 \Psi}{\partial \eta^2} - \eta \Psi = 0. \tag{7.135}$$

See Problem 5 for a series solution.

• Matching Boundary Conditions

The Airy equation has two independent solutions, Ai[η] and Bi[η] (Abramowitz and Stegun, 1968). To satisfy the condition that $\Psi = 0$ as $\eta \to \infty$, we pick

$$\Psi = F\text{Ai}[\eta], \tag{7.136}$$

where coefficient F is determined by the type 3 condition. Now, return to

$$\bar{C} = F e^{\rho/2} p^{-2/3} \text{Ai} \left[(\rho p + 1/4) p^{-2/3} \right] \qquad (7.137)$$

to match the boundary conditions.

Substitute this $\bar{C}$ into equation (7.127) to determine

$$F = \frac{1}{p} \left\{ \frac{\exp[\rho_w/2]}{p^{2/3}} \left(\frac{1}{2} \text{Ai} \left[Y_w \right] - p^{1/3} \text{Ai}' \left[Y_w \right] \right) \right\}^{-1}, \qquad (7.138)$$

where $\text{Ai}' \left[Y_w \right]$ is the derivative of $\text{Ai}[Y_w]$. Therefore,

$$\bar{C}[\rho, p] = \frac{\exp\left[(\rho - \rho_w)/2 \right]}{p} \frac{\text{Ai} \left[Y \right]}{0.5 \text{Ai} \left[Y_w \right] - p^{1/3} \text{Ai}' \left[Y_w \right]}. \qquad (7.139)$$

• Time-Domain Solution

We can use the Stehfest algorithm to invert equation (7.139) for the time-domain solution. Alternatively we can obtain a closed-form solution via the Bromwich integral. The solution (Chen, 1987) is

$$\begin{aligned}
\mathcal{C}[\rho, \tau] &= 1 - \frac{4}{\pi} \exp \left[\frac{\rho - \rho_w}{2} \right] \\
&\quad \cdot \int_0^\infty \frac{\exp[-x^2 \tau]}{x} \frac{\text{Ai}[\eta] f_1 - \text{Bi}[\eta] f_2}{f_1^2 + f_2^2} dx,
\end{aligned} \qquad (7.140)$$

where

$$\begin{aligned}
f_1[x] &= \text{Bi}[\eta_w] + 2x^{2/3} \text{Bi}' [\eta_w], \\
f_2[x] &= \text{Ai}[\eta_w] + 2x^{2/3} \text{Ai}' [\eta_w], \\
\eta &= \frac{1 - 4\rho x^2}{4x^{4/3}}, \quad \eta_w = \frac{1 - 4\rho_w x^2}{4x^{4/3}}.
\end{aligned} \qquad (7.141)$$

Evaluation of $\mathcal{C}[\rho, \tau]$ requires numerical integration of an oscillating integrand. Like the numerical integration of the Hankel transform, the integration is done by summing subintegrals, each of which is integrated between two adjacent roots of $\text{Ai}[\eta] f_1 - \text{Bi}[\eta] f_2 = 0$.

7.6 Simulation by Z-Transform

Recall the z-transform in Section 2.2.5. Output $F[z]$ and input $Q[z]$ are related through

$$F[z] = Q[z]I[z], \tag{7.142}$$

where $I[z]$ is the transfer function

$$I[z] = \sum_{t=0}^{\infty} I_t z^t. \tag{7.143}$$

For data digitally sampled at unit time intervals, we have

$$F_t = \sum_{\tau=0}^{t} Q_\tau I_{t-\tau} \tag{7.144}$$

from which the impulse response can be derived

$$\boxed{I_t = \left(F_t - \sum_{k=1}^{t} Q_k I_{t-k}\right) / Q_0}, \tag{7.145}$$

$$I_0 = F_0/Q_0, \quad t = 1, 2, \ldots$$

Once I_t is determined, the system response to any input function can be easily obtained from equation 7.144, as demonstrated below.

7.6.1 Example 13: Numerical Simulation

In the previous examples we have tried to lump some model parameters together to form a set of dimensionless variables so that a model can be described by a minimum number of independent parameters. In principle only those independent parameters are resolvable through least-squares-based inverse modeling (see Chapter 10). Although the parameters that are lumped together may not be resolvable, they provide tractable physical insight. Hence the following simulations use parameters that are physically measurable rather than the immeasurable dimensionless parameters.

□ Step 1. Acquire digital data. As an example, we select equation (7.21) as output (i.e., $F^t = C_1^t$) due to a step input (i.e., $Q^t = 1$). Here superscript t is used to designate the tth data point to retain the earlier usage of a subscript for the example number. Curve C_1 in Figure 7.10 is the data curve.

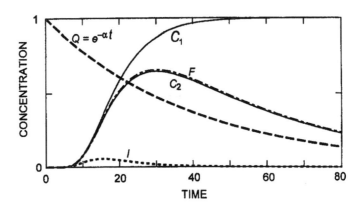

Figure 7.10: Impulse response I is obtained from step response C_1 via the inverse z-transform. Input Q is then convolved with I numerically to yield output F, which deviates from the theoretical output C_2 by a root-mean square error of 0.0055.

□ Step 2. Use inverse z-transform, equation (7.145), to obtain impulse response I^t, which is shown as I in Figure 7.10.

□ Step 3. Convolve input $Q = \exp[-\alpha t]$ with the impulse response I to obtain output F in accordance with equation 7.144.

□ Step 4. Compare the simulated F with the theoretical output C_2 in equation (7.28). Output F follows C_2 closely with a root-mean-square error of 0.0055.

Figure 7.10 is prepared with $\bar{v} = 0.1$, $D = 0.01$, $\alpha = 0.05$, and $x = 1$ for 41 time units (81 points at 0.5 units). Those numbers follow the constraint that $\bar{v}^2 > 4D\alpha$; otherwise, they are somewhat arbitrarily selected. The root-mean-square error of 0.0055 can be made smaller by taking smaller time units.

7.7   Problems, Keys, and Suggested Readings

• Problems

1. Derive the solution for Example 5.

2. If the advective condition in Example 6 is of finite duration, i.e.,

$$-D\frac{\partial C}{\partial x} + \bar{v}C\bigg|_{x=0} = \begin{cases} \bar{v}C_0\exp[-\alpha t], & t \leq t_p \\ 0, & t > t_p \end{cases}, \tag{7.146}$$

derive $C[x,t]$.

3. Given Example 13 as a reference, obtain the responses at $x = 0.25$, 0.5, 1 and 2 due to two pulsed-step inputs. The two pulses are separated by 5 time units and each lasts for a duration of 5 time units. The first pulse has an amplitude of 1 and the second pulse has an amplitude of 1.5.

4. By a change of variables, eliminate the first-derivative term from the following

$$\frac{\partial^2 y}{\partial x^2} + f[x]\frac{\partial y}{\partial x} + g[x]y = 0. \tag{7.147}$$

5. Find a series solution to equation (7.135).

- **Keys**

Key 1

The governing differential equation for a reactive and retardative system is

$$D\frac{\partial^2 C}{\partial x^2} - \bar{v}\frac{\partial C}{\partial x} - \lambda RC = R\frac{\partial C}{\partial t}. \tag{7.148}$$

Divide both sides by R to have

$$\frac{D}{R}\frac{\partial^2 C}{\partial x^2} - \frac{\bar{v}}{R}\frac{\partial C}{\partial x} - \lambda C = \frac{\partial C}{\partial t}. \tag{7.149}$$

Hence the solution can be obtained from the result for a reactive system by replacing D with D/R and $\bar{v}$ with $\bar{v}/R$ (Example 4) except the $\bar{v}$ in the factor of $\exp[\bar{v}x/2D]$,

$$\frac{C}{C_0} = \frac{\exp[\bar{v}x/2D]}{2}\left\{\exp\left[-\frac{\sqrt{\bar{v}^2 + 4D\lambda R}x}{2D}\right]\text{erfc}\left[\frac{x - \sqrt{\bar{v}^2 + 4D\lambda R}t/R}{\sqrt{4Dt/R}}\right]\right.$$
$$\left. + \exp\left[\frac{\sqrt{\bar{v}^2 + 4D\lambda R}x}{2D}\right]\text{erfc}\left[\frac{x + \sqrt{\bar{v}^2 + 4D\lambda R}t/R}{\sqrt{4Dt/R}}\right]\right\} \tag{7.150}$$

for the condition that $C[0,t] = C_0$.

Key 2

In this case, equation (7.44) for $t > t_{\mathrm{p}}$ becomes

$$\frac{\overline{C}[x,p]}{C_0} = \frac{2\overline{v}e^{\overline{v}x/2D}}{p+\alpha}\left(1 - e^{-pt_{\mathrm{p}}}\right)\frac{\exp\left[-\sqrt{\overline{v}^2 + 4DR(p+\lambda)}x/2D\right]}{\overline{v} + \sqrt{\overline{v}^2 + 4DR(p+\lambda)}}. \quad (7.151)$$

The factor $\exp[-pt_{\mathrm{p}}]$ represents a time shift (Basic Rule 3 in Chapter 3).

Therefore, the solution is

$$\frac{C[x,t]}{C_0} = e^{-\alpha t}A_6[x,t], \quad t \le t_{\mathrm{p}},$$

$$\frac{C[x,t]}{C_0} = e^{-\alpha t}A_6[x,t] - e^{-\alpha(t-t_{\mathrm{p}})}A_6[x, t - t_{\mathrm{p}}], \quad t > t_{\mathrm{p}}, \quad (7.152)$$

where $A_6[x,t]$ is defined in equation (7.59).

Note that our solution coincides with an equivalent solution, equation (7.70), which is obtainable by superposition. This is the nature of an exponential decay-like input function. Such type of superposition may not work for other input functions. It is prudent to formally solve a problem up to a stage that is equivalent to equation (7.151), then decide whether superposition is applicable.

Key 3

Figure 7.11 depicts the responses to the two pulsed-step inputs. The two pulses are represented by the light dash box cars marked by distance 0. It illustrates how dispersion modifies the shape and the peak amplitude of the two pulses as the solute migrates. At a distance beyond 2 in this example, the two pulses are indistinguishable. Although the transport velocity is ill-defined in case of dispersion, one can estimate the migration speed of the loci of peak concentration (a phase velocity) which is different from $\overline{v}$.

Key 4

Let

$$y = u[x]v[x], \quad (7.153)$$

then the differential equation becomes

$$\frac{\partial^2 v}{\partial x^2} + \left(2\frac{\partial u/\partial x}{u} + f\right)\frac{\partial v}{\partial x} + \frac{\partial^2 u/\partial x^2 + f\partial u/\partial x + gu}{u}v = 0. \quad (7.154)$$

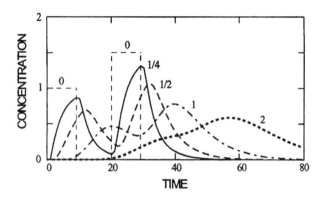

Figure 7.11: Responses to two pulsed inputs of solute (light dash boxcars marked by 0). Numbers are distances from the source location at $x = 0$.

Now, let

$$2\frac{\partial u/\partial x}{u} + f = 0 \tag{7.155}$$

to get

$$u = \exp\left[-2\int f dx\right], \tag{7.156}$$

and the new differential equation in v does not have the first derivative term.

Key 5

Let us rewrite equation (7.135) as

$$\frac{\partial^2 y}{\partial x^2} - xy = 0. \tag{7.157}$$

Let y be represented by the series

$$y = \sum_{n=0}^{\infty} d_n x^{n+s}, \tag{7.158}$$

where s is a real number. Substitute the series into the differential equation to get

$$\sum_{n=0}^{\infty} d_n(n+s)(n+s-1)x^{n+s-2} - \sum_{n=0}^{\infty} d_n x^{n+s+1} = 0 \tag{7.159}$$

for the determination of coefficient d_n.

Now we want to force the coefficient for each power of x in equation (7.159) to vanish so that the differential equation is satisfied for any x. The coefficients for the first five terms of x are given below

$$
\begin{aligned}
x^{s-2} : & \quad d_0 s(s-1) = 0, \\
x^{s-1} : & \quad d_1(s+1)s = 0, \\
x^{s} : & \quad d_2(s+2)(s+1) = 0, \\
x^{s+1} : & \quad d_3(s+3)(s+2) = d_0, \\
x^{s+2} : & \quad d_4(s+4)(s+3) = d_1.
\end{aligned}
\tag{7.160}
$$

We assume that $d_0 \neq 0$; therefore, the choice of

$$
s = 0 \quad \text{or} \quad s = 1
\tag{7.161}
$$

will yield two independent solutions to the Airy equation.

From the above listing we deduce that

$$
\begin{aligned}
d_{3m} &= \frac{1}{(s+3m)(s+3m-1)} d_{3(m-1)}, \\
d_n &= 0 \quad \text{if} \quad n \neq 3m, \quad m = 1,2,3,\ldots
\end{aligned}
\tag{7.162}
$$

For example, $d_1 = d_2 = 0$ and $d_3 = d_0/(s+3)(s+2)$ for $m = 1$.

For the choice of $s = 0$, the series is

$$
\begin{aligned}
f[x] &= d_0 + d_3 x^3 + d_6 x^6 + \ldots \\
&= d_0 + \frac{x^3}{3\cdot 2}d_0 + \frac{x^6}{6\cdot 5}d_3 + \frac{x^9}{9\cdot 8}d_6 + \ldots \\
&= d_0\left\{1 + \frac{1}{3!}x^3 + \frac{1\cdot 4}{6!}x^6 + \frac{1\cdot 4\cdot 7}{9!}x^9 + \ldots\right\}.
\end{aligned}
\tag{7.163}
$$

Similarly, the choice of $s = 1$ yields

$$
\begin{aligned}
g[x] &= d_0 x + \frac{x^4}{4\cdot 3}d_0 + \frac{x^7}{7\cdot 6}d_3 + \frac{x^{10}}{10\cdot 9}d_6 + \ldots \\
&= d_0\left\{x + \frac{2}{4!}x^4 + \frac{2\cdot 5}{7!}x^7 + \frac{2\cdot 5\cdot 8}{10!}x^{10} + \ldots\right\}.
\end{aligned}
\tag{7.164}
$$

Since d_0 is an arbitrary constant, we can set $d_0 = 1$. Any linear combination of the two resultant series

$$
y[x] = Ef[x] + Fg[x],
\tag{7.165}
$$

is still a solution to the Airy equation, where E and F are arbitrary constants. According to Abramowitz and Stegun (1968), the two independent Airy functions

$$
\begin{aligned}
\mathrm{Ai}[x] &= c_1 f[x] - c_2 g[x], \\
\mathrm{Bi}[x] &= \sqrt{3}\left(c_1 f[x] + c_2 g[x]\right), \\
c_1 &= 3^{-2/3}\Gamma[2/3], \quad c_2 = 3^{-1/3}\Gamma[1/3], \quad d_0 = 1
\end{aligned}
\tag{7.166}
$$

are made from the combinations of $f[x]$ and $g[x]$. The choice of the coefficients makes $\mathrm{Ai}[x]$ bounded for $x \geq 0$ and vanish as $x \to \infty$, and $\mathrm{Bi}[x]$ unbounded as $x \to \infty$.

• **Suggested Readings**

For transport in soils we recommend Jury and Roth's (1990) *Transfer Functions and Solute Movement through Soil*. For geological problems, we suggest *Reactive Transport in Porous Media* edited by Lichtener, Steefel, and Oelkers (1996).

7.8   Notations

Symbol	Definition	SI Unit	Dimension
a, b	Width of strip source	m	L
Ai, Bi	Airy functions	-	-
C_i, C_0	Concentration for the ith example	$kg\,m^{-3}$	ML^{-3}
C^{I}	Concentration, impulse response	$kg\,m^{-3}$	ML^{-3}
$\overline{C}$	Laplace transform of C	$kg\,m^{-3}s$	$\mathrm{ML}^{-3}\mathrm{T}$
D	Disperson coefficient or molecular diffusion coefficient	$m^2 s^{-1}$	$\mathrm{L}^2\mathrm{T}^{-1}$
D_{ij}, D_x, D_y	Dispersion coefficient	$m^2 s^{-1}$	$\mathrm{L}^2\mathrm{T}^{-1}$
D_x^y, D_x^z	$= \sqrt{D_y/D_x}, \quad = \sqrt{D_z/D_x}$	-	-
K_{d}	Distribution coefficient	$kg^{-1}\,m^3$	$\mathrm{M}^{-1}\mathrm{L}^3$
G	Solute source	$m^3 s^{-1}$	$\mathrm{L}^3\mathrm{T}^{-1}$
p	Laplace transform parameter	s^{-1}	T^{-1}
Q	Injection rate	$m^3 s^{-1}$	$\mathrm{L}^3\mathrm{T}^{-1}$
r_{w}	Well radius	m	L
R	Retardation factor	-	-
S_{c}	Sorbate concentration	-	-
t	Time	s	T
t_{p}	Pulse duration	s	T
U	$\sqrt{4DR\beta}$	$m\,s^{-1}$	LT^{-1}
$v, \overline{v}$	$v = $ Darcy's velocity, $\overline{v} = v/\phi$	$m\,s^{-1}$	LT^{-1}
$v_{\mathrm{in}}, v_{\mathrm{out}}$	Influx, efflux rate of solute	$m\,s^{-1}$	LT^{-1}
x, y	Distance	m	L
X, Y	$= x - \overline{v}_x t - x'$, etc.	m	L
$\mathcal{C}$	Dimensionless concentration		
$\mathcal{G}_0$	Equation (7.24)		
$\mathcal{L}$	Laplace transform operator		
α	Input solute decay constant also used as dispersivity	s^{-1} m	T^{-1} L
α_{ij}	Dispersivity	m^{-1}	L^{-1}
β	$\lambda - \alpha + \overline{v}^2/4DR$		
λ	Solute decay constant		
ρ	Dimensionless radial distance	-	-
ρ_{w}	Dimensionless well radius	-	-
ρ_{dry}	Dry bulk density	$kg\,m^{-3}$	ML^{-3}
ϕ	Effective porosity	-	-

Chapter 8

SOLVING Ax = b

We will review elementary matrix operations, find eigenvectors, and invert matrices for solving a system of linear equations. The system of equations appears in finite element analyses and in inverse problems for parameter determination. The notations are as follows:

Symbol	Description	Definition
$\mathbf{A}$	Bold, upper case	Matrix
$\mathbf{a}$	Bold, lower case	Column vector
$\mathbf{A}^T$	Superscript T	Transpose
A_{ij}, a_i	Subscript	Entry of a matrix or vector
$\mathbf{A}^k$	Superscript	Iteration step k
$\mathbf{A}^{-1}$		Inverse of square matrix $\mathbf{A}$

Other symbols are self-explanatory. In a product of two or more symbols, repetitive indices mean summation over all repeated indices, for example,

$$B_{ik}C_{kj} = \sum_k B_{ik}C_{kj}. \qquad (8.1)$$

8.1   Elementary Matrix Operations

• **Review of Matrix Operations**

□ Multiplication: $\mathbf{A} = \mathbf{BC}$ means $A_{ij} = B_{ik}C_{kj}$. The number of columns in matrix $\mathbf{B}$ must equal the number of rows in matrix $\mathbf{C}$ for the multiplication to make sense.

279

☐ Commutation: $\mathbf{BC} \neq \mathbf{CB}$. Premultiplication differs from postmultiplication except for diagonal matrices.

☐ Sequence of transpose or inverse: $(\mathbf{AB})^T = \mathbf{B}^T \mathbf{A}^T$ and $(\mathbf{AB})^{-1} = \mathbf{B}^{-1}\mathbf{A}^{-1}$.

☐ Transpose of vector difference: $(\mathbf{d} - \mathbf{g})^T = \mathbf{d}^T - \mathbf{g}^T$. This is valid only if $\mathbf{d}$ and $\mathbf{g}$ are column vectors.

Proof:

$$
(\mathbf{d} - \mathbf{g})^T = \left\{ \begin{array}{c} d_1 - g_1 \\ d_2 - g_2 \\ ... \\ d_n - g_n \end{array} \right\}^T
$$

$$
= \left[\begin{array}{cccc} d_1 - g_1 & d_2 - g_2 & .. & d_n - g_n \end{array} \right]
$$

$$
= \left[\begin{array}{cccc} d_1 & d_2 & ... & d_n \end{array} \right] - \left[\begin{array}{cccc} g_1 & g_2 & ... & g_n \end{array} \right]. \tag{8.2}
$$

Therefore

$$
\boxed{(\mathbf{d} - \mathbf{g})^T = \mathbf{d}^T - \mathbf{g}^T}. \tag{8.3}
$$

☐ Scalar product: $S = (\mathbf{d} - \mathbf{g})^T \mathbf{C}_d^{-1} (\mathbf{d} - \mathbf{g})$.
Let $\mathbf{C}_d$ be a symmetric matrix.

$$
\begin{aligned}
S &= \left(\mathbf{d}^T - \mathbf{g}^T \right) \mathbf{C}_d^{-1} (\mathbf{d} - \mathbf{g}) \\
&= \mathbf{d}^T \mathbf{C}_d^{-1}\mathbf{d} - \mathbf{d}^T \mathbf{C}_d^{-1}\mathbf{g} - \mathbf{g}^T \mathbf{C}_d^{-1}\mathbf{d} + \mathbf{g}^T \mathbf{C}_d^{-1}\mathbf{g}. \tag{8.4}
\end{aligned}
$$

If a is a scalar, then $a = a^T$. Because each term in S is a scalar, we see that

$$
\mathbf{d}^T \mathbf{C}_d^{-1}\mathbf{g} = \mathbf{g}^T \mathbf{C}_d^{-1}\mathbf{d}, \quad \mathbf{g}^T \mathbf{C}_d^{-1}\mathbf{g} = \left(\mathbf{g}^T \mathbf{C}_d^{-1}\mathbf{g} \right)^T. \tag{8.5}
$$

The scalar product becomes

$$
\boxed{S = \mathbf{d}^T \mathbf{C}_d^{-1}\mathbf{d} - 2\mathbf{g}^T \mathbf{C}_d^{-1}\mathbf{d} + \mathbf{g}^T \mathbf{C}_d^{-1}\mathbf{g}}. \tag{8.6}
$$

☐ Differentiation of a scalar product: Let $\mathbf{g}$ be a function of $\mathbf{p}$, but $\mathbf{d}$ and $\mathbf{C}_d$ are independent of $\mathbf{p}$. Assume that $\mathbf{p}$ has a dimension of M and $\mathbf{g}$ has a dimension of N. Then $\partial S/\partial \mathbf{p}$ is a column vector of dimension M,

$$
\frac{\partial S}{\partial \mathbf{p}} = -2\mathbf{G}^T \mathbf{C}_d^{-1}\mathbf{d} + 2\mathbf{G}^T \mathbf{C}_d^{-1}\mathbf{g} \tag{8.7}
$$

or

$$\boxed{\frac{\partial S}{\partial \mathbf{p}} = -2\mathbf{G}^T \mathbf{C}_d^{-1} (\mathbf{d} - \mathbf{g})} , \tag{8.8}$$

where

$$\mathbf{G}^T \equiv \frac{\partial}{\partial \mathbf{p}} \mathbf{g}^T = \left\{ \begin{array}{c} \frac{\partial}{\partial p_1} \\ \frac{\partial}{\partial p_2} \\ .. \\ \frac{\partial}{\partial p_M} \end{array} \right\} \left[\begin{array}{cccc} g_1 & g_2 & .. & g_N \end{array} \right], \quad G_{ij}^T = \frac{\partial g_j}{\partial p_i}. \tag{8.9}$$

• Definition

Square matrix $\mathbf{A}$ is *symmetric* if $\mathbf{A} = \mathbf{A}^T$ (antisymmetric if $\mathbf{A} = -\mathbf{A}^T$, and in this case $A_{ii} = 0$). *Adjoint* $\mathbf{A}^\dagger$ is the complex conjugate of transpose $\mathbf{A}$, i.e., $\mathbf{A}^\dagger = \mathbf{A}^{*T}$.

Square matrix $\mathbf{A}$ is *Hermitian* if $\mathbf{A}^\dagger = \mathbf{A}$. Square matrix $\mathbf{A}$ is orthogonal if $\mathbf{A}\mathbf{A}^T = \mathbf{I} = \mathbf{A}^T\mathbf{A}$; it is *unitary* if $\mathbf{A}\mathbf{A}^\dagger = \mathbf{A}^\dagger\mathbf{A} = \mathbf{I}$. A matrix is normal if $\mathbf{A}\mathbf{A}^\dagger = \mathbf{A}^\dagger\mathbf{A}$. For an orthogonal matrix, $\mathbf{A}^{-1} = \mathbf{A}^T$.

8.2 Eigenproblems

8.2.1 Eigenmatrix

If a linear operator can be represented by an $N \times N$ square matrix $\mathbf{A}$, the operation of $\mathbf{A}$ on vector $\mathbf{x}$ may generate a vector that is proportional to $\mathbf{x}$

$$\boxed{\mathbf{A}\mathbf{x} = \lambda\mathbf{x}} \tag{8.10}$$

or

$$(\mathbf{A} - \lambda\mathbf{I})\mathbf{x} = \mathbf{0}. \tag{8.11}$$

This is a linear system of N simultaneous equations. The system is homogeneous and has a nontrivial solution $\mathbf{x}$ only if the determinant vanishes

$$\det|\mathbf{A} - \lambda\mathbf{I}| = 0, \tag{8.12}$$

which is commonly known as the *secular* equation or *characteristic* equation.

The determinant has a dimension of $N \times N$. Thus it is an N-degree polynomial in λ. The roots λ_j ($j = 1, 2, ...N$) of this polynomial are the *eigenvalues* of $\mathbf{A}$.

Eigenvalues are real for a Hermitian matrix. If $\lambda_j > 0$ for all j, $\mathbf{A}$ is positive definite; if $\lambda_j \geq 0$, $\mathbf{A}$ is semipositive definite. Some eigenvalues of a real, nonsymmetric matrix may be complex; a complex but a non-Hermitian matrix usually has complex eigenvalues.

For each eigenvalue λ_j, there is a corresponding *eigenvector* $\mathbf{x}_j$. Eigenvectors for distinct eigenvalues are orthogonal (orthonormal if normalized), i.e., the scalar product

$$\mathbf{x}_i^T \mathbf{x}_j = \begin{cases} 1 & \text{if } i = j, \\ 0 & \text{if } i \neq j. \end{cases} \tag{8.13}$$

Many computer codes are available for eigenproblems (e.g., Press et al., 1992, Maple®, Mathcad®, Mathematica®, Matlab®).

If the secular equation has a repeated root, the repeated eigenvalue is *degenerative*. The eigenvectors associated with a degenerative eigenvalue can be made orthogonal by the Schmidt (or Gram-Schmidt) orthogonalization procedure.

A Hermitian matrix $\mathbf{A}$ can be diagonalized by a unitary transform

$$\boxed{\mathbf{A}_D = \mathbf{X}^{-1}\mathbf{A}\mathbf{X}}, \tag{8.14}$$

where $\mathbf{X}$ is an $N \times N$ eigenmatrix with the column vectors that are the eigenvectors $\mathbf{x}_k$, and $\mathbf{A}_D$ is the diagonalized matrix whose diagonal entries are the eigenvalues. Note that

$$\mathbf{X}^T\mathbf{X} = \mathbf{I} = \mathbf{X}\mathbf{X}^T. \tag{8.15}$$

• **Remark**

If $\mathbf{A}$ represents a 3×3 conductivity matrix (tensor), its eigenvalues (λ) are the principal conductivities, and the associated eigenvectors ($\mathbf{x}$) give the orientation of the principal axes of the conductivity ellipsoid. A symmetric $\mathbf{A}$ means that the flow properties are bidirectional; the properties remain the same irrespective of flow direction.

8.2.2 Least-Squares Method

• Symmetric Matrix

The eigenproblem $\mathbf{Au} = \lambda\mathbf{u}$ can be rewritten for symmetric matrix $\mathbf{A}$ as

$$\mathbf{AU} = \mathbf{UD}, \tag{8.16}$$

where $\mathbf{D}$ is an $N \times N$ diagonal matrix with the diagonal entries that are the eigenvalues λ, and $\mathbf{U}$ is the eigenmatrix.

Postmultiplying equation (8.16) by $\mathbf{U}^T$ and noting the orthonormal property of eigenvectors yields

$$\boxed{\mathbf{A} = \mathbf{UDU}^T}. \tag{8.17}$$

So the solution to $\mathbf{Ax} = \mathbf{b}$ is

$$\mathbf{x} = \mathbf{A}^{-1}\mathbf{b} = \mathbf{UD}^{-1}\mathbf{U}^T\mathbf{b}. \tag{8.18}$$

If any eigenvalue is zero, $\mathbf{D}^{-1}$ becomes singular and the inverse of $\mathbf{A}$ does not exist.

• A General Coefficient Matrix

A system of N linear equations and M unknowns

$$\boxed{\underset{N \times M}{\mathbf{B}}\,\underset{M \times 1}{\mathbf{x}} = \mathbf{y}} \tag{8.19}$$

represents an overdetermined or overconstrained problem if $N > M$. Because $\mathbf{B}$ is not a square matrix, direct inversion of the matrix is not possible. It can be solved by premultiplying it with $\mathbf{B}^T$ to obtain

$$\underset{M \times 1}{\mathbf{x}} = \underset{M \times M}{\left(\mathbf{B}^T\mathbf{B}\right)^{-1}}\underset{M \times N}{\mathbf{B}^T}\underset{N \times 1}{\mathbf{y}}. \tag{8.20}$$

• Least-Squares (LSQ)

Now, let us examine the least-squares method. One can minimize the error e for a trial solution $\mathbf{x}^t$,

$$\mathbf{e} = \mathbf{Bx}^t - \mathbf{y}, \tag{8.21}$$

by minimizing the sum of the squared errors $\mathbf{e}^T\mathbf{e}$ to yield the desired parameter vector (see Section 10.1.1)

$$\boxed{\mathbf{x}^t = \left(\mathbf{B}^T\mathbf{B}\right)^{-1}\mathbf{B}^T\mathbf{y}}. \tag{8.22}$$

Since $\mathbf{B}^T\mathbf{B}$ is an $M \times M$ square matrix, its inverse can be achieved through various equation solving techniques.

8.3 $\mathbf{x} = \mathbf{A}^{-1}\mathbf{b}$

Given a set of N simultaneous equations

$$\boxed{\mathbf{Ax} = \mathbf{b}}, \tag{8.23}$$

we are to determine M unknown entries of $\mathbf{x}$. The right–hand side vector $\mathbf{b}$ has a dimension of N. Coefficient matrix $\mathbf{A}$ has a dimension of $N \times M$. If $N > M$, the problem is overdetermined (greater number of equations, data points, or constraints than number of unknowns); if $N < M$, it is underdetermined.

If $N = M$, then premultiply both sides by $\mathbf{A}^{-1}$ to yield

$$\mathbf{x} = \mathbf{A}^{-1}\mathbf{b}. \tag{8.24}$$

Direct matrix inversion is rarely used, however, for solving a large system of linear equations. What follows are a few common methods for solving simultaneous equations in finite element analyses or linear inverse problems, which are addressed in the next two chapters. These methods are readily available in many commercial software packages.

8.3.1   Gaussian Elimination and Backsubstitution

Consider an example of $N = M = 4$ to illustrate the method of Gaussian elimination and back substitution,

$$\begin{bmatrix} A_{11} & A_{12} & A_{13} & A_{14} \\ A_{21} & A_{22} & A_{23} & A_{24} \\ A_{31} & A_{32} & A_{33} & A_{34} \\ A_{41} & A_{42} & A_{43} & A_{44} \end{bmatrix} \begin{Bmatrix} x_1 \\ x_2 \\ x_3 \\ x_4 \end{Bmatrix} = \begin{Bmatrix} b_1 \\ b_2 \\ b_3 \\ b_4 \end{Bmatrix}. \tag{8.25}$$

The strategy is to transform $\mathbf{A}$ into an upper triangular matrix, then use backward substitution to recover $\mathbf{x}$.

- ## Elimination Procedure

☐ 1. Divide the first rows of **A** and **b** by A_{11}

$$A_{1j}^1 = \frac{A_{1j}}{A_{11}}, \quad b_j^1 = \frac{b_j}{A_{11}} \tag{8.26}$$

to make $A_{11}^1 = 1$. Superscript k indicates that the value has been upgraded k times.

☐ 2. Multiply the revised first row by $-A_{i1}$ and add the results to the ith rows for $i = 2, 3, 4$ so that all entries in the first column of **A** are zero except the first row

$$\begin{bmatrix} 1 & A_{12}^1 & A_{13}^1 & A_{14}^1 \\ 0 & A_{22}^1 & A_{23}^1 & A_{24}^1 \\ 0 & A_{32}^1 & A_{33}^1 & A_{34}^1 \\ 0 & A_{42}^1 & A_{43}^1 & A_{44}^1 \end{bmatrix} \begin{Bmatrix} x_1 \\ x_2 \\ x_3 \\ x_4 \end{Bmatrix} = \begin{Bmatrix} b_1^1 \\ b_2^1 \\ b_3^1 \\ b_4^1 \end{Bmatrix}. \tag{8.27}$$

☐ 3. Repeat the elimination procedure for each of the remaining rows to get

$$\begin{bmatrix} 1 & A_{12}^1 & A_{13}^1 & A_{14}^1 \\ 0 & 1 & A_{23}^2 & A_{24}^2 \\ 0 & 0 & 1 & A_{34}^3 \\ 0 & 0 & 0 & A_{44}^3 \end{bmatrix} \begin{Bmatrix} x_1 \\ x_2 \\ x_3 \\ x_4 \end{Bmatrix} = \begin{Bmatrix} b_1^1 \\ b_2^2 \\ b_3^3 \\ b_4^3 \end{Bmatrix}. \tag{8.28}$$

- ## Back Substitution

Starting from the last row, we get

$$x_4 = b_4^3 / A_{44}^3 \tag{8.29}$$

as a seed for the recursive relation

$$x_i = b_i^i - \sum_{j=i+1}^{N} A_{ij}^i x_j, \quad i = N-1, N-2, ..., 1. \tag{8.30}$$

The Gaussian elimination and back substitution procedure is straightforward. One drawback is that **b** needs to be known in advance. Division by A_{ii}^i at the ith elimination step may be troublesome if A_{ii}^i is zero or near zero. Proper pivoting by swapping rows and columns may be necessary to have nonzero A_{ii}^i. In finite element analyses the procedure is usually stable because the diagonal entry of a coefficient matrix is generally larger than other entries in the same row. For stability of matrix inversion, small positive damping factors have frequently been added to the diagonal entries of the to-be-inverted matrix. (See Chapter 10.)

8.3.2   LU Decomposition

• Method

Suppose that the $N \times N$ matrix $\mathbf{A}$ can be resolved into a product of one lower ($\mathbf{L}$) and one upper ($\mathbf{U}$) triangular matrix

$$\mathbf{A} = \mathbf{LU}. \qquad (8.31)$$

Then

$$\begin{aligned} \mathbf{Ax} &= \mathbf{LUx} = \mathbf{Ly} = \mathbf{b}, \\ \mathbf{y} &= \mathbf{Ux}. \end{aligned} \qquad (8.32)$$

Through forward substitution, one gets the intermediate solution

$$\mathbf{y} = \mathbf{L}^{-1}\mathbf{b}; \qquad (8.33)$$

and through back substitution, one gets the final answer

$$\boxed{\mathbf{x} = \mathbf{U}^{-1}\mathbf{y}}. \qquad (8.34)$$

• Example

The procedures for *LU* decomposition can be demonstrated with a 3×3 matrix

$$\begin{bmatrix} A_{11} & A_{12} & A_{13} \\ A_{21} & A_{22} & A_{23} \\ A_{31} & A_{32} & A_{33} \end{bmatrix} = \begin{bmatrix} L_{11} & 0 & 0 \\ L_{21} & L_{22} & 0 \\ L_{31} & L_{32} & L_{33} \end{bmatrix} \begin{bmatrix} U_{11} & U_{12} & U_{13} \\ 0 & U_{22} & U_{23} \\ 0 & 0 & U_{33} \end{bmatrix}. \qquad (8.35)$$

$\mathbf{L}$ and $\mathbf{U}$ are obtained from the matrix multiplication

$$A_{ij} = L_{ik}U_{kj} = L_{i1}U_{1j} + L_{i2}U_{2j} + ... + L_{iN}U_{Nj} \qquad (8.36)$$

with the understanding that

$$
\begin{aligned}
L_{ik} &= 0 \quad \text{if} \quad k > i, \\
U_{kj} &= 0 \quad \text{if} \quad k < j.
\end{aligned}
\tag{8.37}
$$

There are N^2 equations and $N^2 + N$ unknowns (in $\mathbf{L}$ and $\mathbf{U}$). So we should place additional N arbitrary constraints. Let those constraints be

$$
L_{kk} = 1, \quad k = 1, 2, ..., N
\tag{8.38}
$$

and the problem is solvable for $\mathbf{L}$ and $\mathbf{U}$. The decomposition starts with $i = j = 1$ for

$$
A_{11} = L_{11}U_{11} \quad \text{or} \quad U_{11} = A_{11}.
\tag{8.39}
$$

For each of $j = 2, 3, .., N$, perform the pair of recursive relations (Crout's algorithm)

$$
\begin{aligned}
U_{i,j} &= A_{ij} - \sum_{k=1}^{i-1} L_{ik}U_{kj}, \quad i = 1, ..., j; \\
L_{ij} &= \frac{1}{U_{ij}} \left(A_{ij} - \sum_{k=1}^{j-1} L_{ik}U_{kj} \right), \quad i = j+1, ..., N.
\end{aligned}
\tag{8.40}
$$

Unlike the Gaussian method in which $\mathbf{A}$ is modified during forward elimination, LU decomposition remains intact during forward and backward substitutions. Hence, one decomposition can be used repeatedly for different $\mathbf{b}$s.

8.3.3 Iteration Methods

There are many iteration schemes for solving a system of equations. We begin with an example to show how iteration may work.

• An Example

Find the square root of a real number r, i.e., $\sqrt{r} = ?$

Let $s^2 = r$, then adding s^2 to both sides gives

$$
2s^2 = r + s^2.
\tag{8.41}
$$

Divide the expression by s to obtain the square root

$$s = \frac{s + r/s}{2}. \tag{8.42}$$

However, this equation cannot be solved because the unknown s appears on both sides.

To iterate for a solution, let

$$s^{k+1} = \frac{s^k + r/s^k}{2}, \quad k = 1, 2, ..., \tag{8.43}$$

where superscript k denotes step of iteration.

Now, take $r = 9$ as an example (i.e., what is $\sqrt{9}$?). Let the initial guess of the square root of 9 be $s^1 = 10$. The approximations to the square root at successive steps of iteration are $s^2 = 5.45$, $s^3 = 3.55$, $s^4 = 3.043$, and $s^5 = 3.0003$. Thus an accuracy of 0.01% for the square root of 9 is achieved at the 4th step from an outrageous guess value of 10.

The number of steps needed for a given accuracy for any iteration scheme depends on how close the initial guess is. Convergence of iterations must be monitored during numerical computation.

- **Jacobi Iteration**

Let

$$\mathbf{A} = \mathbf{U} + \mathbf{D} + \mathbf{L}, \tag{8.44}$$

where $\mathbf{U}$, $\mathbf{D}$, and $\mathbf{L}$ are, respectively, the upper triangular matrix, the diagonal matrix, and the lower triangular matrix of $\mathbf{A}$. For example, the partition of a 4×4 matrix $\mathbf{A}$ gives

$$\mathbf{Ax} = \begin{bmatrix} D_{11} & U_{12} & U_{13} & U_{14} \\ L_{21} & D_{22} & U_{23} & U_{24} \\ L_{31} & L_{32} & D_{33} & U_{34} \\ L_{41} & L_{42} & L_{43} & D_{44} \end{bmatrix} \begin{Bmatrix} x_1 \\ x_2 \\ x_3 \\ x_4 \end{Bmatrix} = \begin{Bmatrix} b_1 \\ b_2 \\ b_3 \\ b_4 \end{Bmatrix} = \mathbf{b}. \tag{8.45}$$

Rewrite the linear equation $\mathbf{Ax} = \mathbf{b}$ as

$$\mathbf{Dx} = -(\mathbf{U} + \mathbf{L})\mathbf{x} + \mathbf{b} \tag{8.46}$$

and the Jacobi iteration scheme is

$$\boxed{\mathbf{x}^{k+1} = \mathbf{D}^{-1}\left[-(\mathbf{U} + \mathbf{L})\mathbf{x}^k + \mathbf{b}\right]}. \tag{8.47}$$

Since $\mathbf{D}$ is a diagonal matrix, the inversion can be easily performed. The entries of $\mathbf{D}^{-1}$ are simply the inverse of the corresponding entries of $\mathbf{D}$.

• Gauss-Seidel Iteration

Rewrite the linear equation as

$$(\mathbf{D} + \mathbf{L})\,\mathbf{x} = -\mathbf{U}\mathbf{x} + \mathbf{b} \tag{8.48}$$

and set the iteration scheme

$$\boxed{\mathbf{x}^{k+1} = (\mathbf{D} + \mathbf{L})^{-1}\left(-\mathbf{U}\mathbf{x}^{k'} + \mathbf{b}\right)}. \tag{8.49}$$

Since $(\mathbf{D} + \mathbf{L})$ is a lower triangular matrix, $\mathbf{x}^{k+1}$ can be obtained through forward substitution (see Problem 3).

The major difference between the two schemes is that all $\mathbf{x}^k$ are used to obtain $\mathbf{x}^{k+1}$ in the Jacobi scheme but in the Gauss-Seidel scheme, newly acquired $x_1^{k+1}, x_2^{k+1}, .., x_{i-1}^{k+1}$ are also included for the computation of x_i^{k+1} For this reason, the step k' in $\mathbf{x}^{k'}$ is primed.

8.4 Singular Value Decomposition

If matrix $\mathbf{A}$ is singular or near singular numerically, the inverse $\mathbf{A}^{-1}$ does not exist or the condition is so poor that the solution may not be meaning-ful. An ill-conditioned matrix may be encountered in least-squares inverse problems. The technique of singular value decomposition (SVD) can help us to diagnose singularity and frequently, but not always, achieve a useful numerical solution.

• Nonsquare Matrix and Eigenproblems

We seek an alternative solution to $\mathbf{B}\mathbf{x} = \mathbf{y}$ by decomposing $\mathbf{B}$ in a manner analogous to equation (8.17) to get a stable inversion of matrix $\mathbf{B}^T\mathbf{B}$. The $N \times M$ matrix $\mathbf{B}$ can be premultiplied or postmultiplied by $\mathbf{B}^T$ to transform it into two square matrices. Both of the resulting square matrices can be formulated to yield two eigenproblems

$$\underset{N \times N}{\mathbf{B}\mathbf{B}^T}\,\mathbf{u}_k = \alpha_k \underset{N \times 1}{\mathbf{u}_k}, \tag{8.50}$$

$$\underset{M \times M}{\mathbf{B}^T\mathbf{B}}\,\mathbf{v}_j = \beta_j \underset{M \times 1}{\mathbf{v}_j}, \tag{8.51}$$

where α and β are the eigenvalues of $\mathbf{BB}^T$ and $\mathbf{B}^T\mathbf{B}$, respectively, while $\mathbf{u}_k$ and $\mathbf{v}_j$ are their corresponding eigenvectors. As indicated by the number of entries for each vector, $\mathbf{u}$ is related to data and $\mathbf{v}$ is related to parameters.

• **Relation between Eigenvectors u and v**

The two sets of eigenvectors ($\mathbf{u}$ and $\mathbf{v}$) are related through $\mathbf{B}$. Let us assume that

$$\underset{M\times N}{\mathbf{B}^T}\,\underset{N\times 1}{\mathbf{u}_j} = \lambda_j\,\underset{M\times 1}{\mathbf{v}_j}\,, \qquad (8.52)$$

$$\underset{N\times M}{\mathbf{B}}\,\underset{M\times 1}{\mathbf{v}_k} = \lambda_k'\,\underset{N\times 1}{\mathbf{u}_k}\,, \qquad (8.53)$$

where $j = 1, ..., M$ and $k = 1, ..., N$. We are to show that $\lambda_j = \lambda_k'$ if $j = k$ for k up to M if $M < N$, and for nonzero λ_j and λ_k'.

Premultiply equation (8.52) by $\mathbf{v}_j^T$ to get

$$\underset{1\times M}{\mathbf{v}_j^T}\,\underbrace{\mathbf{B}^T\mathbf{u}_j}_{M\times 1}= \mathbf{v}_j^T\lambda_j\mathbf{v}_j = \lambda_j. \qquad (8.54)$$

Then, take the transpose of equation (8.53) and postmultiply the result by $\mathbf{u}_k$ to obtain

$$\underbrace{\mathbf{v}_k^T\mathbf{B}^T}_{1\times N}\,\underset{N\times 1}{\mathbf{u}_k} = \lambda_k'\mathbf{u}_k^T\mathbf{u}_k = \lambda_k'. \qquad (8.55)$$

The left-hand sides of the above two equations are equal for all $j = k$ and $\lambda_k' \neq 0$ or $\lambda_j \neq 0$. Therefore $\lambda_j = \lambda_k'$ (note that there are $N - M$ undefined λ_j if $N > M$).

• **Singular Values and Eigenvalues**

Next we show that $\alpha = \beta = \lambda^2$. From equation (8.52), we obtain

$$\underset{N\times N}{\underbrace{\mathbf{BB}^T}}\,\underset{N\times 1}{\mathbf{u}_j} = \lambda_j\,\underset{N\times M}{\mathbf{B}}\,\underset{M\times 1}{\mathbf{v}_j}\,, \qquad (8.56)$$

that is

$$\alpha_j\mathbf{u}_j = \lambda_j\lambda_j\mathbf{u}_j.$$

Therefore

$$\boxed{\alpha = \lambda^2}\,. \qquad (8.57)$$

The λ's are *singular values* of $\mathbf{B}$ (note that $\lambda \neq 0$). The eigenvalues of $\mathbf{BB}^T$ or $\mathbf{B}^T\mathbf{B}$ are the squares of the corresponding singular values of $\mathbf{B}$.

Here we face a paradox: the number of eigenvalues α is N whereas the number of eigenvalues β is M. We cannot claim that $\alpha = \beta$ unless $N = M$. To fix it, let us make an eigenmatrix from equation (8.53)

$$\underset{N\times M}{\mathbf{B}}\ \underset{M\times M}{\mathbf{V}} = \underset{N\times M}{\mathbf{U}^s}\ \underset{M\times M}{\mathbf{\Lambda}}, \tag{8.58}$$

where $\mathbf{\Lambda}$ is the diagonal matrix that has entries made of λs arranged in descending order of absolute values, and $\mathbf{U}^s$ is a subset of the $N \times N$ matrix $\mathbf{U}$, comprising the first M columns of $\mathbf{U}$.

Now $\mathbf{B}$ can be decomposed by postmultiplying the above relation with $\mathbf{V}^T$

$$\mathbf{BVV}^T = \mathbf{U}^s\mathbf{\Lambda V}^T. \tag{8.59}$$

Since $\mathbf{VV}^T = \mathbf{I}$ theoretically, we obtain

$$\boxed{\underset{N\times M}{\mathbf{B}} = \underset{N\times M}{\mathbf{U}^s}\ \underset{M\times M}{\mathbf{\Lambda}}\ \underset{M\times M}{\mathbf{V}}^T}. \tag{8.60}$$

This is the *singular value decomposition* (SVD) of $\mathbf{B}$ as used by Press et al. (1992) or *spectral decomposition* of $\mathbf{B}$.

Alternatively we can preserve $\mathbf{U}$ and change $\mathbf{\Lambda}$ to make the eigenmatrix from equation (8.53) as

$$\underset{N\times M}{\mathbf{B}}\ \underset{M\times M}{\mathbf{V}} = \underset{N\times N}{\mathbf{U}}\ \underset{N\times M}{\mathbf{\Lambda}^a}, \tag{8.61}$$

where $\mathbf{\Lambda}^a$ is augmented by adding null row vectors to $\mathbf{\Lambda}$. In this case, $\mathbf{\Lambda}^a$ is not a diagonal matrix. The SVD of $\mathbf{B}$ is accordingly

$$\boxed{\mathbf{B} = \underset{N\times N}{\mathbf{U}}\ \underset{N\times M}{\mathbf{\Lambda}^a}\ \underset{M\times M}{\mathbf{V}}^T}. \tag{8.62}$$

This formula is perhaps more popular with symbolic programming such as Maple®. See Example 1 for a demonstration of the equality between the two forms of SVD.

292   CHAPTER 8. SOLVING AX = B

• Solution by SVD

Substituting the decomposed $\mathbf{B}$ into the least-squares solution

$$
\begin{aligned}
\mathbf{x} &= \left(\mathbf{B}^T\mathbf{B}\right)^{-1}\mathbf{B}^T\mathbf{y} \\
&= \left(\mathbf{V}\Lambda\mathbf{U}^{sT}\mathbf{U}^s\Lambda\mathbf{V}^T\right)^{-1}\left(\mathbf{V}\Lambda\mathbf{U}^{sT}\right)\mathbf{y}.
\end{aligned}
\tag{8.63}
$$

Noting that $\mathbf{U}^{-1} = \mathbf{U}^T$, we get

$$
\boxed{\mathbf{x} = \mathbf{V}\Lambda^{-1}\mathbf{U}^{sT}\mathbf{y}}.
\tag{8.64}
$$

Alternatively if equation (8.62) is used,

$$
\boxed{\mathbf{x} = \mathbf{V}\Lambda^{a-1}\mathbf{U}^T\mathbf{y}}.
\tag{8.65}
$$

• Resolution

SVD also gives resolution on parameters and data. Replacing $\mathbf{y}$ by $\mathbf{Bx}$ in equation (8.64) yields the computed

$$
\mathbf{x}^t = \mathbf{V}\Lambda^{-1}\mathbf{U}^{sT}\left(\mathbf{U}^s\Lambda\mathbf{V}^T\right)\mathbf{x}
\tag{8.66}
$$

or

$$
\boxed{\mathbf{x}^t = \mathbf{V}\mathbf{V}^T\mathbf{x}}.
\tag{8.67}
$$

If $\mathbf{V}\mathbf{V}^T = \mathbf{I}$ numerically, then $\mathbf{x}^t = \mathbf{x}$ and we achieve a perfect parameter resolution. Thus the $M \times M$ matrix $\mathbf{V}\mathbf{V}^T$ is the *parameter resolution matrix*.

According to the prediction by the least-squares parameters in equations (8.60) and (8.64), one obtains

$$
\mathbf{y}^t = \mathbf{B}\mathbf{x}^t = \mathbf{U}^s\Lambda\mathbf{V}^T\left(\mathbf{V}\Lambda^{-1}\mathbf{U}^{sT}\right)\mathbf{y}
\tag{8.68}
$$

or

$$
\boxed{\mathbf{y}^t = \mathbf{U}^s\mathbf{U}^{sT}\mathbf{y}}.
\tag{8.69}
$$

This $N \times N$ *data resolution matrix* $\mathbf{U}^s\mathbf{U}^{sT}$ is not always equal to $\mathbf{I}$. However, the relation

$$
\mathbf{U}^{sT}\mathbf{U}^s = \mathbf{I}
\tag{8.70}
$$

was used in equation (8.67) (see Example 1 for a demonstration).

• Summary

Coefficient matrix $\mathbf{A}$ of the system of $\mathbf{Ax} = \mathbf{b}$ has M eigenvalues if $\mathbf{A}$ is an $M \times M$ matrix. A nonsquare matrix $\mathbf{B}$ does not have eigenvalues. Hence, if $\mathbf{B}$ is $N \times M$ in dimension $(M < N)$, then $\mathbf{B}$ has M singular values, which are equal to the square roots of the eigenvalues for $\mathbf{B}^T\mathbf{B}$.

For problem $\mathbf{Ax} = \mathbf{b}$,

$$\boxed{\mathbf{A} = \mathbf{U}\mathbf{\Lambda}\mathbf{V}^T} \tag{8.71}$$

If $\mathbf{A}$ is real and symmetric, then $\mathbf{U} = \mathbf{V}$ and the eigenvalues and singular values (λ) of $\mathbf{A}$ are the same, and

$$\mathbf{A}^{-1} = \mathbf{V}\mathbf{\Lambda}^{-1}\mathbf{U}^T; \tag{8.72}$$

otherwise $\mathbf{U} \neq \mathbf{V}$.

• Condition Number

Singularity in $\mathbf{A}$ occurs when one or more eigenvalues are zero. Matrix $\mathbf{A}$ is ill-conditioned if the ratio $\lambda_{\min}/\lambda_{\max}$ is less than a computer's floating point precision (e.g., 10^{-6} for single precision, Press et al., 1992). As used in Maple[R], the condition number is defined as

$$\boxed{N_c = \max|\lambda| / \min|\lambda|}. \tag{8.73}$$

The number of nonzero λ is the rank (R) of matrix $\mathbf{A}$. If $R = M$, then all parameter space $\mathbf{x}$ can be mapped by $\mathbf{A}$ into the $\mathbf{y}$ space $(\mathbf{Ax} = \mathbf{y})$ and $\mathbf{x}$ can be obtained.

If $R < M$, there are $M - R$ null eigenvalues (eigenvectors can still be constructed from null eigenvalues). Some $\mathbf{x}$ cannot be mapped into $\mathbf{y}$ (some $\mathbf{Ax} = \mathbf{0}$ occur and the uncertainty in the determination of $\mathbf{x}$ arises). The number of zero eigenvalues is the number of linearly independent vectors. See Problem 1 for the usage of a matrix's condition number and see Example 2 for the number of linearly independent vectors or variables.

A FORTRAN program code for SVD can be found in Press et al. (1992). In case of singularity, replace each ∞ in $\mathbf{\Lambda}^{-1}$ by 0 to achieve a minimum length solution ($|\mathbf{x}|$). See Example 2 and Problem 2 for how to use SVD for finding a solution from noisy input and a near-singular matrix $\mathbf{A}$.

8.5   Examples

All examples and problems in this chapter are solved with Maple$^{\circledR}$ which is included in a Tex-based Scientific WorkPlace$^{\circledR}$ word processor with which this book was prepared. A reader needs similar software to perform cut-and-paste in the following examples.

8.5.1   Example 1: SVD

• **Problem**

Find the singular value decomposition of

$$\mathbf{B} = \begin{bmatrix} 3 & 2 \\ -1 & 3 \\ 4 & 1 \end{bmatrix}. \tag{8.74}$$

• **Solution**

  □ 1. Obtain

$$\mathbf{B}\mathbf{B}^T = \begin{bmatrix} 3 & 2 \\ -1 & 3 \\ 4 & 1 \end{bmatrix} \begin{bmatrix} 3 & -1 & 4 \\ 2 & 3 & 1 \end{bmatrix} = \begin{bmatrix} 13.0 & 3.0 & 14.0 \\ 3.0 & 10.0 & -1.0 \\ 14.0 & -1.0 & 17.0 \end{bmatrix}. \tag{8.75}$$

Its eigenvalues (α) are 29.22, 10.78, and 1.9901×10^{-8}. Since one of the eigenvalues is essentially 0, matrices $\mathbf{B}\mathbf{B}^T$ and $\mathbf{B}$ are ill-conditioned.

The eigenmatrix $\mathbf{U}$ arranged in descending order of eigenvalues is

$$\mathbf{U} = \begin{bmatrix} -0.65882 & -0.17161 & -0.7324 \\ -6.3834 \times 10^{-2} & -0.95737 & 0.28172 \\ -0.74959 & 0.23236 & 0.61978 \end{bmatrix} \tag{8.76}$$

(e.g., the first column is the eigenvector for eigenvalue 29.22).

  □ 2. Obtain

$$\mathbf{B}^T\mathbf{B} = \begin{bmatrix} 3 & -1 & 4 \\ 2 & 3 & 1 \end{bmatrix} \begin{bmatrix} 3 & 2 \\ -1 & 3 \\ 4 & 1 \end{bmatrix} = \begin{bmatrix} 26.0 & 7.0 \\ 7.0 & 14.0 \end{bmatrix}.$$

This gives an eigenmatrix of

$$\mathbf{V} = \begin{bmatrix} -0.90852 & 0.41786 \\ -0.41786 & -0.90851 \end{bmatrix} \tag{8.77}$$

for eigenvalues (β) of 29.22 and 10.78. Note that $\alpha = \beta$ as expected, except for an unmatchable value of α (i.e., the third value).

□ 3. The singular values (λ) are 5.4055 and 3.2834 according to the relation $\lambda = \sqrt{\beta}$. Thus, equation (8.62) yields

$$\mathbf{B} = \mathbf{U}\mathbf{\Lambda}^a\mathbf{V}^T = \begin{bmatrix} -0.65882 & -0.17161 & -0.73 \\ -6.3834 \times 10^{-2} & -0.95737 & 0.28 \\ -0.74959 & 0.23236 & 0.62 \end{bmatrix}$$

$$\cdot \begin{bmatrix} 5.4055 & 0 \\ 0 & 3.2834 \\ 0 & 0 \end{bmatrix} \begin{bmatrix} -.90851 & -.41786 \\ .41786 & -.90851 \end{bmatrix} \tag{8.78}$$

($\mathbf{\Lambda}^a$ is constructed by augmenting $\mathbf{\Lambda}$ with a null row vector).

This is the Maple® form of SVD in equation (8.62). Note that in this expression, the third column of $\mathbf{U}$ (corresponding to $\alpha = 0$) contributes nothing to the product of $\mathbf{U}\mathbf{\Lambda}^a\mathbf{V}^T$. So, effectively, we can delete the third column of $\mathbf{U}$ and the third row of $\mathbf{\Lambda}^a$ to have the alternative form

$$\mathbf{B} = \begin{bmatrix} -.65882 & -.17161 \\ -6.3834 \times 10^{-2} & -.95737 \\ -.74959 & .23236 \end{bmatrix} \begin{bmatrix} 5.4055 & 0 \\ 0 & 3.2834 \end{bmatrix}$$

$$\cdot \begin{bmatrix} -.90851 & -.41786 \\ .41786 & -.90851 \end{bmatrix} = \begin{bmatrix} 3.0 & 2.0 \\ -1.0 & 3.0 \\ 4.0 & 1.0 \end{bmatrix}$$

$$= \mathbf{U}^s\mathbf{\Lambda}\mathbf{V}^T, \tag{8.79}$$

which is equation (8.60) and recovers $\mathbf{B}$ as expected. In this alternative expression the dimension of $\mathbf{\Lambda}$ was reduced to the rank of $\mathbf{B}$. The two expressions of SVD reproduce $\mathbf{B}$ to the same precision.

□ 4. Check the orthogonality of parameter resolution matrix $\mathbf{V}$

$$\mathbf{V}\mathbf{V}^T = \begin{bmatrix} 1.0 & 4.1786 \times 10^{-6} \\ 0 & 1.0 \end{bmatrix} \tag{8.80}$$

and eigenmatrix $\mathbf{U}$

$$\underset{3\times 3}{\mathbf{U}}\,\underset{3\times 3}{\mathbf{U}^T}= \begin{bmatrix} .9999 & 1.7654 \times 10^{-5} & 4.2712 \times 10^{-5} \\ 1.7654 \times 10^{-5} & 1.0 & -7.435 \times 10^{-7} \\ 4.2712 \times 10^{-5} & -7.435 \times 10^{-7} & 1.0 \end{bmatrix}. \qquad (8.81)$$

Both are essentially identity matrices.

However, the data resolution matrix for equation (8.69)

$$\underset{3\times 2}{\mathbf{U}^s}\,\underset{2\times 3}{\mathbf{U}^{sT}}= \begin{bmatrix} .46349 & .20635 & .45397 \\ .20635 & .92063 & -.17461 \\ .45397 & -.17461 & .61588 \end{bmatrix}. \qquad (8.82)$$

is highly dispersed to the off-diagonal entries. On the other hand,

$$\underset{2\times 3}{\mathbf{U}^{sT}}\,\underset{3\times 2}{\mathbf{U}^s}= \begin{bmatrix} 1 & -1.876 \times 10^{-6} \\ -1.876 \times 10^{-6} & 1 \end{bmatrix} \qquad (8.83)$$

is essentially an identity matrix. This example is consistent with the derivation of equation (8.67).

8.5.2   Example 2: Ill-Conditioned Matrix

• **Problem**

Solve the simultaneous equations

$$\begin{aligned} x_1 + x_3 &= 2, \\ x_2 &= 1, \\ x_1 + x_3 &= 3. \end{aligned} \qquad (8.84)$$

• **Solution**

This set of simultaneous equations is ill-conditioned. It cannot be solved except for x_2. Rewrite the set of equations in matrix form

$$\mathbf{Ax} = \mathbf{b} \qquad (8.85)$$

$$\mathbf{A} = \begin{bmatrix} 1 & 0 & 1 \\ 0 & 1 & 0 \\ 1 & 0 & 1 \end{bmatrix}, \qquad \mathbf{b} = \begin{bmatrix} 2 \\ 1 \\ 3 \end{bmatrix}. \qquad (8.86)$$

The eigenvalues of **A** are $2, 1$ and 0.

The zero eigenvalue poses a problem for inverting **A**. Since there is one occurrence of zero eigenvalue, there corresponds one independent variable, x_2. Because this x_2 is unrelated to x_1 and x_3, we can exclude x_2 from the solution processes. However, we will retain it for the sake of using SVD.

The eigenvalues (α) for $\mathbf{A}\mathbf{A}^T$ (or β for $\mathbf{A}^T\mathbf{A}$) are $4, 1$ and 0. So the singular values λ for **A** are 2, 1, and 0 because $\lambda = \sqrt{\alpha}$. They are the same as the eigenvalues of **A**.

Let us try to solve the problem by means of SVD.

☐ 1. Make SVD of **A**

$$\mathbf{A} = \mathbf{U}\mathbf{\Lambda}\mathbf{V}^T, \tag{8.87}$$

where

$$\mathbf{U} = \begin{bmatrix} -.70711 & 0 & .70711 \\ 0 & -1.0 & 0 \\ -.70711 & 0 & -.70711 \end{bmatrix}, \qquad \mathbf{\Lambda} = \begin{bmatrix} 2.0 & 0 & 0 \\ 0 & 1.0 & 0 \\ 0 & 0 & 0 \end{bmatrix}. \tag{8.88}$$

Because **A** is real and symmetric, $\mathbf{V} = \mathbf{U}$. Since one entry of $\mathbf{\Lambda}$ is zero, matrix **A** is singular.

☐ 2. Obtain

$$\begin{aligned} \mathbf{x} &= \mathbf{A}^{-1}\mathbf{b} \\ &= \mathbf{V}\mathbf{\Lambda}^{-1}\mathbf{U}^T\mathbf{b}. \end{aligned} \tag{8.89}$$

Noting that the inverse of $\mathbf{\Lambda}$

$$\mathbf{\Lambda}^{-1} = \begin{bmatrix} 1/2.0 & 0 & 0 \\ 0 & 1/1.0 & 0 \\ 0 & 0 & \infty \end{bmatrix}, \tag{8.90}$$

is singular, there is no solution in general. However, as suggested by Press et al. (1992), replacing ∞ with 0 will give a reasonable solution

$$\begin{aligned} \mathbf{x} &= \mathbf{U}\mathbf{\Lambda}^{-1}\mathbf{U}^T\mathbf{b} \\ &= \begin{bmatrix} -.70711 & 0 & .70711 \\ 0 & -1.0 & 0 \\ -.70711 & 0 & -.70711 \end{bmatrix} \begin{bmatrix} 0.5 & 0 & 0 \\ 0 & 1.0 & 0 \\ 0 & 0 & 0 \end{bmatrix} \\ &\quad \cdot \begin{bmatrix} -.70711 & 0 & -.70711 \\ 0 & -1.0 & 0 \\ .70711 & 0 & -.70711 \end{bmatrix} \begin{bmatrix} 2 \\ 1 \\ 3 \end{bmatrix} = \begin{bmatrix} 1.25 \\ 1.0 \\ 1.25 \end{bmatrix}. \end{aligned} \tag{8.91}$$

Thus $x_1 + x_3 = 1.25 + 1.25 = 2.5$ is a compromised solution for the sum of x_1 and x_3 to be both 2 and 3, a conflicting demand.

8.5.3 Example 3: Ill-Conditioned Matrix (continued)

If the second equation in the set of simultaneous equations in Example 2 is

$$x_2 + 2x_3 = 1, \tag{8.92}$$

the coefficient matrix is

$$\mathbf{A} = \begin{bmatrix} 1 & 0 & 1 \\ 0 & 1 & 2 \\ 1 & 0 & 1 \end{bmatrix}. \tag{8.93}$$

The SVD of $\mathbf{A}$ is

$$\begin{bmatrix} -.4544 & -.5418 & -.7071 \\ -.7662 & .6426 & 0 \\ -.4544 & -.5418 & .7071 \end{bmatrix} \begin{bmatrix} 2.7152 & 0 & 0 \\ 0 & 1.2758 & 0 \\ 0 & 0 & 0 \end{bmatrix}$$

$$\cdot \begin{bmatrix} -.3347 & -.2822 & -.8991 \\ -.8493 & .5037 & .1581 \\ .4083 & .8165 & -.4083 \end{bmatrix}. \tag{8.94}$$

Since one of the eigenvalues vanishes, it is still ill-conditioned. However, SVD gives a solution

$$\mathbf{x}^T = \begin{bmatrix} 1.75 & -.49999 & .75001 \end{bmatrix}. \tag{8.95}$$

Again, $x_1 + x_3 = 2.5$ is a compromised solution to the conflict that the sum of x_1 and x_3 is both 2 and 3.

8.5.4 Example 4: A Well-Behaved Matrix

If the coefficient matrix $\mathbf{A}$ in Example 2 is modified as

$$\mathbf{A} = \begin{bmatrix} 1 & 0 & 1 \\ 0 & 1 & 2 \\ 1 & -1 & 1 \end{bmatrix}, \tag{8.96}$$

all three components of $\mathbf{x}$ are related (none is independent). The SVD of $\mathbf{A}$ is

$$\begin{bmatrix} -.4886 & .2737 & -.8285 \\ -.7511 & -.6151 & .2398 \\ -.4440 & .7394 & .5061 \end{bmatrix} \begin{bmatrix} 2.6253 & 0 & 0 \\ 0 & 1.7054 & 0 \\ 0 & 0 & .4467 \end{bmatrix}$$

$$\cdot \begin{bmatrix} -.3552 & -.1170 & -.9274 \\ .5940 & -.7943 & -.1274 \\ -.7218 & -.5962 & .3516 \end{bmatrix}. \tag{8.97}$$

At the condition number of $N_c = 2.6253/0.4467 = 5.9$, the solution is

$$\mathbf{x}^T = \begin{bmatrix} 1 & -1 & 1 \end{bmatrix}. \tag{8.98}$$

8.6 Problems, Keys, and Suggested Readings

- **Problems**

   1. Determine the eigenvectors and SVD of $\mathbf{A} = \begin{bmatrix} 4 & 2 \\ 2 & 3 \end{bmatrix}$.
   2. Discuss and solve

$$\begin{aligned} x + y &= 2.0 \\ x + 1.001\,y &= 2.1 \end{aligned} \tag{8.99}$$

by the methods of Gaussian elimination and SVD.

   3. Given a set of 4 simultaneous equations $\mathbf{Ax} = \mathbf{b}$, use the Gauss-Seidel iteration to write an algorithm for solving $\mathbf{x}$.

- **Keys**

Key 1

   Eigenvector: For a 2×2 matrix, we will obtain the eigenvector with a calculator before using SVD. The eigenproblem is

$$\mathbf{Ax} = \lambda\mathbf{x} \quad \text{or} \quad (\mathbf{A} - \lambda\mathbf{I})\mathbf{x} = 0. \tag{8.100}$$

From the determinant

$$\det \mathbf{A} = \begin{vmatrix} 4 - \lambda & 2 \\ 2 & 3 - \lambda \end{vmatrix}, \tag{8.101}$$

we get the polynomial

$$(4 - \lambda)(3 - \lambda) - 4 = 8 - 7\lambda + \lambda^2 = 0. \tag{8.102}$$

It has two roots, $(\lambda_1 \quad \lambda_2) = (5.5616 \quad 1.4384)$, which are the eigenvalues.

For eigenvalue $\lambda_1 = 5.5616$, we have two equations

$$\begin{bmatrix} 4 - 5.5616 & 2 \\ 2 & 3 - 5.5616 \end{bmatrix} \begin{bmatrix} x_1^{(1)} \\ x_2^{(1)} \end{bmatrix} = 0 \qquad (8.103)$$

or

$$\begin{aligned} -1.5616\, x_1^{(1)} + 2\, x_2^{(1)} &= 0, \\ 2\, x_1^{(1)} - 2.5616\, x_2^{(1)} &= 0. \end{aligned} \qquad (8.104)$$

Since the determinant is zero, the two are not independent of each other and we can determine the ratio of $x_1^{(1)}$ to $x_2^{(1)}$. That is, $x_1^{(1)} = 1.2807 x_2^{(1)}$, a line passing through the origin with a positive slope (Figure 8.1), or

$$\mathbf{x}^{(1)} = \begin{bmatrix} 1.2807 \\ 1.0000 \end{bmatrix}. \qquad (8.105)$$

Normalization gives the first eigenvector

$$\mathbf{x}^{(1)} = \begin{bmatrix} 0.7882 \\ 0.6154 \end{bmatrix}. \qquad (8.106)$$

Similarly, the second eigenvalue yields

$$x_1^{(2)} = -0.7808 x_2^{(2)} \qquad (8.107)$$

or a normalized eigenvector

$$\mathbf{x}^{(2)} = \begin{bmatrix} -0.6154 \\ 0.7882 \end{bmatrix}. \qquad (8.108)$$

The eigenmatrix for the two eigenvectors is

$$\mathbf{X} = \begin{bmatrix} \mathbf{x}^{(1)} & \mathbf{x}^{(2)} \end{bmatrix} = \begin{bmatrix} 0.7882 & -0.6154 \\ 0.6154 & 0.7882 \end{bmatrix}. \qquad (8.109)$$

The two eigenvectors form the new orthogonal basis vectors and the two eigenvalues constitute the diagonal entries of a new diagonal matrix. If the original $\mathbf{A}$ represents an anisotropic conductivity tensor (matrix), the diagonal entries of the new diagonal matrix are the principal conductivities along the principal axes represented by the new basis vectors

$$\mathbf{A}_{\text{new}} = \begin{bmatrix} 5.56 & 0. \\ 0. & 1.44 \end{bmatrix}. \qquad (8.110)$$

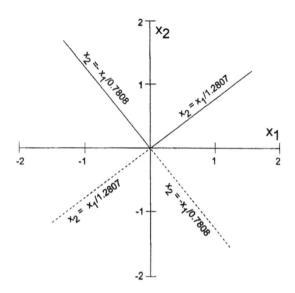

Figure 8.1: Paired eigenvectors. Dotted pair is from the SVD solution.

SVD Solution: First, form

$$\mathbf{AA}^T = \begin{bmatrix} 4 & 2 \\ 2 & 3 \end{bmatrix} \begin{bmatrix} 4 & 2 \\ 2 & 3 \end{bmatrix}^T = \begin{bmatrix} 20 & 14 \\ 14 & 13 \end{bmatrix} \tag{8.111}$$

to find eigenmatrix

$$\mathbf{U} = \begin{bmatrix} -.78821 & .61541 \\ -.61541 & -.78821 \end{bmatrix} \tag{8.112}$$

for eigenvalues α of 30.931 and 2.0691.

Next, form

$$\mathbf{A}^T\mathbf{A} = \begin{bmatrix} 4 & 2 \\ 2 & 3 \end{bmatrix}^T \begin{bmatrix} 4 & 2 \\ 2 & 3 \end{bmatrix} = \begin{bmatrix} 20 & 14 \\ 14 & 13 \end{bmatrix} \tag{8.113}$$

to find eigenmatrix $\mathbf{V}$. Because $\mathbf{A}$ is symmetric, $\beta = \alpha$ and $\mathbf{V} = \mathbf{U}$. Note that $\mathbf{U} = \mathbf{X}$ of equation (8.109) except for the sign for orientation of basis vectors. This difference is immaterial.

The singular values ($\lambda = \sqrt{\alpha}$) are 5.5616 and 1.4384, which are the same as the eigenvalues for a symmetric $\mathbf{A}$. The SVD of $\mathbf{A}$ is therefore

$$\mathbf{A} \quad = \mathbf{U}\mathbf{\Lambda}\mathbf{V}^T = \begin{bmatrix} -.78821 & .61541 \\ -.61541 & -.78821 \end{bmatrix}$$

$$\cdot \begin{bmatrix} 5.5616 & 0 \\ 0 & 1.4384 \end{bmatrix} \begin{bmatrix} -.78821 & -.61541 \\ .61541 & -.78821 \end{bmatrix} \cdot \qquad (8.114)$$

The condition number N_c of matrix **A** is defined as the ratio of maximum to the minimum singular values,

$$N_c = \frac{\max |\lambda|}{\min |\lambda|}. \qquad (8.115)$$

A number near 1 implies it is stable in finding parameters or unknown **x**. For our example, the condition number is $5.5616/1.4384 = 3.8664$. Large N_c means that the solution is sensitive to errors in observation **b** or input, which in conjunction with the theory (model) determine **A**. Note that the ratio $\lambda_{min}/\lambda_{max}$ has also been used as a measure of matrix condition.

Key 2

SVD: Recast the problem as

$$\mathbf{Ax} = \mathbf{b}, \qquad \mathbf{A} = \begin{bmatrix} 1 & 1 \\ 1 & 1.001 \end{bmatrix}, \qquad \mathbf{b} = \begin{bmatrix} 2.0 \\ 2.1 \end{bmatrix}. \qquad (8.116)$$

The SVD of coefficient matrix **A** is

$$\mathbf{A} = \mathbf{U\Lambda V}^T = \begin{bmatrix} -.70693 & -.70728 \\ -.70728 & .70693 \end{bmatrix}$$

$$\cdot \begin{bmatrix} 2.0005 & 0 \\ 0 & 4.9988 \times 10^{-4} \end{bmatrix} \begin{bmatrix} -.70693 & -.70728 \\ -.70728 & .70693 \end{bmatrix}. \qquad (8.117)$$

Its condition number is

$$N_c = \frac{2.0005}{4.9988 \times 10^{-4}} = 4002.0. \qquad (8.118)$$

Such a large condition number indicates that the solution is sensitive to error.

Gaussian Elimination: Using the Gaussian elimination method yields

$$\mathbf{x} = \begin{bmatrix} -98.0 \\ 100.0 \end{bmatrix}. \qquad (8.119)$$

Now if **b** has an error of ± 0.1 (5%), say in an extreme situation

$$\mathbf{b}' = \begin{bmatrix} 2.1 \\ 2.0 \end{bmatrix}, \tag{8.120}$$

our **x** becomes

$$\mathbf{x}' = \begin{bmatrix} 102.1 \\ -100.0 \end{bmatrix}. \tag{8.121}$$

The solution flip-flops, indicating the system is very sensitive to input noise or error in vector **b**.

SVD Solution: Now, try the SVD inversion

$$\mathbf{x} = \mathbf{A}^{-1}\mathbf{b} = \mathbf{V}\boldsymbol{\Lambda}^{-1}\mathbf{U}^T\mathbf{b}. \tag{8.122}$$

Because

$$\boldsymbol{\Lambda}^{-1} = \begin{bmatrix} 1/2.0005 & 0 \\ 0 & 1/4.9988 \times 10^{-4} \end{bmatrix} \tag{8.123}$$

is near singular, remember the black-box operation (Press et al., 1992) by replacing ∞ with 0 to yield

$$\boldsymbol{\Lambda}^{-1} = \begin{bmatrix} 1/2.0005 & 0 \\ 0 & 0 \end{bmatrix}. \tag{8.124}$$

It is easy to find

$$\mathbf{V} = \begin{bmatrix} -.70693 & -.70728 \\ -.70728 & .70693 \end{bmatrix}, \tag{8.125}$$

$$\mathbf{U}^T = \begin{bmatrix} -.70693 & -.70728 \\ -.70728 & .70693 \end{bmatrix}. \tag{8.126}$$

Finally, perform

$$\mathbf{x} = \mathbf{V}\boldsymbol{\Lambda}^{-1}\mathbf{U}^T\mathbf{b} = \begin{bmatrix} 1.0245 \\ 1.0250 \end{bmatrix} \tag{8.127}$$

and

$$\mathbf{x}' = \mathbf{V}\boldsymbol{\Lambda}^{-1}\mathbf{U}^T\mathbf{b}' = \begin{bmatrix} 1.0245 \\ 1.0250 \end{bmatrix}. \tag{8.128}$$

Thus, we have obtained the same answer for **b** and **b**' by SVD in contrast to the unstable solutions obtained with the Gaussian elimination method. Also, the two methods yield solutions that are different by a factor of about 100. We conclude that SVD is much less sensitive to noise. At least, it does not flip-flop.

- **Suggested Readings**

Almost any textbook on linear algebra covers the topics summarized in this chapter. *Numerical Recipes* by Press et al. (1992) is highly recommended. For inversion of large matrix, see George and Liu's (1981) *Computer Solution of Large Sparse Positive Definite System.*

Chapter 9

FINITE ELEMENT ANALYSIS

Finite element analysis is a numerical technique for solving differential equations. It involves discretization of a region of interest into many meshes or elements. The junction points of two or more elements are nodes (some nodes can be located at the boundary between two elements or inside an element). The dependent variable of the differential equation (e.g., hydraulic head) is represented by a set of values at the nodes. Within or on the boundary of an element, the values of the dependent variable are interpolated from those nodal values. The task of finite element analysis is to determine the nodal values. This is achieved by substituting the interpolation functions into the differential equation, minimizing the weighted errors due to approximation, and solving the resulted set of simultaneous equations.

We begin the finite element analysis by introducing an axi-symmetric groundwater flow problem — a numerical solution for the equivalent Theis well function— except that the well can have a nonzero diameter and hydraulic properties can vary radially.

9.1 1D Finite Element Method

9.1.1 Formulation of a Problem

For axi-symmetric and depth-independent groundwater flow, the flow depends on radial distance r only. The region of interest between r_{inner} and

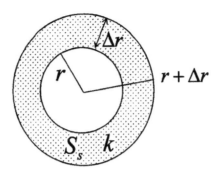

Figure 9.1: A ring element between r and $r + \Delta r$. Properties are uniform within each ring.

r_outer is discretized into n ring elements. Each ring element is homogeneous but the hydraulic properties may differ from one element to the next.

Now, consider the mass balance in one ring element between r and $r + \Delta r$ (Figure 9.1). This ring element is our control volume. Conservation of mass requires the net outflow from the ring element be equal to the change in mass storage

$$\frac{\partial}{\partial r}\left[-\left(2\pi r b\right) k\frac{\partial h}{\partial r}\right] \Delta r = -S_s \frac{\partial h}{\partial t}\left[\pi \left(r + \Delta r\right)^2 b - \pi r^2 b\right], \qquad (9.1)$$

where h is the hydraulic head $[m]$, k is the hydraulic conductivity $[ms^{-1}]$, b is the axial height $[m]$, S_s is the specific storage $[m^{-1}]$, and t is the time $[s]$. The term in the brackets on the left-hand side of the mass balance equation is the Darcy flow across the cylindrical surface of the ring element. Its derivative multiplied by Δr is the net outflow, which is accompanied by the decrease of water storage in the ring volume as indicated by the negative sign on the right-hand side of the equation.

Simplify the mass balance equation and ignore $(\Delta r)^2$ term to yield

$$\frac{1}{r}\frac{\partial}{\partial r}\left(kr\frac{\partial h}{\partial r}\right) - S_s\frac{\partial h}{\partial t} = 0 \qquad (9.2)$$

in which the hydraulic conductivity and specific storage can be functions of position. We will solve this equation together with appropriate initial and boundary conditions by finite element analysis.

9.1.2 Galerkin Weighted Residual

• Weighted Residual

Let the hydraulic head h be approximated by $\widetilde{h}$, the finding of which is the main task of finite element analysis. In general, $\widetilde{h}$ will not satisfy the differential equation. In other words, substitution of $\widetilde{h}$ into equation (9.2) results in an error,

$$\text{Error} = k\left(\frac{\partial^2 \widetilde{h}}{\partial r^2} + \frac{1}{r}\frac{\partial \widetilde{h}}{\partial r}\right) - S_s \frac{\partial \widetilde{h}}{\partial t} \neq 0. \tag{9.3}$$

For an acceptable $\widetilde{h}$, this error (residual) should be made as small as possible.

A common practice in finite element analysis is to weight the error over an entire element before minimizing the residual. At node i,

$$\text{Weighted Error}_i = W_i \times \text{Error}_i, \tag{9.4}$$

where $W_i[r]$ is a weighting function. This weighted error is integrated over element e to which node i belongs, yielding the weighted residual at node i,

$$R_i^e = \int\limits_0^{2\pi} \int\limits_{r_i}^{r_j} \int\limits_{z_1}^{z_2} W_i \left[k\left(\frac{\partial^2 \widetilde{h}}{\partial r^2} + \frac{1}{r}\frac{\partial \widetilde{h}}{\partial r}\right) - S_s \frac{\partial \widetilde{h}}{\partial t}\right] r\,dz\,dr\,d\phi, \tag{9.5}$$

where r_i and r_j are the radial distances at the inner and outer nodes of element e as designated by superscript e.

We choose a weighting function such that

$$\begin{aligned} W_i[r] &= \begin{cases} 1 & \text{at } r = r_i \\ 0 & \text{at } r = r_j \end{cases}, \\ 0 &< W_i[r] < 1, \quad r_i < r < r_j. \end{aligned} \tag{9.6}$$

Away from the boundary nodes, an explicit representation of the weighting function, $W_i[r]$, is yet to be determined.

Hereafter, the tilde in $\widetilde{h}$ is omitted for notational simplicity. Also, because of axi-symmetry, ϕ and z dependencies will be omitted.

• Reduction of the Second-Order Derivative

The hydraulic head and normal component of hydraulic flux (in this case, $-k\partial h/\partial r$) are continuous at any interior boundary. Across the boundary of dissimilar materials, however, the normal derivative is discontinuous and the second-order differential $\partial^2 h/\partial r^2$ is not defined. To avoid the difficulty of integrating over a material boundary, the second-order differential is reduced to the first order through integration by parts

$$
R_i^e = kW_i r \frac{\partial h}{\partial r}\bigg|_{r_i}^{r_j}
$$

$$
-k \int_{r_i}^{r_j} \frac{\partial h}{\partial r} \frac{\partial}{\partial r}(W_i r)\, dr + k \int_{r_i}^{r_j} W_i \frac{\partial h}{\partial r}\, dr - S_s \int_{r_i}^{r_j} W_i \frac{\partial h}{\partial t} r\, dr \quad (9.7)
$$

in which the material properties k and S_s are understood to be homogeneous within element e.

Carrying out the differentiation of the second term on the right-hand side of the above equation to nullify the third term, and accounting for the nature of the weighting function at the boundary nodes, we obtain the weighted residual at node i of element e,

$$
-R_i^e = k \int_{r_i}^{r_j} \frac{\partial W_i}{\partial r} \frac{\partial h}{\partial r} r\, dr + S_s \int_{r_i}^{r_j} W_i \frac{\partial h}{\partial t} r\, dr + k\, r \frac{\partial h}{\partial r}\bigg|_{r_i}. \quad (9.8)
$$

• Linear Interpolation

To proceed further, we need to define $h[r]$ and $W_i[r]$. First, interpolate $h[r]$ in terms of nodal values h_i and h_j, i.e.,

$$
h = N_i h_i + N_j h_j, \quad \text{or}
$$

$$
\boxed{h = \begin{bmatrix} N_i & N_j \end{bmatrix} \begin{Bmatrix} h_i \\ h_j \end{Bmatrix},} \quad (9.9)
$$

where N_i and N_j are interpolation functions, and a pair of square brackets designates a row vector, and a pair of braces denotes a column vector. The interpolation functions are fixed at the end nodes,

$$
N_i = \begin{cases} 1 & \text{at } r = r_i \\ 0 & \text{at } r = r_j \end{cases}, \quad N_j = \begin{cases} 0 & \text{at } r = r_i \\ 1 & \text{at } r = r_j \end{cases} \quad (9.10)
$$

such that $h[r_i] = h_i$ and $h[r_j] = h_j$. Between the endpoints, the N_i and N_j are yet to be defined.

Use of the interpolation functions yields

$$
\begin{aligned}
-R_i^e &= k \int_{r_i}^{r_j} \frac{\partial W_i}{\partial r} \frac{\partial}{\partial r} \left[\begin{array}{cc} N_i & N_j \end{array}\right] r\, dr \left\{\begin{array}{c} h_i \\ h_j \end{array}\right\} \\
&\quad + S_s \int_{r_i}^{r_j} W_i \left[\begin{array}{cc} N_i & N_j \end{array}\right] r\, dr \left\{\begin{array}{c} \partial h_i/\partial t \\ \partial h_j/\partial t \end{array}\right\} + k\, r\frac{\partial h}{\partial r}\bigg|_i, \quad (9.11)
\end{aligned}
$$

in which the nodal values and their time derivatives are factored out of the integral signs because they are fixed at the nodes. The flux term $kr\partial h/\partial r$ is either specified at the exterior boundary or cancelled at the interior boundary, so there is no need to interpolate the flux term.

Similarly, the residual at node j of element e can be found. Express the two nodal residuals as a column vector

$$
\begin{aligned}
\left\{\begin{array}{c} -R_i^e \\ -R_j^e \end{array}\right\} &= k\bar{r} \left[\begin{array}{cc} \int_{r_i}^{r_j} \frac{\partial W_i}{\partial r}\frac{\partial N_i}{\partial r} dr & \int_{r_i}^{r_j} \frac{\partial W_i}{\partial r}\frac{\partial N_j}{\partial r} dr \\ \int_{r_i}^{r_j} \frac{\partial W_j}{\partial r}\frac{\partial N_i}{\partial r} dr & \int_{r_i}^{r_j} \frac{\partial W_j}{\partial r}\frac{\partial N_j}{\partial r} dr \end{array}\right] \left\{\begin{array}{c} h_i \\ h_j \end{array}\right\} \\
&\quad + S_s\bar{r} \left[\begin{array}{cc} \int_{r_i}^{r_j} W_i N_i\, dr & \int_{r_i}^{r_j} W_i N_j\, dr \\ \int_{r_i}^{r_j} W_j N_i\, dr & \int_{r_i}^{r_j} W_j N_j\, dr \end{array}\right] \left\{\begin{array}{c} \partial h_i/\partial t \\ \partial h_j/\partial t \end{array}\right\} \\
&\quad + k \left\{\begin{array}{c} + r\frac{\partial h}{\partial r}\big|_i \\ - r\frac{\partial h}{\partial r}\big|_j \end{array}\right\}, \quad (9.12)
\end{aligned}
$$

in which we have used the mean value,

$$
\bar{r} = \frac{r_i + r_j}{2} \quad \text{for small} \quad \Delta r = r_j - r_i, \quad (9.13)
$$

to approximate r in equation (9.12) to simplify the integration. Note that the two entries in the flux vector have different signs, one for influx and the other for efflux.

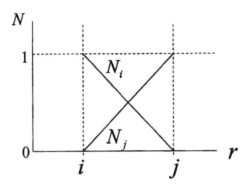

Figure 9.2: Interpolation functions $N_i[r]$ and $N_j[r]$.

Explicit expressions for W and N between the end nodes are in order now. First, let us make

$$W_i = N_i \quad \text{and} \quad W_j = N_j. \tag{9.14}$$

Next, make N a linear function of r (Figure 9.2)

$$N_i = \frac{-r + r_j}{\Delta r}, \quad N_j = \frac{r - r_i}{\Delta r}. \tag{9.15}$$

The residual obtained from the identical weighting and interpolation functions is known as the *Galerkin weighted residual.*

As defined here, N_i and N_j also serve as local coordinates with each being defined between 0 and 1. They are the *shape functions* or the natural coordinates of the element. When the interpolation and shape functions are identical, the finite element scheme is *isoparametric.*

9.1.3   Elementary Matrices

The integrals in equation (9.12) can be integrated exactly. They are of the following forms:

$$\int_{r_i}^{r_j} \frac{\partial N_i}{\partial r} \frac{\partial N_i}{\partial r} dr = \frac{1}{\Delta r}, \quad \int_{r_i}^{r_j} \frac{\partial N_i}{\partial r} \frac{\partial N_j}{\partial r} dr = -\frac{1}{\Delta r}$$

$$\int_{r_i}^{r_j} N_i N_i \, dr = \frac{\Delta r}{3}, \quad \int_{r_i}^{r_j} N_i N_j \, dr = \frac{\Delta r}{6}. \tag{9.16}$$

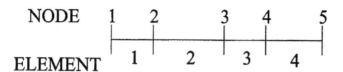

Figure 9.3: Element and node numbers for a one-dimensional problem.

Substitution of these relations into equation (9.12) yields

$$\left\{ \begin{array}{c} -R_i^e \\ -R_j^e \end{array} \right\} = [k]^e \left\{ \begin{array}{c} h_i \\ h_j \end{array} \right\}$$

$$+ [c]^e \left\{ \begin{array}{c} \partial h_i/\partial t \\ \partial h_j/\partial t \end{array} \right\} + \left\{ \begin{array}{c} +k \left. r\frac{\partial h}{\partial r} \right|_i \\ -k \left. r\frac{\partial h}{\partial r} \right|_j \end{array} \right\}, \qquad (9.17)$$

where $[k]^e$ is the elementary conductance matrix

$$[k]^e = \frac{k\bar{r}}{\Delta r} \left[\begin{array}{cc} 1 & -1 \\ -1 & 1 \end{array} \right] \qquad (9.18)$$

and $[c]^e$ is the elementary capacitance matrix

$$[c]^e = \frac{S_s \bar{r} \Delta r}{6} \left[\begin{array}{cc} 2 & 1 \\ 1 & 2 \end{array} \right]. \qquad (9.19)$$

Both are symmetric matrices.

9.1.4 Finite-Element Equation

Node i is shared by two adjacent elements. Both elements contribute to the residual at node i. The residual at node i is thus the sum of the residuals at the common node i of the two elements

$$R_i = \sum_e R_i^e . \qquad (9.20)$$

Essentially it sums all elements e having i as a common node (there are only two elements for 1D case). Hereafter nodes i or j will refer to global nodal numbers rather than local elementary nodes.

Let us consider an example of four elements and five nodes (Figure 9.3). For element 1 the residuals are

$$
\left\{ \begin{array}{c} -R_1 \\ -R_2 \end{array} \right\}^{(1)} = \left(\frac{k\bar{r}}{\Delta r} \right)^{(1)} \left[\begin{array}{cc} 1 & -1 \\ -1 & 1 \end{array} \right] \left\{ \begin{array}{c} h_1 \\ h_2 \end{array} \right\} + \left(\frac{S_s\bar{r}\Delta r}{6} \right)^{(1)}
$$
$$
\cdot \left[\begin{array}{cc} 2 & 1 \\ 1 & 2 \end{array} \right] \left\{ \begin{array}{c} \frac{\partial h_1}{\partial t} \\ \frac{\partial h_2}{\partial t} \end{array} \right\} + \left[\begin{array}{c} +k\,r\frac{\partial h}{\partial r}\big|_1 \\ -k\,r\frac{\partial h}{\partial r}\big|_2 \end{array} \right]^{(1)} . \tag{9.21}
$$

This column vector of residuals can be augmented with null vectors to represent other nodes. For example, adding the null row vector for node 3 yields

$$
\left\{ \begin{array}{c} -R_1 \\ -R_2 \\ -R_3 \end{array} \right\}^{(1)} = \left(\frac{k\bar{r}}{\Delta r} \right)^{(1)} \left[\begin{array}{ccc} 1 & -1 & 0 \\ -1 & 1 & 0 \\ 0 & 0 & 0 \end{array} \right] \left\{ \begin{array}{c} h_1 \\ h_2 \\ h_3 \end{array} \right\} + \left(\frac{S_s\bar{r}\Delta r}{6} \right)^{(1)}
$$
$$
\cdot \left[\begin{array}{ccc} 2 & 1 & 0 \\ 1 & 2 & 0 \\ 0 & 0 & 0 \end{array} \right] \left\{ \begin{array}{c} \frac{\partial h_1}{\partial t} \\ \frac{\partial h_2}{\partial t} \\ \frac{\partial h_3}{\partial t} \end{array} \right\} + \left[\begin{array}{c} +k\,r\frac{\partial h}{\partial r}\big|_1 \\ -k\,r\frac{\partial h}{\partial r}\big|_2 \\ 0 \end{array} \right]^{(1)} . \tag{9.22}
$$

Similarly, the residuals for element 2 are

$$
\left\{ \begin{array}{c} -R_1 \\ -R_2 \\ -R_3 \end{array} \right\}^{(2)} = \left(\frac{k\bar{r}}{\Delta r} \right)^{(2)} \left[\begin{array}{ccc} 0 & 0 & 0 \\ 0 & 1 & -1 \\ 0 & -1 & 1 \end{array} \right] \left\{ \begin{array}{c} h_1 \\ h_2 \\ h_3 \end{array} \right\} + \left(\frac{S_s\bar{r}\Delta r}{6} \right)^{(2)}
$$
$$
\cdot \left[\begin{array}{ccc} 0 & 0 & 0 \\ 0 & 2 & 1 \\ 0 & 1 & 2 \end{array} \right] \left\{ \begin{array}{c} \frac{\partial h_1}{\partial t} \\ \frac{\partial h_2}{\partial t} \\ \frac{\partial h_3}{\partial t} \end{array} \right\} + \left[\begin{array}{c} 0 \\ +k\,r\frac{\partial h}{\partial r}\big|_2 \\ -k\,r\frac{\partial h}{\partial r}\big|_3 \end{array} \right]^{(2)} . \tag{9.23}
$$

Adding the two column vectors of residuals for elements 1 and 2 together yields

$$
\left\{ \begin{array}{c} -R_1 \\ -R_2 \\ -R_3 \end{array} \right\} = \left[\begin{array}{ccc} (k\bar{r}/\Delta r)_1 & -(k\bar{r}/\Delta r)_1 & 0 \\ -(k\bar{r}/\Delta r)_1 & (k\bar{r}/\Delta r)_1 + (k\bar{r}/\Delta r)_2 & -(k\bar{r}/\Delta r)_2 \\ 0 & -(k\bar{r}/\Delta r)_2 & (k\bar{r}/\Delta r)_2 \end{array} \right] \left\{ \begin{array}{c} h_1 \\ h_2 \\ h_3 \end{array} \right\}
$$
$$
+ \frac{1}{6} \left[\begin{array}{ccc} 2(S_s\bar{r}\Delta r)_1 & (S_s\bar{r}\Delta r)_1 & 0 \\ (S_s\bar{r}\Delta r)_1 & 2(S_s\bar{r}\Delta r)_1 + 2(S_s\bar{r}\Delta r)_2 & (S_s\bar{r}\Delta r)_2 \\ 0 & (S_s\bar{r}\Delta r)_2 & 2(S_s\bar{r}\Delta r)_2 \end{array} \right]
$$
$$
\cdot \left\{ \begin{array}{c} \frac{\partial h_1}{\partial t} \\ \frac{\partial h_2}{\partial t} \\ \frac{\partial h_3}{\partial t} \end{array} \right\} + \left[\begin{array}{c} +k\,r\frac{\partial h}{\partial r}\big|_1 \\ -k\,r\frac{\partial h}{\partial r}\big|_1 + k\,r\frac{\partial h}{\partial r}\big|_2 \\ -k\,r\frac{\partial h}{\partial r}\big|_3 \end{array} \right] . \tag{9.24}
$$

So the residual at common node 2 is

$$-R_2 = -R_2^{(1)} - R_2^{(2)}, \qquad (9.25)$$

while the residual at nodes 1 and 3 remain the same. Note that $-k\, r\partial h/\partial r|_1 + k\, r\partial h/\partial r|_2$ vanishes because of the continuity in normal flux.

By deduction, the global conductance matrix of our example of four elements and five nodes is

$$[K] = \begin{bmatrix} k_{11}^1 & -k_{12}^1 & 0 & 0 & 0 \\ -k_{21}^1 & \boxed{\begin{array}{c} k_{22}^1 \\ +k_{22}^2 \end{array}} & -k_{23}^2 & 0 & 0 \\ 0 & -k_{32}^2 & \boxed{\begin{array}{c} k_{33}^2 \\ +k_{33}^3 \end{array}} & -k_{34}^3 & 0 \\ 0 & 0 & -k_{43}^3 & \boxed{\begin{array}{c} k_{44}^3 \\ +k_{44}^4 \end{array}} & -k_{45}^4 \\ 0 & 0 & 0 & -k_{54}^4 & k_{55}^4 \end{bmatrix}, \qquad (9.26)$$

where the superscripts refer to element numbers and the subscripts refer to global nodal numbers. Similarly, the global capacitance matrix $[C]$ can be constructed.

Note that, in this example, $[K]$ is symmetric and the nonzero entries are banded (three bands). It is sparse if a matrix is dominated by zero entries.

In summary, the global conductance and capacitance matrices for all nodes and elements are, respectively,

$$K_{ij} = \sum_e k_{ij}^e, \quad C_{ij} = \sum_e c_{ij}^e. \qquad (9.27)$$

Setting the residual at every global node to vanish yields the finite element equation

$$\boxed{[C]\left\{\frac{\partial h}{\partial t}\right\} + [K]\{h\} = \{f\}}. \qquad (9.28)$$

The right-hand column vector is

$$\{f\} = -\left\{\begin{array}{c} \left(kr\frac{\partial h}{\partial r}\right)\big|_1 \\ 0 \\ 0 \\ 0 \\ -\left(kr\frac{\partial h}{\partial r}\right)\big|_n \end{array}\right\} = \left\{\begin{array}{c} (\text{in flux})|_1 \\ 0 \\ 0 \\ 0 \\ (\text{in flux})|_n \end{array}\right\} \qquad (9.29)$$

in which the fluxes at the internal nodes cancel each other as shown in equation (9.24). Note that the sign of the boundary fluxes $\{f\}$ represents the influxes into the boundary elements.

9.1.5   Differential Equation in Time

Equation (9.28) is a system of first-order differential equations in time. It can be solved by the finite difference method or the finite element method. For notational simplicity, the symbols {} and [] will be dropped in the following sections. Instead, bold capital letters will designate matrices (e.g., $\mathbf{C}$ for $[C]$) and bold lower case letters designate column vectors (e.g., $\mathbf{h}$). In matrix notation, the finite element equation is

$$\boxed{\mathbf{C}\frac{\partial \mathbf{h}}{\partial t} + \mathbf{Kh} = \mathbf{f}}. \qquad (9.30)$$

- **Finite Difference**

    Explicit Scheme: Use the Taylor series expansion to obtain $\mathbf{h}^{m+1}$ at the $(m+1)$th time step from values at the mth step

$$\mathbf{h}^{m+1} = \mathbf{h}^m + \frac{\partial \mathbf{h}^m}{\partial t}\Delta t + \frac{\partial^2 \mathbf{h}^m}{\partial t^2}\frac{(\Delta t)^2}{2!} + \dots \qquad (9.31)$$

Put the first-order approximation

$$\frac{\partial \mathbf{h}^m}{\partial t} \approx \frac{\mathbf{h}^{m+1} - \mathbf{h}^m}{\Delta t} \qquad (9.32)$$

into equation (9.30) and evaluate $\mathbf{h}$ and $\mathbf{f}$ at time $m\Delta t$

$$\mathbf{h}^{m+1} = \mathbf{h}^m - \Delta t\, \mathbf{C}^{-1}(\mathbf{Kh}^m - \mathbf{f}^m). \qquad (9.33)$$

This is an example of the explicit-scheme forward finite-difference method because all needed values to evaluate $\mathbf{h}^{m+1}$ are explicitly known at time $m\Delta t$. For numerical stability in time stepping, the time step must be less than a critical value

$$\Delta t < \Delta t_{\text{critical}}, \qquad (9.34)$$

which is proportional to the square of the smallest Δr.

Implicit Scheme: Use the Taylor series backwards

$$\mathbf{h}^m = \mathbf{h}^{m+1} - \frac{\partial \mathbf{h}^{m+1}}{\partial t} \Delta t + \frac{\partial^2 \mathbf{h}^{m+1}}{\partial t^2} \frac{(\Delta t)^2}{2!} - \cdots \tag{9.35}$$

to obtain

$$\frac{\partial \mathbf{h}^{m+1}}{\partial t} \approx \frac{\mathbf{h}^{m+1} - \mathbf{h}^m}{\Delta t}. \tag{9.36}$$

Evaluating $\mathbf{h}$ and $\mathbf{f}$ at time $m\Delta t$, the backward finite difference method gives

$$\mathbf{h}^{m+1} = (\mathbf{C} + \Delta t \mathbf{K})^{-1} \left(\mathbf{C}\mathbf{h}^m + \Delta t \mathbf{f}^{m+1} \right). \tag{9.37}$$

This implicit scheme is unconditionally stable for any Δt.

Weighted Scheme: Multiply the explicit solution by $(1 - \omega)\mathbf{C}$ and the implicit solution by $\omega(\mathbf{C} + \mathbf{K}\Delta t)$, then add the two products to yield

$$\boxed{\mathbf{h}^{m+1} = (\mathbf{C} + \omega \Delta t \mathbf{K})^{-1} \left\{ [\mathbf{C} - (1 - \omega) \Delta t \mathbf{K}] \, \mathbf{h}^m + [(1 - \omega) \, \mathbf{f}^m + \omega \mathbf{f}^{m+1}] \, \Delta t \right\}}, \tag{9.38}$$

where the weighting factor ω is less than one.

If $\omega = 0$, equation (9.38) is the explicit solution; and if $\omega = 1$, it is implicit. If $\omega = 1/2$, one gets the central difference or *Crank-Nicholson* solution

$$\mathbf{h}^{m+1} = \left(\mathbf{C} + \frac{\Delta t}{2}\mathbf{K} \right)^{-1} \left[\left(\mathbf{C} - \frac{\Delta t}{2}\mathbf{K} \right) \mathbf{h}^m + \frac{\Delta t}{2} \left(\mathbf{f}^{m+1} + \mathbf{f}^m \right) \right]. \tag{9.39}$$

• **Finite Element in Time**

Consider a temporal finite-element scheme that is based on linear interpolation within the time interval $0 \leq t \leq \Delta t$

$$\mathbf{h} = \begin{bmatrix} M_0 & M_1 \end{bmatrix} \left\{ \begin{array}{c} \mathbf{h}^0 \\ \mathbf{h}^1 \end{array} \right\}, \quad M_0 = \frac{\Delta t - t}{\Delta t}, \quad M_1 = \frac{t}{\Delta t}. \tag{9.40}$$

Apply the Galerkin weighted residual method at time 1 (the end of a chosen time interval)

$$R_1 = \int_0^{\Delta t} M_1 \left(\mathbf{C}\frac{\partial \mathbf{h}}{\partial t} + \mathbf{K}\mathbf{h} - \mathbf{f} \right) dt. \tag{9.41}$$

Substituting the relation

$$\frac{\partial \mathbf{h}}{\partial t} = \frac{1}{\Delta t} \begin{bmatrix} -1 & 1 \end{bmatrix} \left\{ \begin{array}{c} \mathbf{h}^0 \\ \mathbf{h}^1 \end{array} \right\} = \frac{\mathbf{h}^1 - \mathbf{h}^0}{\Delta t} \tag{9.42}$$

into R_1 yields

$$R_1 = \frac{1}{(\Delta t)^2} \int_0^{\Delta t} \left(t\mathbf{C} \begin{bmatrix} -1 & 1 \end{bmatrix} + t\mathbf{K} \begin{bmatrix} \Delta t - t & t \end{bmatrix} \right) dt \left\{ \begin{matrix} \mathbf{h}^0 \\ \mathbf{h}^1 \end{matrix} \right\} - \frac{1}{\Delta t} \int_0^{\Delta t} t\mathbf{f}\, dt$$

$$= \begin{bmatrix} -\frac{\mathbf{C}}{2} + \frac{\mathbf{K}\Delta t}{6} & \frac{\mathbf{C}}{2} + \frac{\mathbf{K}\Delta t}{3} \end{bmatrix} \left\{ \begin{matrix} \mathbf{h}^0 \\ \mathbf{h}^1 \end{matrix} \right\} - \frac{1}{\Delta t} \int_0^{\Delta t} t\mathbf{f}\, dt. \tag{9.43}$$

Letting this residual be zero yields

$$\boxed{\mathbf{h}^1 = \left(\mathbf{C} + \frac{2\Delta t}{3}\mathbf{K} \right)^{-1} \left[\left(\mathbf{C} - \frac{\Delta t}{3}\mathbf{K} \right) \mathbf{h}^0 + \frac{2}{\Delta t} \int_0^{\Delta t} t\mathbf{f}\, dt \right]} . \tag{9.44}$$

This is equivalent to setting $\omega = 2/3$ in equation (9.38) provided that $\mathbf{f}$ is independent of time. The finite element scheme in time is also unconditionally stable.

9.1.6   Lumped Finite-Element Formulation

The elementary capacitance matrix

$$[c]^e = \frac{S_s \bar{r} \Delta r}{6} \begin{bmatrix} 2 & 1 \\ 1 & 2 \end{bmatrix} \tag{9.45}$$

is a consistent formulation. An alternative representation is to make

$$N_i N_j = \begin{cases} 1/n & \text{if } i = j, \\ 0 & \text{if } i \neq j, \end{cases} \tag{9.46}$$

where n is the number of nodes in an element. In this case the elementary capacitance matrix becomes

$$[c]^e = \frac{S_s \bar{r} \Delta r}{2} \begin{bmatrix} 1 & 0 \\ 0 & 1 \end{bmatrix}. \tag{9.47}$$

This is known as the lumped finite-element formulation. The sum of all entries for the lumped matrix equals that of the consistent matrix.

Because the lumped $[c]^e$ is diagonal, its resulting global capacitance matrix is also diagonal. The lumped formulation has the advantage of less numerical oscillation during matrix inversion as compared to the consistent formulation. Also, if the lumped matrix is used, the inversion of matrix $\mathbf{C}$ in the explicit scheme is essentially a division by C_{ii} only.

9.1.7 Nature of Coefficient Matrix

Equation (9.38) is a set of linear simultaneous equations

$$
\begin{aligned}
\mathbf{Ax} &= \mathbf{b}, \\
\mathbf{x} &= \mathbf{h}^1, \quad \mathbf{b} = \mathbf{Bh}^0 + \left[(1 - \omega)\, \mathbf{f}^0 + \omega \mathbf{f}^1 \right] \Delta t, \\
\mathbf{A} &= \mathbf{C} + \omega \Delta t \mathbf{K}, \quad \mathbf{B} = \mathbf{C} - (1 - \omega)\, \Delta t \mathbf{K}.
\end{aligned}
\tag{9.48}
$$

As noted earlier, capacitance matrix $\mathbf{C}$ and conductance matrix $\mathbf{K}$ are symmetric, banded and sparse; and so are the resultant matrices $\mathbf{A}$ and $\mathbf{B}$. For each row of $\mathbf{K}$, say row i for node i, the nonzero entries in the matrix start from node $(i - 1)$ on the left of node i and end at node $(i + 1)$ on the right. The bandwidth is one plus the difference between the highest and lowest nodal numbers (for nodes that are tied to node i). In this one-dimensional example, the bandwidth is 3. Because of the symmetric banded structure, only the upper nonzero band plus the diagonal entries of a matrix need to be stored in a computer. For example, all zero entries in equation (9.26) can be excluded from storage.

Presumably global matrices are assembled from one element to the next. To save computer memory and improve execution efficiency, global matrices $\mathbf{K}$, $\mathbf{C}$, and $\mathbf{B}$ are rarely assembled as full matrices. Usually these three matrices are directly compiled to form the column vector $\mathbf{b}$. Coefficient matrix $\mathbf{A}$ is the only one that is assembled in practice. (Some codes do not assemble the entire entries of $\mathbf{A}$, either.) Taking advantage of the symmetric and banded nature, the square matrix $\mathbf{A}$ or the banded part of $\mathbf{A}$ is commonly stored as a rectangular matrix to reduce the demand of computer memory.

9.1.8 Initial and Boundary Conditions

Equation (9.44) is a set of n simultaneous equations for a finite element system consisting of n nodes. Given initial condition $\mathbf{h}^0$ at n nodal points, application of equation (9.44) or an equivalent one in the finite difference scheme yields nodal values $\mathbf{h}^1$ at time $1\Delta t$. Repetitive application results in $\mathbf{h}^m$ at the desired time step $m\Delta t$. However, equation solving should be done only after the boundary conditions are implemented.

Neumann Condition: A specified boundary flux is usually imposed through $\mathbf{f}$ during compilation of the global matrices. At the boundary where flux is not specified, the entry of $\mathbf{f}$ at that boundary node is zero.

Dirichlet Condition: Boundary values h_{1b} and h_{nb} at nodes 1 and n respectively are incorporated directly into $\mathbf{h}^1$. In the 4-element problem,

$$\{\mathbf{h}^1\}^T = \begin{bmatrix} h_{1b}^1 & h_2^1 & h_3^1 & h_4^1 & h_{5b}^1 \end{bmatrix}. \tag{9.49}$$

The Dirichlet conditions are imposed during equation solving and the manner of implementation depends on how the equations are solved.

As an example, consider a Dirichlet point at one interior node (e.g., the drawdown measured at an observation well). Let the head at node 3 at time t in a five-node system be h_3^*, then

$$\begin{bmatrix} A_{11} & A_{12} & & & \\ A_{21} & A_{22} & A_{23} & & \\ & A_{32} & A_{33} & A_{34} & \\ & & A_{43} & A_{44} & A_{45} \\ & & & A_{54} & A_{55} \end{bmatrix} \begin{Bmatrix} h_1^1 \\ h_2^1 \\ h_3^* \\ h_4^1 \\ h_5^1 \end{Bmatrix} = \begin{Bmatrix} b_1 \\ b_2 \\ b_3 \\ b_4 \\ b_5 \end{Bmatrix}. \tag{9.50}$$

Since h_3^* is given, its contribution can be calculated,

$$\begin{bmatrix} A_{11} & A_{12} & & & \\ A_{21} & A_{22} & 0 & & \\ 0 & 0 & 0 & & \\ & 0 & A_{44} & A_{45} \\ & & A_{54} & A_{55} \end{bmatrix} \begin{Bmatrix} h_1^1 \\ h_2^1 \\ 0 \\ h_4^1 \\ h_5^1 \end{Bmatrix} = \begin{Bmatrix} b_1 \\ b_2 - A_{23}h_3^* \\ 0 \\ b_4 - A_{43}b_3^* \\ b_5 \end{Bmatrix}, \tag{9.51}$$

and column 3 and row 3 are deleted from consideration. The remainder is a set of four simultaneous equations.

9.1.9 Source Term

In the presence of a water source or sink H, the mass balance equation is modified

$$S_s \frac{\partial h}{\partial t} - \frac{1}{r}\frac{\partial}{\partial r}\left(kr\frac{\partial h}{\partial r}\right) = H, \tag{9.52}$$

where H has a unit of $[m^3 m^{-3} s^{-1}]$. It is positive for a source and negative for a sink.

Assuming that water production is uniform throughout the ring element, the weighted residual should contain an extra term for the source

$$R_{iH} = \int_{r_i}^{r_j} N_i H r \, dr = \frac{H\bar{r}\Delta r}{2} = R_{jH}. \tag{9.53}$$

Comparing equation (9.52) with equation (9.28), it is clear that R_{iH} should be added to the term of influx vector f.

If H is a ring source at r_p (i.e., $H = H\delta[r - r_p]$),

$$
R_{iH} = \int_{r_i}^{r_j} N_i\, H\delta[r - r_p]\, r\, dr = H N_i r|_{r=r_p} = H \frac{-r_p + r_j}{\Delta r} r_p,
$$

$$
R_{jH} = H \frac{r_p - r_i}{\Delta r} r_p. \tag{9.54}
$$

If the ring source is located at the nodes, then

$$
\begin{aligned}
R_{iH} &= H r_i, \quad R_{jH} = 0 \quad \text{if} \quad r_p = r_i, \\
R_{iH} &= 0, \quad R_{jH} = H r_j \quad \text{if} \quad r_p = r_j.
\end{aligned} \tag{9.55}
$$

Note that the residual for the ring source has a hidden dimension of $[m^2 s^{-1}]$ despite the apparent dimension of $[m^1 s^{-1}]$.

9.1.10 Numerical Instability and Oscillation

In addition to the inherent error in the finite-element approximation to the differential equation, there are errors associated with hardware and software. Because of the finite precision in digital representation of variables, round-off error occurs during computation. The error could be cumulative during time stepping and could result in a solution that diverges more and more from the true solution. Improper choice of element size and time step can thus lead to numerical oscillation around the true solution. Those potential sources of errors should be mitigated.

• Numerical Instability

Numerical instability refers to the amplification of errors when equation solving proceeds from one time step to the next. Rewrite equation (9.38)

$$
\mathbf{h}^{m+1} = (\mathbf{C} + \omega \Delta t \mathbf{K})^{-1} [\mathbf{C} - (1 - \omega)\,\Delta t \mathbf{K}]\, \mathbf{h}^m + \dots \tag{9.56}
$$

as a recurrence relation for the nodal hydraulic head $\mathbf{h}$ at a given choice of ω and Δt. The error propagates through the same recurrence relation during time stepping. An initial error is amplified or damped by the same factor that transforms $\mathbf{h}^m$ into $\mathbf{h}^{m+1}$.

Approximation by a Scalar: For the time being, let us treat $\mathbf{C}$ and $\mathbf{K}$ as scalar to take a heuristic view on how to mitigate the error amplification. We demand

$$-1 < \frac{C - (1 - \omega)\,\Delta t K}{C + \omega \Delta t K} < 1 \tag{9.57}$$

so that the error propagation can be attenuated.

Multiplying the above relation by $C + \omega \Delta t K$ (which is positive), adding $C - \omega \Delta t K$, and dividing the result by $-2\Delta t K$ yields

$$\boxed{\omega > \frac{1}{2} - \frac{C}{\Delta t K} > -\frac{C}{\Delta t K}}\,. \tag{9.58}$$

Since C and K are positive, the stability is guaranteed if $\omega \geq 1/2$. Thus all the time schemes considered so far, except the explicit forward finite difference, are numerically stable during time stepping.

Eigenvector: The heuristic view is not elegant. Let's have an alternative (Bathe and Wilson, 1976). Rewrite the finite element equation (9.30)

$$\mathbf{C}\,\dot{\mathbf{h}} + \mathbf{K}\mathbf{h} = \mathbf{f}, \quad \dot{\mathbf{h}} = \partial \mathbf{h}/\partial t. \tag{9.59}$$

Assume that $\mathbf{h}$ can be represented by the product of a time-independent square matrix $\mathbf{\Phi}$ and a time-dependent column vector $\mathbf{x}$

$$\mathbf{h} = \mathbf{\Phi}\mathbf{x}. \tag{9.60}$$

The finite element equation becomes

$$\mathbf{C}\mathbf{\Phi}\,\dot{\mathbf{x}} + \mathbf{K}\mathbf{\Phi}\mathbf{x} = \mathbf{f}. \tag{9.61}$$

Premultiplying this relation by the transpose of $\mathbf{\Phi}$,

$$\mathbf{\Phi}^T \mathbf{C}\mathbf{\Phi}\,\dot{\mathbf{x}} + \mathbf{\Phi}^T \mathbf{K}\mathbf{\Phi}\mathbf{x} = \mathbf{\Phi}^T \mathbf{f}. \tag{9.62}$$

So far we have introduced two variables $\mathbf{\Phi}$ and $\mathbf{x}$ to represent one variable $\mathbf{h}$. One of the two must be constrained. Let us constrain the choice of $\mathbf{\Phi}$ by

$$\mathbf{\Phi}^T \mathbf{C}\mathbf{\Phi} = \mathbf{I}\,, \tag{9.63}$$

where $\mathbf{I}$ is an identity matrix.

Then let us determine the nature of $\mathbf{\Phi}^T\mathbf{K}\mathbf{\Phi}$ via another designation

$$\mathbf{\Lambda} = \mathbf{\Phi}^T\mathbf{K}\mathbf{\Phi} = \mathbf{\Phi}^T\mathbf{C}\mathbf{\Phi}\mathbf{\Lambda} \tag{9.64}$$

or equivalently

$$\mathbf{K}\mathbf{\Phi} = \mathbf{C}\mathbf{\Phi}\mathbf{\Lambda}. \tag{9.65}$$

The new matrix $\mathbf{\Lambda}$ is a diagonal matrix consisting of the entries that are eigenvalues λ for the generalized eigenproblem

$$\mathbf{K}\phi = \lambda\mathbf{C}\phi. \tag{9.66}$$

(If $\mathbf{C} = \mathbf{I}$, it is a typical eigenproblem.) Eigenvectors ϕ constitute the columns of eigenmatrix $\mathbf{\Phi}$. This choice of $\mathbf{\Phi}$ is $\mathbf{C}$-orthonormalized.

Substituting the constraint into equation (9.62) gives

$$\dot{\mathbf{x}} + \mathbf{\Lambda}\mathbf{x} = \mathbf{\Phi}^T\mathbf{f}. \tag{9.67}$$

This relation is a collection of the component equations

$$\frac{dx}{dt} + \lambda x = b, \tag{9.68}$$

where b is a row entry of $\mathbf{\Phi}^T\mathbf{f}$. Comparing this relation with equation (9.28) and referring to equation (9.38) yields

$$x^{m+1} = (1 + \omega\lambda\Delta t)^{-1}[1 - (1 - \omega)\lambda\Delta t]x^m + ... \tag{9.69}$$

Impose the condition that

$$-1 < \frac{1 - (1 - \omega)\lambda\Delta t}{1 + \omega\lambda\Delta t} < 1 \tag{9.70}$$

to damp the error propagation during time stepping. Then, rewrite the constraint as

$$\omega > \frac{1}{2} - \frac{1}{\lambda\Delta t} > -\frac{1}{\lambda\Delta t}. \tag{9.71}$$

Here the eigenvalue λ is positive because matrix $\mathbf{K}$ is symmetric. Thus we have reached the same conclusion as the heuristic approach that for stable time-stepping,

$$\omega \geq \frac{1}{2}. \tag{9.72}$$

• **Numerical Oscillation**

Numerical oscillation can occur when a large time step Δt is used. The step should be as large as possible but not large enough to overshoot or undershoot the expected (true) results. An optimal Δt is a function of element geometry, shape function, material properties, and choice of ω. We will defer this topic to later discussion. Here, it suffices to use a rule of thumb

$$\Delta t \le \frac{L^2}{4D\left(1-\omega\right)}, \quad \omega \ne 1, \tag{9.73}$$

where L is the length of the shortest element for a one-dimensional case (otherwise L^2 is the area of the smallest element) and D is the hydraulic diffusivity defined as the ratio of the hydraulic conductivity k to the specific storage S_s (i.e., $D = k/S_s$).

9.2   2D Finite Element Method

Now we proceed to two-dimensional problems described by the governing differential equation

$$\frac{\partial}{\partial x}\left(k_x\frac{\partial h}{\partial x}\right) + \frac{\partial}{\partial y}\left(k_y\frac{\partial h}{\partial y}\right) = S_s\frac{\partial h}{\partial t} \tag{9.74}$$

where h is the hydraulic head $[m]$, k_x and k_y are anisotropic hydraulic conductivity $[ms^{-1}]$, S_s is the specific storage $[m^2m^{-3}]$, and t is the time $[s]$. This represents a transient flow in the cross-sectional view or plan view of saturated porous media.

9.2.1   Procedures

The procedures for 2D-finite element analysis follow closely the 1D procedures.

   □ 1. Discretize the region of interest into triangular elements. Other element geometry is feasible but can be decomposed into triangular elements. This uses the advantage that exact integration can be made for triangular elements with linear interpolation functions.

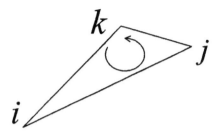

Figure 9.4: A triangular element, nodes numbered counterclockwise.

□ 2. Adopt the method of Galerkin weighted residual. At node i in triangular element e, the residual is

$$R_i^e = \iint_e N_i \left[\frac{\partial}{\partial x} \left(k_x \frac{\partial h}{\partial x} \right) + \frac{\partial}{\partial y} \left(k_y \frac{\partial h}{\partial y} \right) - S_s \frac{\partial h}{\partial t} \right] dx dy. \tag{9.75}$$

The weighting function and interpolation function are identical in this scheme.

□ 3. Use a linear interpolation function such that $h[x, y]$ is a linear combination of nodal values $\begin{bmatrix} h_i & h_j & h_k \end{bmatrix}^T$ at the three corner nodes (i, j, k) of a triangular element (Figure 9.4)

$$h = \begin{bmatrix} N_i & N_j & N_k \end{bmatrix} \left\{ \begin{array}{c} h_i \\ h_j \\ h_k \end{array} \right\} = N_i h_i + N_j h_j + N_k h_k. \tag{9.76}$$

Note that nodes i, j, and k are ordered in a counterclockwise sense.

□ 4. Let the interpolation function $N_i[x, y]$ for h be the same as the interpolation function for coordinates in terms of local nodal coordinates, i.e.,

$$\begin{aligned} x &= N_i x_i + N_j x_j + N_k x_k, \\ y &= N_i y_i + N_j y_j + N_k y_k. \end{aligned} \tag{9.77}$$

There are three unknowns (N_i, N_j, and N_k) and two equations. So, place another constraint

$$1 = N_i + N_j + N_k. \tag{9.78}$$

In matrix form, the three simultaneous equations become

$$\begin{bmatrix} 1 & 1 & 1 \\ x_i & x_j & x_k \\ y_i & y_j & y_k \end{bmatrix} \left\{ \begin{array}{c} N_i \\ N_j \\ N_k \end{array} \right\} = \left\{ \begin{array}{c} 1 \\ x \\ y \end{array} \right\}. \tag{9.79}$$

□ 5. Solve the above three simultaneous equations to obtain

$$N_i = \frac{1}{2A}\left(a_i + b_i x + c_i y\right),$$

$$A = \text{area of element}$$

$$= \frac{1}{2}\begin{vmatrix} 1 & 1 & 1 \\ x_i & x_j & x_k \\ y_i & y_j & y_k \end{vmatrix} = \frac{1}{2}\left(x_j y_k + x_k y_i + x_i y_j - x_k y_j - x_i y_k - x_j y_i\right),$$

$$a_i = x_j y_k - x_k y_j, \quad b_i = y_j - y_k, \quad c_i = x_k - x_j. \tag{9.80}$$

Similarly, other components (e.g., N_j, c_j) can be obtained by cyclic permutation of i, j, and k. For example, $c_j = x_i - x_k$. In this case, local coordinates (area coordinates) and interpolation functions are identical — *isoparametric* elements.

□ 6. Integrate the weighted residual by parts to change the second-order derivative into a first-order derivative,

$$\begin{aligned} R_i^e = &\int N_i \left(k_x \frac{\partial h}{\partial x} + k_y \frac{\partial h}{\partial y}\right)\bigg|_n \, d\ell \\ &- \iint\limits_e \left(k_x \frac{\partial N_i}{\partial x}\frac{\partial h}{\partial x} + k_y \frac{\partial N_i}{\partial y}\frac{\partial h}{\partial y} + N_i S_s \frac{\partial h}{\partial t}\right) dx\,dy, \end{aligned} \tag{9.81}$$

where the first integral is around the boundary of the element and subscript n means outward normal across the boundary. This relation has resulted from the application of the divergence theorem in two dimensions. For the time being, let's designate the first integral as f_i^e.

□ 7. Implement the interpolation function

$$\begin{aligned} R_i^e = f_i^e - &\iint\limits_e \left(k_x \frac{\partial N_i}{\partial x}\frac{\partial}{\partial x}\left[\begin{matrix} N_i & N_j & N_k \end{matrix}\right] + k_y \frac{\partial N_i}{\partial y}\frac{\partial}{\partial y}\left[\begin{matrix} N_i & N_j & N_k \end{matrix}\right]\right. \\ &\left. + N_i S_s \left[\begin{matrix} N_i & N_j & N_k \end{matrix}\right]\frac{\partial}{\partial t}\right) dx\,dy \left\{\begin{matrix} h_i \\ h_j \\ h_k \end{matrix}\right\}. \end{aligned} \tag{9.82}$$

□ 8. Simplify the above relation using the following relations

$$\iint\limits_e N_i^p N_j^q N_k^r dx\,dy = \frac{p!\,q!\,r!}{(p+q+r+2)!}2A \tag{9.83}$$

(See Problem 1 for proof.) For example,

$$\iint_e \frac{\partial N_i}{\partial x}\frac{\partial N_j}{\partial x}dxdy = \frac{b_ib_j}{4A}, \quad \iint_e \frac{\partial N_i}{\partial y}\frac{\partial N_j}{\partial y}dxdy = \frac{c_ic_j}{4A},$$

$$\iint_e (N_i)^2\,dxdy = \frac{A}{6}, \quad \iint_e N_iN_jdxdy = \frac{A}{12}. \tag{9.84}$$

The residuals at the three corner nodes of element e are

$$\left\{ \begin{array}{c} R_i \\ R_j \\ R_k \end{array} \right\}^e = \left\{ \begin{array}{c} f_i \\ f_j \\ f_k \end{array} \right\}^e - [\mathbf{k}]^e\,\{\mathbf{h}\}^e - [\mathbf{c}]^e\left\{ \frac{\partial\mathbf{h}}{\partial t} \right\}^e, \tag{9.85}$$

where

$$[\mathbf{k}]^e = \frac{k_x^e}{4A}\left[\begin{array}{ccc} b_ib_i & b_ib_j & b_ib_k \\ b_jb_i & b_jb_j & b_jb_k \\ b_kb_i & b_kb_j & b_kb_k \end{array} \right] + \frac{k_y^e}{4A}\left[\begin{array}{ccc} c_ic_i & c_ic_j & c_ic_k \\ c_jc_i & c_jc_j & c_jc_k \\ c_kc_i & c_kc_j & c_kc_k \end{array} \right], \tag{9.86}$$

$$[\mathbf{c}]^e = \frac{S_s^e A}{12}\left[\begin{array}{ccc} 2 & 1 & 1 \\ 1 & 2 & 1 \\ 1 & 1 & 2 \end{array} \right], \quad \{\mathbf{h}\}^e = \left\{ \begin{array}{c} h_i \\ h_j \\ h_k \end{array} \right\}^e. \tag{9.87}$$

Note that the symbol c has been used for two different meanings: single subscript for interpolation function and double subscripts for capacitance matrices.

☐ 9. Change the local nodal numbers to global ones and sum up the residuals in all elements that share a common global node i,

$$R_i = \sum_e R_i^e = \sum_e f_i^e - \sum_e \left[[\mathbf{k}]\,\{\mathbf{h}\} + [\mathbf{c}]\left\{ \frac{\partial\mathbf{h}}{\partial t} \right\} \right]_i^e. \tag{9.88}$$

☐ 10. Make the residual at each node zero to obtain

$$\mathbf{C}\frac{\partial\mathbf{h}}{\partial t} + \mathbf{Kh} = \mathbf{f}, \tag{9.89}$$

$$K_{ij} = \sum_e k_{ij}^e, \quad C_{ij} = \sum_e c_{ij}^e,$$

$$f_i = \sum_e \int N_i\left(k_x\frac{\partial h}{\partial x} + k_y\frac{\partial h}{\partial y} \right)\bigg|_n d\ell.$$

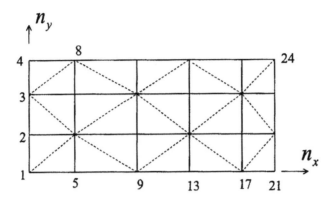

Figure 9.5: Sequence of node numbering for $n_y < n_x$ starts from the origin upward along n_y-axis, then continues upward along each consecutive column. Numbering of element will not affect the size of coefficient matrix.

Note that the summation for f_i is effective for the external boundary where the fluxes are specified. The contributions to $\mathbf{f}$ along the interior boundaries of elements cancel one another because the outward normal $\hat{n}$ has opposite signs across the boundary between two elements.

☐ 11. Impose a constant head boundary condition, if appropriate, and solve the first-order differential equation by means of equation (9.38)

$$
\begin{aligned}
\mathbf{h}^{m+1} \;=\; & (\mathbf{C} + \omega \Delta t \mathbf{K})^{-1} \\
& \cdot \left\{ \left[\mathbf{C} - (1-\omega)\,\Delta t \mathbf{K}\right]\mathbf{h}^m + \left[(1-\omega)\,\mathbf{f}^m + \omega \mathbf{f}^{m+1}\right]\Delta t \right\} \quad (9.90)
\end{aligned}
$$

for nodal $\mathbf{h}^{m+1}$.

9.2.2   Remarks

• Symmetric and Banded Matrix

The conductivity matrix $\mathbf{K}$ and capacitance matrix $\mathbf{C}$ are symmetric, sparse, and banded. For each row i (corresponding to global node i) of $\mathbf{K}$ or $\mathbf{C}$, the nonzero entries K_{ij} or C_{ij} occur only where nodes i and j have direct tie lines. Hence, the number of nonzero entries is equal to the number of triangular elements that join together at node i plus one (number of tie lines plus the node i itself). Along each row, the column range of nonzero entries spans from $j_{\min}$ to $j_{\max}$ with $j_{\min}$ being the least nodal numbers and $j_{\max}$ being the

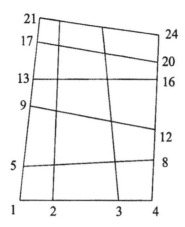

Figure 9.6: Sequence of node numbering for $n_y > n_x$ starts from the lower-left corner rightward, then continues rightward along each consecutive row.

greatest nodal numbers around node i. The bandwidth is the maximum of $[2\,(j_{\max} - j_{\min}) + 1]$ among all rows. A good programming practice is to keep the difference $(j_{\max} - j_{\min})$ for all rows as small as possible.

As the example shown in Figure 9.5, a rectangular region is divided into $n_x \times n_y$ rectangular elements ($n_x = 5$ and $n_y = 3$) and each rectangular element is further split into two triangular elements. The total number of nodes or unknown nodal values is $(n_x+1) \times (n_y+1)$ $(= 24)$. The dimension of $\mathbf{K}$ is $[(n_x + 1) \times (n_y + 1)]^2$ $(= 576)$. Since the row number n_y is less than the column number n_x (i.e., $n_y < n_x$), the numbering begins along the n_y-axis to minimize the maximum difference in node numbers around a given node.

The numbering sequence in Figure 9.5 shows that a node can be tied to as many as 8 other nodes. For example, node 14 is tied to nodes 9, 10, 11, 13, 15, 17, 18, and 19. The maximum difference in nodal number around any node is 10. Corresponding to node 14, row 14 in the 24×24 square matrix $\mathbf{K}$ has a bandwidth of 11, which is $2(n_y + 1) + 3$ for $n_y = 3$. The entries outside this band are zero. Within the band, there are 9 nonzero entries from columns 9 through 19 and zero entries occur at nodes 12 and 16 that are not tied to node 14. The half-bandwidth plus the diagonal is $n_y + 1 + 2$.

The zero entries outside the band can be excluded from the memory storage. Hence, it is common to store the square matrix $\mathbf{K}$ as a rectangular matrix. The storage requirement for the symmetric band is equal to the total number of nodal points multiplied by the half-bandwidth, i.e.,

$$(n_y + 3) \left[(n_x + 1) \times (n_y + 1) \right]$$

($= 144$). This represents only a fraction,

$$\frac{(n_y + 3) \left[(n_x + 1) \times (n_y + 1) \right]}{\left[(n_x + 1) \times (n_y + 1) \right] \left[(n_x + 1) \times (n_y + 1) \right]},$$

($= 0.25 = 6/24 = 144/576$) of the total entries of **K**. The saving in storage increases with increasing difference between n_x and n_y. However, the reduction in storage is exercised at the expense of more programming effort.

For the case that $n_y > n_x$, the numbering starts from the origin rightward along the n_x-axis (Figure 9.6). The storage saving for the two examples is the same but their node numbering sequences are different. The numbering sequence of elements will not affect the size of the coefficient matrix.

- **Axi-Symmetric Problems**

For an axi-symmetric problem (i.e., h independent of azimuth), $dxdydz$ can be replaced by $2\pi r dr dz$ for evaluating the residual R_i^e. The integration is usually not exact. However, if r in the $2\pi r dr dz$ is replaced by the average of the global nodal r-coordinates,

$$\bar{r}^e = \frac{1}{3} \left(r_i^e + r_j^e + r_k^e \right), \tag{9.91}$$

the $\bar{r}^e$ can be factored out of the integral sign.

In so doing, the integrands look exactly like those for the cases in rectangular coordinates and exact integration can be obtained by multiplying $\bar{r}^e$ to each term of the above 2D formulation at the elementary level before assembling the global matrices. This approximation is acceptable if the element size is sufficiently small.

9.3   Transport Equations

9.3.1   Advective Heat Transfer

Now consider advective heat transfer in steady groundwater flow

$$\frac{\partial}{\partial x} \left(k_x \frac{\partial T}{\partial x} \right) + \frac{\partial}{\partial y} \left(k_y \frac{\partial T}{\partial y} \right) - (\rho c_v)_w \mathbf{v} \cdot \nabla T = \rho c_v \frac{\partial T}{\partial t}, \tag{9.92}$$

where T is the temperature, k is the anisotropic thermal conductivity (k_x, k_y), $\mathbf{v}$ is the steady groundwater flow velocity (Darcy's velocity), $(\rho c_v)_w$ is the volumetric heat capacity of water, and ρc_v is the volumetric heat capacity of the fluid-rock mixture. Comparison of this relation with equation (9.74) indicates that the only difference is the addition of an advective term.

Using the linear interpolation

$$T = \begin{bmatrix} N_i & N_j & N_k \end{bmatrix} \left\{ \begin{array}{c} T_i \\ T_j \\ T_k \end{array} \right\}, \tag{9.93}$$

the weighted residual at node i in element e due to advection is

$$
\begin{aligned}
-R_{vi}^e &= (\rho c_v)_w \iint_e N_i \mathbf{v} \cdot \nabla T \, dx dy \\
&= (\rho c_v)_w \iint_e N_i \left(v_x \frac{\partial T}{\partial x} + v_y \frac{\partial T}{\partial y} \right) dx dy = \frac{(\rho c_v)_w}{2A} \\
&\quad \cdot \iint_e N_i \left(v_x \begin{bmatrix} b_i & b_j & b_k \end{bmatrix} + v_y \begin{bmatrix} c_i & c_j & c_k \end{bmatrix} \right) dx dy \left\{ \begin{array}{c} T_i \\ T_j \\ T_k \end{array} \right\}^e \\
&= \frac{(\rho c_v)_w}{6} \left(v_x \begin{bmatrix} b_i & b_j & b_k \end{bmatrix} + v_y \begin{bmatrix} c_i & c_j & c_k \end{bmatrix} \right)^e \left\{ \begin{array}{c} T_i \\ T_j \\ T_k \end{array} \right\}^e. \tag{9.94}
\end{aligned}
$$

Extending this relation to other nodes results in an elementary matrix for advection,

$$\mathbf{k}_v^e = \frac{(\rho c_v)_w}{6} \begin{bmatrix} v_x b_i + v_y c_i & v_x b_j + v_y c_j & v_x b_k + v_y c_k \\ v_x b_i + v_y c_i & v_x b_j + v_y c_j & v_x b_k + v_y c_k \\ v_x b_i + v_y c_i & v_x b_j + v_y c_j & v_x b_k + v_y c_k \end{bmatrix}^e. \tag{9.95}$$

Because of its association with the nodal temperatures, $\mathbf{k}_v^e$ should be added to the elementary conductivity matrix and then assembled into a global advection-conduction matrix. Because $\mathbf{k}_v^e$ is not symmetric, the resulting advection-conduction matrix $\mathbf{K}$ is not symmetric either. The entries of the asymmetric and banded $\mathbf{K}$ must be stored and an equation solver for asymmetric coefficient matrix must be employed.

9.3.2 Solute Transport

Solute transport in steady groundwater flow is governed by

$$\frac{\partial(\theta C^s)}{\partial t} = \frac{\partial}{\partial x}\left[D_x\frac{\partial(\theta C^s)}{\partial x}\right] + \frac{\partial}{\partial y}\left[D_y\frac{\partial(\theta C^s)}{\partial y}\right]$$

$$-\mathbf{v}\cdot\nabla(C^s) - \frac{\partial}{\partial t}\left(\rho_b K_d C^s\right) - \lambda\left(\theta + \rho_b K_d\right)C^s, \quad (9.96)$$

where C^s is the solute concentration $[kg\,m^{-3}]$, θ is the volumetric water content of the porous medium [-], D_x and D_y are the dispersion coefficients $[m^2 s^{-1}]$, $\mathbf{v}$ is Darcy's velocity $[m\,s^{-1}]$, ρ_b is the bulk density of dry porous medium $[kg\,m^{-3}]$, K_d is the distribution coefficient $[kg^{-1}m^3]$, λ is the solute decay constant $[s^{-1}]$, and t is the time $[s]$. Note that θ is the effective porosity in the case of saturated flow.

• Dispersion-Adsorption

Comparison with equation (9.92) indicates that the finite element equation must include terms corresponding to the residual for adsorption or absorption. Let

$$C^s = \begin{bmatrix} N_i & N_j & N_k \end{bmatrix} \left\{\begin{array}{c} C^s_i \\ C^s_j \\ C^s_k \end{array}\right\} \quad (9.97)$$

and K_d, ρ_b, and θ be constant within element e. The weighted residual for the adsorption and the first-order decay terms is

$$-R^e_{ai} = \iint\limits_e N_i\left(\frac{\partial}{\partial t}\left(\rho_b K_d C^s\right) + \lambda\left(\theta + \rho_b K_d\right)C^s\right)dxdy$$

$$= \rho_b K_d \iint\limits_e N_i\frac{\partial C^s}{\partial t}\,dxdy + \lambda\left(\theta + \rho_b K_d\right)\iint\limits_e N_i C^s\,dxdy$$

$$= \rho_b K_d \iint\limits_e \begin{bmatrix} N_i N_i & N_i N_j & N_i N_k \end{bmatrix} dxdy \left\{\begin{array}{c} \partial C^s_i/\partial t \\ \partial C^s_j/\partial t \\ \partial C^s_k/\partial t \end{array}\right\} + \lambda\left(\theta + \rho_b K_d\right)$$

$$\cdot \iint\limits_e \begin{bmatrix} N_i N_i & N_i N_j & N_i N_k \end{bmatrix} dxdy \left\{\begin{array}{c} C^s_i \\ C^s_j \\ C^s_k \end{array}\right\}. \quad (9.98)$$

Thus, R^e_{ai} contributes to both **K** and **C**. For the $\partial C^s / \partial t$ term (equivalent to $\partial h / \partial t$ or $\partial T / \partial t$ terms) or the capacitance matrix **C**, add the adsorption term

$$
\mathbf{c}^e_a = \frac{(\rho_b K_d A)^e}{12}
\begin{bmatrix}
2 & 1 & 1 \\
1 & 2 & 1 \\
1 & 1 & 2
\end{bmatrix}
\tag{9.99}
$$

to form an adsorption-capacitance matrix **C**. For the C^s term (equivalent to h or T terms) or advection-dispersion matrix **K**, add the adsorption term

$$
\mathbf{k}^e_a = \frac{[\lambda (\theta + \rho_b K_d) A]^e}{12}
\begin{bmatrix}
2 & 1 & 1 \\
1 & 2 & 1 \\
1 & 1 & 2
\end{bmatrix}.
\tag{9.100}
$$

The net result is an adsorption-advection-dispersion matrix **K**, which is asymmetric even though $\mathbf{k}^e_a$ is symmetric.

Note that we have used **C** for capacitance matrix and C^s for solute concentration. The distinction should be clear from the context.

For unsaturated flow, hydraulic properties and moisture content depend on the hydraulic head. The problem becomes nonlinear and usually requires iteration to solve. Appropriate functional relations between properties and hydraulic head (constitutive relations) must also be incorporated in order to evaluate the residual.

• Upstream Weighting

Numerical oscillations may occur when steep gradients across temperature or concentration fronts are simulated. The oscillation is usually suppressed by giving more weight to the upstream node in the advective term. For example, replacing the linear interpolation function by the modified nonlinear function (Simunek et al., 1994; Yeh and Tripathi, 1990)

$$
N'_i = N_i - 3\alpha_k N_j N_i + 3\alpha_j N_k N_i,
\tag{9.101}
$$

where the weighting factor (Christie et al., 1976)

$$
\alpha_i = \coth\left(\frac{vL}{2D}\right) - \frac{2D}{vL}
\tag{9.102}
$$

with L being the length of the element boundary opposing node i. Implementation of nonlinear N'_i, etc. for $\mathbf{v} \cdot \nabla T$ or $\mathbf{v} \cdot \nabla(C^s)$ makes the finite-element formulation more complicated.

• Numerical Oscillation and Dispersion

Numerical oscillation and dispersion can be reduced by proper sizing of spatial and temporal grids (Simunek et al., 1994). First, for an optimal choice of characteristic length l of an element, compare the relative influence of advective flux $v_i C^s / \theta$ and dispersive flux $D_{ii} C^s / l$. Keep the grid Peclet number (the ratio of the advective to diffusive fluxes) under 5,

$$P_{e_i^e} = \frac{v_i C^s / \theta}{D_{ii} C^s / l} = \frac{v_i l}{\theta D_{ii}} \le 5. \qquad (9.103)$$

A tolerable small oscillation can still be maintained for $P_{e_i^e}$ up to 10.

Second, pick time step Δt from the Courant number

$$C_{r_i^e} = \frac{v_i / \theta}{l / (\Delta t / R)} = \frac{v_i \Delta t}{\theta R l} \le 1, \quad R = 1 + \rho_b K_d / \theta, \qquad (9.104)$$

where R is the retardation factor, [-]. $C_{r_i^e}$ is a measure of the relative velocity of groundwater flow and solute transport. An alternative choice of Δt is constrained by

$$P_{e_i^e} \cdot C_{r_i^e} \le 2 \quad \text{or} \quad \left(\frac{v_i}{\theta}\right)^2 \frac{\Delta t}{D_{ii} R} \le 2. \qquad (9.105)$$

A third option is to add an artificial longitudinal dispersivity

$$\overline{D}_L = \frac{|v| \Delta t}{2\theta R} - D_L - \frac{\theta D_w \tau}{|v|} \qquad (9.106)$$

as introduced by Perrochet and Berod (1993) and implemented by Simunek et al. (1994), where D_w is the ionic or molecular dispersion coefficient in free water $[m^2 s^{-1}]$ and τ is the tortuosity [-].

• Cauchy (or mixed) Boundary Condition

In addition to the Dirichlet condition (specified boundary concentration) and Neumann condition (specified boundary solute flux), a third type of condition which prescribes both concentration and flux may be imposed,

$$-\theta \widehat{n} \cdot D \nabla C^s + \mathbf{v} \cdot \widehat{n} C^s = \mathbf{v} \cdot \widehat{n} C^{s\prime}, \qquad (9.107)$$

where $\widehat{n}$ is the outward normal and $C^{s\prime}$ is the concentration of incoming fluid. If the fluid flows out of the region, then

$$C^s = C^{s\prime} \quad \text{and} \quad \theta \widehat{n} \cdot D \nabla C^s = 0. \qquad (9.108)$$

Consider the case that $C^s \neq C^{s'}$ and rewrite the Cauchy condition

$$\theta \widehat{n} \cdot D \nabla C^s = \mathbf{v} \cdot \widehat{n} \left(C^s - C^{s'} \right). \tag{9.109}$$

Substitute this relation into the equivalent of equation (9.89)

$$f_i = \sum_e \int N_i \left(D_x \frac{\partial C^s}{\partial x} + D_y \frac{\partial C^s}{\partial y} \right) \Big|_n d\ell \tag{9.110}$$

to yield

$$
\begin{aligned}
f_{id}^e &= \sum_e \int N_i \left(\mathbf{v} \cdot \widehat{n} \left(C^s - C^{s'} \right) \right) d\ell \\
&= \sum_e \mathbf{v} \cdot \widehat{n} \int N_i \begin{bmatrix} N_i & N_j & N_k \end{bmatrix} \begin{Bmatrix} C_i^s \\ C_j^s \\ C_k^s \end{Bmatrix} d\ell \\
&\quad - \sum_e \mathbf{v} \cdot \widehat{n} \int N_i C^{s'} d\ell.
\end{aligned} \tag{9.111}
$$

Extending to other boundary nodes yields

$$
\begin{aligned}
\mathbf{f}_d^e &= \sum_e \mathbf{v} \cdot \widehat{n} \int \begin{bmatrix} N_i N_i & N_i N_j & N_i N_k \\ N_j N_i & N_j N_j & N_j N_k \\ N_k N_i & N_k N_j & N_k N_k \end{bmatrix} \begin{Bmatrix} C_i^s \\ C_j^s \\ C_k^s \end{Bmatrix} d\ell \\
&\quad - \sum_e \mathbf{v} \cdot \widehat{n} \int \begin{bmatrix} N_i \\ N_j \\ N_k \end{bmatrix} C^{s'} d\ell.
\end{aligned} \tag{9.112}
$$

The first term in $\mathbf{f}_d^e$ is symmetric and the entries are added directly into the adsorption-advection-dispersion matrix $\mathbf{K}$ because they are related to the nodal concentrations; the contributions from the second term are subtracted from the $\mathbf{f}$.

9.4 Problems, Keys, and Suggested Readings

• Problems

1. Show that

$$\iint N_i^p N_j^q N_k^r dx dy = \frac{p! q! r!}{(p+q+r+2)!} 2A \tag{9.113}$$

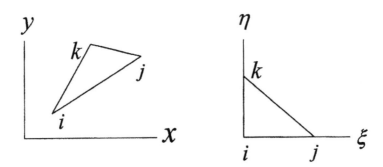

Figure 9.7: Transformation of a triangular element in the xy-plane into a right isosceles in the $\xi\eta$-plane.

for a linear triangular element, where p, q, and r are integers and A is the area of triangle.

• **Keys**

Key 1

For ease of integration, we make a linear transformation from an arbitrarily shaped triangle in the xy-plane into a right isosceles triangle in the $\xi\eta$-plane (Figure 9.7)

$$
\begin{aligned}
x &= x[\xi, \eta], \\
y &= y[\xi, \eta].
\end{aligned} \tag{9.114}
$$

The corner nodes are $(0,0)$, $(1,0)$ and $(0,1)$ of (ξ_i, η_i), (ξ_j, η_j) and (ξ_k, η_k), respectively. The interpolation (shape) functions in the right isosceles triangle are

$$
\psi_i = 1 - \xi - \eta, \quad \psi_j = \xi, \quad \psi_k = \eta \tag{9.115}
$$

such that

$$
\begin{aligned}
\xi &= \psi_i \xi_i + \psi_j \xi_j + \psi_k \xi_k = \psi_j, \\
\eta &= \psi_i \eta_i + \psi_j \eta_j + \psi_k \eta_k = \psi_k, \\
0 &\le \xi \le 1 \quad \text{and} \quad 0 \le \eta \le 1.
\end{aligned} \tag{9.116}
$$

Using linear interpolation for both N and ψ, the transform relations are

$$
x = x_i \psi_i + x_j \psi_j + x_k \psi_k,
$$

$$y = y_i\psi_i + y_j\psi_j + y_k\psi_k. \tag{9.117}$$

The differentials dx and dy are related to $d\xi$ and $d\eta$ through a Jacobian transformation matrix

$$\begin{bmatrix} dx \\ dy \end{bmatrix} = \begin{bmatrix} \frac{\partial x}{\partial \xi} & \frac{\partial x}{\partial \eta} \\ \frac{\partial y}{\partial \xi} & \frac{\partial y}{\partial \eta} \end{bmatrix} \begin{bmatrix} d\xi \\ d\eta \end{bmatrix}. \tag{9.118}$$

The Jacobian determinant is

$$\begin{aligned}
\det J &= \begin{vmatrix} \frac{\partial x}{\partial \xi} & \frac{\partial x}{\partial \eta} \\ \frac{\partial y}{\partial \xi} & \frac{\partial y}{\partial \eta} \end{vmatrix} \\
&= (-x_i + x_j)(-y_i + y_k) - (-x_i + x_k)(-y_i + y_j) \\
&= (x_j y_k - x_k y_j) + (x_k y_i - x_i y_k) + (x_i y_j - x_j y_i) \\
&= 2A, \tag{9.119}
\end{aligned}$$

where A is the area of the triangle in the xy-plane

$$A = \begin{vmatrix} 1 & 1 & 1 \\ x_i & x_j & x_k \\ y_i & y_j & y_k \end{vmatrix}. \tag{9.120}$$

The relation of an area element is

$$dxdy = \det J \, d\xi d\eta. \tag{9.121}$$

Let the desired integral be designated W. Then

$$W = \iint \psi_i^p \psi_j^q \psi_k^r \det J \, d\xi d\eta = 2A \int_0^1 \xi^q d\xi \int_0^{1-\xi} (1 - \xi - \eta)^p \, \eta^r d\eta. \tag{9.122}$$

Integration by parts yields

$$W = 2A \int_0^1 \xi^q d\xi \left\{ (1 - \xi - \eta)^p \frac{\eta^{r+1}}{r+1} \bigg|_0^{1-\xi} + p \int_0^{1-\xi} (1 - \xi - \eta)^{p-1} \frac{\eta^{r+1}}{r+1} d\eta \right\}. \tag{9.123}$$

The first term in the braces vanishes.

Repeating integration by parts yields

$$
\begin{aligned}
W &= 2A \int_0^1 \xi^q d\xi \left\{ p! \frac{\eta^{r+p+1}}{(r+1)...(r+p+1)} \Big|_0^{1-\xi} \right\} \\
&= \frac{2Ap!r!}{(r+p+1)!} \int_0^1 \xi^q (1-\xi)^{r+p+1} d\xi,
\end{aligned}
\tag{9.124}
$$

and eventually

$$
\begin{aligned}
W &= \frac{2Ap!r!}{(r+p+1)!} \left\{ (p+r+1) \int_0^1 (1-\xi)^{r+p} \frac{\xi^{q+1}}{q+1} d\xi \right\} \\
&= \frac{2Ap!r!}{(r+p+1)!} \left\{ (p+r+1)! \int_0^1 \frac{\xi^{r+p+q+1}}{(q+1)...(r+p+q+1)} d\xi \right\} \\
&= \frac{r!p!q!}{(r+p+q+2)!} 2A.
\end{aligned}
\tag{9.125}
$$

• Suggested Readings

Many good textbooks on finite element analyses are available for engineering applications. Bathe and Wilson's (1976) *Numerical Methods in Finite Element Analysis* and Burnet's (1987) *Finite Element Analysis* are quite useful. For hydrogeological applications, we suggest Huyakorn and Pinder's (1983) *Computational Methods in Subsurface Flow* and Istok's (1989) *Groundwater Modeling by the Finite Element Method.* Lists of commercial software are available through the International Groundwater Modeling Center (igwmc@mines.colorado.edu). Some of them are public-domain programs sponsored by the U. S. Government.

Chapter 10

INVERSE PROBLEMS

In constructing a mathematical model, the problem of interest is usually generalized to produce a conceptual model, which is then formulated by relating model-defining parameters. The next step is to quantify a parameterized model in view of data and constraints. Given a set of parameter values for certain theories, hypotheses, or models, one can predict the outcome. If the predictions do not satisfactorily match the observations, the parameter values are modified iteratively to improve fitting, provided that the theoretical basis or model is valid. A model is an approximation to a real-world problem. Hence, models of a given situation are frequently revised to better match the data, to account for new observations, or to meet the improved understanding of the problem. Model revisions are usually accompanied by addition or deletion of parameters or modification of functional relations.

This chapter describes how to determine model parameter values. Models are assumed to be valid; the only unknowns are parameter values that define the models. Our modeling is restricted to determining parameter values and using these values to predict model behavior. Conceptualization of a problem and parameterization of a model are beyond the scope here.

Prediction based on a given set of parameter values is forward modeling. Determination of parameter values from observed data is inverse modeling. Forward modeling is an integral part of inverse modeling because inversion requires minimizing the discrepancy between prediction and observation. Inversion can be achieved in two ways. One, a modeler iteratively modifies parameter values until attaining best match. Such forward modeling is done usually with the aid of visual display on a monitor. The other, an inverse algorithm is adopted to automatically or semi-automatically obtain the parameter values from the observed data and an initial set of trial parameter values. The algorithm also provides an estimate of parameter uncertainty and resolution. This chapter is concerned with the second approach and introduces some inversion algorithms that are based on the least-squares method.

A well-posed inverse problem requires demonstration of the *existence* of the problem, the *uniqueness* in solution of the problem, the *stability* in algorithm used to achieve a solution, and the *efficiency* of delivering a final product. Generally, in view of observed data and our understanding of a real-world system, a problem is presumed to exist. For example, feeling strong ground motion suggests an earthquake or an explosion has happened. Detection of a contaminant plume in groundwater suggests contamination must have happened in the past. The question is how to relate the observation to earthquake mechanism and wave propagation or how to decipher the migration history of contaminant. A cause generally has an effect. Can an effect result from different causes? Is it unique in theory or model? Even if it is, have we counted and resolved all parameters that define a model? Inverse uniqueness has two levels of questions: the model itself and the model-defining parameters. The latter is related to the stability of a solution algorithm. How sensitive are parameters to uncertainty of observed data? Are the errors amplified during inversion? Is the inversion algorithm efficient in terms of ease of usage and cost of running the program? This chapter deals only with how to obtain parameter values for an accepted model from observed data.

10.1   Linear Inversion

Let the model parameter **p** and the output data **d** be related by

$$\boxed{\mathbf{d} = \mathbf{Gp}}, \tag{10.1}$$

where **G** is a generating function for a given system, a theoretical relation between model parameters and predictions. If **G** is independent of **p**, the system is linear and **G** is a linear operator. Numerically, **d** is represented by a column vector of n entries and **p** is another column vector of m entries; **G** is then an n by m matrix representing the *kernel* of the linear system. The task of inversion is to find **p** from observation $\mathbf{d}^{obs}$ for a given **G**. Let us begin with two examples to demonstrate the linear operation.

10.1.1 Example 1: A Linearizable System

• Least-Squares Method

Drawdown Δh at distance r from a Theis well which is being pumped at a constant rate of Q is given approximately by Jacob's semilog solution for large time t,

$$
\begin{aligned}
\Delta h \;\approx\; & \frac{Q}{4\pi T_{\mathrm{h}}} \left(-0.5772 - \ln \left[\frac{r^2 S}{4 T_{\mathrm{h}} t} \right] \right) \\
= \; & \frac{Q}{4\pi T_{\mathrm{h}}} \left(-0.5772 - \ln \left[\frac{r^2 S}{4 T_{\mathrm{h}}} \right] + \ln[t] \right),
\end{aligned}
\tag{10.2}
$$

where T_{h} is the transmissivity and S is the storativity. The goal of the inversion is to determine T_{h} and S. The relation between drawdown Δh and time t is logarithmic, not linear. However, if $\ln[t]$ is treated as an independent variable, the relation between Δh and $\ln[t]$ is linear.

Let the above relation be represented by

$$
y = a + bx,
\tag{10.3}
$$

where

$$
\begin{aligned}
y \;\equiv\; & \Delta h, \quad x \equiv \ln[t] \\
a \equiv \; & \frac{Q}{4\pi T_{\mathrm{h}}} \left(-0.5772 - \ln \left[\frac{r^2 S}{4 T_{\mathrm{h}}} \right] \right), \quad b \equiv \frac{Q}{4\pi T_{\mathrm{h}}}.
\end{aligned}
\tag{10.4}
$$

We are to determine two parameters a and b from the observed y at given x and, once they are determined, to predict y for any x.

Let there be n pairs of observations $(x, \, y^{\mathrm{obs}})$. The error or misfit between each pair of the prediction and observation is

$$
\delta_i = y_i - y_i^{\mathrm{obs}}, \quad i = 1, 2, ..., n.
\tag{10.5}
$$

We are to minimize the sum of the squares of errors

$$
\delta^2 = \sum_{i=1}^{n} \delta_i^2 = \sum_{i=1}^{n} \left(y_i - y_i^{\mathrm{obs}} \right)^2.
\tag{10.6}
$$

The minimum occurs where the partial derivatives of δ^2 with respect to parameters a and b vanish:

$$\frac{\partial \delta^2}{\partial a} = 2 \sum_{i=1}^{n} \left(y_i - y_i^{\text{obs}} \right) = 0,$$

$$\frac{\partial \delta^2}{\partial b} = 2 \sum_{i=1}^{n} \left(y_i - y_i^{\text{obs}} \right) x_i = 0. \tag{10.7}$$

Rewrite the above two *normal equations*

$$an + b \sum_{i=1}^{n} x_i = \sum_{i=1}^{n} y_i^{\text{obs}},$$

$$a \sum_{i=1}^{n} x_i + b \sum_{i=1}^{n} x_i^2 = \sum_{i=1}^{n} x_i y_i^{\text{obs}}. \tag{10.8}$$

Solving this system of two simultaneous equations with *Cramer's rule* yields

$$a = \frac{1}{D} \left(\sum_{i=1}^{n} x_i^2 \sum_{i=1}^{n} y_i^{\text{obs}} - \sum_{i=1}^{n} x_i \sum_{i=1}^{n} x_i y_i^{\text{obs}} \right),$$

$$b = \frac{1}{D} \left(n \sum_{i=1}^{n} x_i y_i^{\text{obs}} - \sum_{i=1}^{n} x_i \sum_{i=1}^{n} y_i^{\text{obs}} \right), \tag{10.9}$$

$$D = n \sum_{i=1}^{n} x_i^2 - \left(\sum_{i=1}^{n} x_i \right)^2, \quad D \neq 0.$$

This is the so-called least-squares method. We have assumed that $n > m = 2$. It is an *overdetermined* inverse problem in the sense of the least-squares method. Note that the y^{obs} and x are assumed to be free of error for the above derivation. The fact that argument t of the logarithmic function (in $x = \ln t$) cannot have a dimensional unit is also ignored if compatible units are made to render $r^2 S / 4t T_{\text{h}}$ dimensionless.

The transmissivity T_{h} is determined from the given Q and the computed slope b; and the storativity S is then computed from T_{h} and intercept a.

• Vector Representation

For notational simplicity, let us change the preceding series representations to vector form,

$$\mathbf{d} \equiv \mathbf{y} = \begin{bmatrix} y_1 & y_2 & \cdots & y_n \end{bmatrix}^T,$$

$$\mathbf{p} = \left\{ \begin{array}{c} p_1 \\ p_2 \end{array} \right\} \equiv \left\{ \begin{array}{c} a \\ b \end{array} \right\}, \tag{10.10}$$

where [...] is a row vector, {...} is a column vector, and superscript T designates transpose of a vector or matrix. Rewrite the relation $y_i = a + bx_i$ in matrix notation

$$\left\{ \begin{array}{c} d_1 \\ d_2 \\ .. \\ d_n \end{array} \right\} = \begin{bmatrix} 1 & x_1 \\ 1 & x_2 \\ .. & .. \\ 1 & x_n \end{bmatrix} \left\{ \begin{array}{c} p_1 \\ p_2 \end{array} \right\} \tag{10.11}$$

or equivalently, equation (10.1),

$$\mathbf{d} = \mathbf{Gp}, \tag{10.12}$$

where the n by m data matrix $\mathbf{G}$ is the data kernel of linear operator $\mathbf{G}$. The data kernel depends on the input data; it numerically represents the kernel but is not the kernel itself.

Define the misfit or error column vector as

$$\boldsymbol{\varepsilon} = \mathbf{d}^{\mathrm{obs}} - \mathbf{d} = \mathbf{d}^{\mathrm{obs}} - \mathbf{Gp}. \tag{10.13}$$

The sum of the squares of misfits is a scalar

$$\begin{aligned} \delta^2 &= \boldsymbol{\varepsilon}^T \boldsymbol{\varepsilon} = \left(\mathbf{d}^{\mathrm{obs}} - \mathbf{d} \right)^T \left(\mathbf{d}^{\mathrm{obs}} - \mathbf{d} \right) \\ &= \left(\mathbf{d}^{\mathrm{obs}} - \mathbf{Gp} \right)^T \left(\mathbf{d}^{\mathrm{obs}} - \mathbf{Gp} \right). \end{aligned} \tag{10.14}$$

Minimization of δ^2 with respect to $\mathbf{p}$ yields a column vector of m null entries

$$\frac{\partial \delta^2}{\partial \mathbf{p}} = 2\mathbf{G}^T \left(\mathbf{d}^{\mathrm{obs}} - \mathbf{d} \right) = \mathbf{0}, \tag{10.15}$$

which leads to a set of m normal equations,

$$\mathbf{G}^T \mathbf{Gp} = \mathbf{G}^T \mathbf{d}^{\mathrm{obs}}. \tag{10.16}$$

(See Section 8.1 for how to differentiate a vector or a matrix product.)

Now $\mathbf{G}^T\mathbf{G}$ is an $m \times m$ square matrix. Its inverse gives the least-squares parameter estimators

$$\boxed{\mathbf{p} = \left(\mathbf{G}^T\mathbf{G}\right)^{-1}\mathbf{G}^T\mathbf{d}^{\mathrm{obs}}}. \tag{10.17}$$

Note that if the approximation $\mathbf{d}^{\mathrm{obs}} \approx \mathbf{G}\mathbf{p}$ is taken, premultiplication of $\mathbf{G}^T$ could have yielded the $\mathbf{p}$ without going through the step of minimization. See Section 8.2 for using singular value decomposition to analyze the singularity of matrix $\mathbf{G}^T\mathbf{G}$.

10.1.2   Example 2: Partially Linearizable

Depending on the choice of parameters, a system may be linear or nonlinear as illustrated by a simple geophysical problem. The gravitational acceleration $\vec{g}_i$ at point i due to point mass s_j at point j is

$$\vec{g}_i = \gamma\frac{s_j}{r_{ij}^2}\,\widehat{e}_{i\to j}, \tag{10.18}$$

$$r_{ij}^2 = (x_j - x_i)^2 + (y_j - y_i)^2 + (z_j - z_i)^2,$$

where γ is the gravitational constant and the acceleration is in the direction from observation point i toward mass point j $(\widehat{e}_{i\to j})$, and r_{ij} is the distance between the observation and mass points. To avoid the complexity of performing vector operations, let us consider the vertical component of the acceleration only (z is downward positive). If there are m point masses, then the vertical component of acceleration at point i due to all point masses is

$$d_i = \sum_{j=1}^{m} g_i\frac{z_j - z_i}{r_{ij}} = \gamma\sum_{j=1}^{m}\frac{s_j\,(z_j - z_i)}{r_{ij}^3}. \tag{10.19}$$

(This formulation is consistent with the practice in gravity survey that uses a gravimeter to measure the vertical components of gravity only.)

Our inverse problem is to determine the mass s_j and its location (x_j, y_j, z_j) from n measurements d_i^{obs} of vertical acceleration (gravity). Equation (10.19) is linear in mass but nonlinear in location. Hence, as a whole it is nonlinear.

If the locations of mass points are given, the problem becomes linear. The case for two observations $(n = 2)$ and three mass points $(m = 3)$ is

$$\left\{ \begin{array}{c} d_1 \\ d_2 \end{array} \right\} = \left[\begin{array}{ccc} G_{11} & G_{12} & G_{13} \\ G_{21} & G_{22} & G_{23} \end{array} \right] \left\{ \begin{array}{c} s_1 \\ s_2 \\ s_3 \end{array} \right\}, \qquad (10.20)$$

where the data kernel is

$$G_{ij} = \gamma \frac{z_j - z_i}{r_{ij}^3}. \qquad (10.21)$$

Here we have three unknowns (s_1, s_2, s_3) and two equations. The inverse problem is underdetermined $(n < m)$.

Generally an underdetermined problem, $\mathbf{d} = \mathbf{Gp}$, has no solution. However, if the determinant

$$\det \left(\mathbf{G}^T \mathbf{G} \right) = \left| \left| \mathbf{G}^T \mathbf{G} \right| \right| = 0, \qquad (10.22)$$

then relative values of parameters $\mathbf{p}$ can be determined but the determination cannot come from the inversion of a singular $\mathbf{G}^T \mathbf{G}$, as given in equation (10.17).

10.2 Constrained Linear Inversion

10.2.1 Constraint of Parameters

Inversion of a linear system

$$\boxed{ \underset{n \times 1}{\mathbf{d}} = \underset{n \times m}{\mathbf{G}} \underset{m \times 1}{\mathbf{p}} } \qquad (10.23)$$

can be achieved by several ways. Here, by means of least-squares method, we are to find parameter $\mathbf{p}$ from the minimization of

$$\boxed{ \chi^2 = \left(\mathbf{d}^{\mathrm{obs}} - \mathbf{d} \right)^T \mathbf{C}_d^{-1} \left(\mathbf{d}^{\mathrm{obs}} - \mathbf{d} \right) + \left(\mathbf{p}^{\mathrm{prior}} - \mathbf{p} \right)^T \mathbf{C}_p^{-1} \left(\mathbf{p}^{\mathrm{prior}} - \mathbf{p} \right) } \qquad (10.24)$$

(Tarantola, 1987). The first term of χ^2 (chi square, a scalar) represents the misfit between prediction and observation, and the second term represents prior knowledge $\mathbf{p}^{\mathrm{prior}}$ or preprocessing constraint. This χ^2 has been called objective function, merit function, or sum of misfit squared. (Some users may define a χ^2 by multiplying the right-hand side of the above relation with a factor of $1/2$.)

The $n \times n$ matrix $\mathbf{C}_d$ is the covariance matrix of the observed data. Its diagonal entries are the squares of standard deviations and its off-diagonal entries represent correlations of data. If the data are independent of one another, then the data covariance matrix is diagonal. $\mathbf{C}_d$ is a measure of experimental as well as theoretical uncertainties.

Similarly, the $m \times m$ covariance matrix $\mathbf{C}_p$ represents the uncertainty in model parameters and is usually assumed to be diagonal. (All $\mathbf{p}$ are independent of one another; otherwise the least-squares method based on minimization of χ^2 with respect to $\mathbf{p}$ is theoretically invalid.) The correlation or cross-correlation coefficients are

$$r_{ij} = \frac{C_{ij}}{\sqrt{C_{ii}C_{jj}}} \qquad (10.25)$$

for either the data or the parameters.

The inclusion of $\mathbf{C}_d^{-1}$ and $\mathbf{C}_p^{-1}$ in the misfit function renders the misfit and constraint dimensionless so that the two can be added together to give a dimensionless χ^2. Note that the δ^2 in Section 10.1 is permissible only if all entries of $\mathbf{d}$ have the same physical units. Also, the two terms in χ^2 can be viewed, respectively, as the normalized misfit between observation and prediction, and between model and preprocessing parameters.

• **Maximum Likelihood Estimates**

Minimization of χ^2 with respect to the parameter vector yields a set of m simultaneous equations

$$\frac{\partial \chi^2}{\partial \mathbf{p}} = \mathbf{G}^T \mathbf{C}_d^{-1} \left(\mathbf{d}^{\text{obs}} - \mathbf{Gp} \right) + \mathbf{C}_p^{-1} \left(\mathbf{p}^{\text{prior}} - \mathbf{p} \right) = 0. \qquad (10.26)$$

Rearranging the terms yields our first algorithm for linear inversion

$$\boxed{\mathbf{p} = \mathbf{C}_{p'} \left(\mathbf{G}^T \mathbf{C}_d^{-1} \mathbf{d}^{\text{obs}} + \mathbf{C}_p^{-1} \mathbf{p}^{\text{prior}} \right)}, \qquad (10.27)$$

where

$$\mathbf{C}_{p'}^{-1} = \mathbf{C}_p^{-1} + \mathbf{G}^T \mathbf{C}_d^{-1} \mathbf{G} \qquad (10.28)$$

and $\mathbf{C}_{p'}$ is the postprocessed covariance matrix of parameters. $\mathbf{C}_{p'}$ represents parameter uncertainties (diagonal entries) and correlation or interdependence of parameters (off-diagonal entries). This algorithm requires inversion of an $m \times m$ matrix $\mathbf{C}_{p'}$.

Implicit in the above derivation is the assumption that the misfits or errors $\left(\mathbf{d}^{\mathrm{obs}} - \mathbf{Gp}\right)/\mathbf{C}_d^{1/2}$ and $\left(\mathbf{p}^{\mathrm{prior}} - \mathbf{p}\right)/\mathbf{C}_p^{1/2}$ have Gaussian distributions with zero mean and unit standard deviation. Their probability density functions are proportional to

$$\exp\left[-\frac{\left(\mathbf{d}^{\mathrm{obs}} - \mathbf{Gp}\right)^T \left(\mathbf{d}^{\mathrm{obs}} - \mathbf{Gp}\right)}{2\mathbf{C}_d}\right]$$

and

$$\exp\left[-\frac{\left(\mathbf{p}^{\mathrm{prior}} - \mathbf{p}\right)^T \left(\mathbf{p}^{\mathrm{prior}} - \mathbf{p}\right)}{2\mathbf{C}_p}\right],$$

respectively. $\mathbf{C}_d^{1/2}$ is a diagonal matrix consisting of the square roots of the diagonal entries of the diagonal $\mathbf{C}_d$. Thus $\mathbf{C}_d^{1/2}$ represents the standard deviations for individual observations. The joint probability density function of the two is the product of the two probability density functions. Hence, the probability density function as a whole is proportional to $\exp[-\chi^2/2]$.

Minimization of χ^2 is equivalent to maximizing $\exp[-\chi^2/2]$ and the joint probability density function. The results of minimization are the maximum likelihood central estimates of model parameters, which are the least-squares parameters.

It is noted that LSQ (least-squares method) deals with normally distributed random errors or misfits. It does not minimize systematic errors.

• Alternative Formulas

The central estimators in equation (10.27) can be rewritten alternatively. Since

$$
\begin{aligned}
\mathbf{p} &= \mathbf{C}_{p'}\left\{\mathbf{G}^T\mathbf{C}_d^{-1}\mathbf{d}^{\mathrm{obs}} + \left(\mathbf{G}^T\mathbf{C}_d^{-1}\mathbf{G} + \mathbf{C}_p^{-1} - \mathbf{G}^T\mathbf{C}_d^{-1}\mathbf{G}\right)\mathbf{p}^{\mathrm{prior}}\right\} \\
&= \mathbf{C}_{p'}\left\{\mathbf{G}^T\mathbf{C}_d^{-1}\mathbf{d}^{\mathrm{obs}} + \left(\mathbf{C}_{p'}^{-1} - \mathbf{G}^T\mathbf{C}_d^{-1}\mathbf{G}\right)\mathbf{p}^{\mathrm{prior}}\right\} \\
&= \mathbf{C}_{p'}\mathbf{G}^T\mathbf{C}_d^{-1}\mathbf{d}^{\mathrm{obs}} + \left(\mathbf{I} - \mathbf{C}_{p'}\mathbf{G}^T\mathbf{C}_d^{-1}\mathbf{G}\right)\mathbf{p}^{\mathrm{prior}},
\end{aligned}
\tag{10.29}
$$

we obtain the second algorithm

$$\boxed{\mathbf{p} = \mathbf{p}^{\mathrm{prior}} + \mathbf{C}_{p'}\mathbf{G}^T\mathbf{C}_d^{-1}\left(\mathbf{d}^{\mathrm{obs}} - \mathbf{Gp}^{\mathrm{prior}}\right)},\tag{10.30}$$

where $\mathbf{I}$ is the identity matrix.

The preceding two linear-inversion formulas require inverting an $m \times m$ matrix. Both are suitable for solving overdetermined problems. For an underdetermined problem ($n < m$), however, we prefer using a third formula that inverts an $n \times n$ matrix.

Starting from the trivial identity,

$$\mathbf{G}^T \mathbf{C}_d^{-1} \mathbf{G} \mathbf{C}_p \mathbf{G}^T + \mathbf{G}^T = \mathbf{G}^T \mathbf{C}_d^{-1} \mathbf{G} \mathbf{C}_p \mathbf{G}^T + \mathbf{G}^T \quad \text{or}$$
$$\mathbf{G}^T \mathbf{C}_d^{-1} \left(\mathbf{G} \mathbf{C}_p \mathbf{G}^T + \mathbf{C}_d \right) = \left(\mathbf{G}^T \mathbf{C}_d^{-1} \mathbf{G} + \mathbf{C}_p^{-1} \right) \mathbf{C}_p \mathbf{G}^T, \quad (10.31)$$

one obtains

$$\mathbf{C}_{p'} \mathbf{G}^T \mathbf{C}_d^{-1} = \mathbf{C}_p \mathbf{G}^T \mathbf{A}, \quad \mathbf{A} = \mathbf{G} \mathbf{C}_p \mathbf{G}^T + \mathbf{C}_d. \quad (10.32)$$

Substituting this relation into equation (10.30) yields the third algorithm for linear inversion (Tarantola, 1987, p.70)

$$\boxed{\mathbf{p} = \mathbf{p}^{\text{prior}} + \mathbf{C}_p \mathbf{G}^T \mathbf{A}^{-1} \left(\mathbf{d}^{\text{obs}} - \mathbf{G} \mathbf{p}^{\text{prior}} \right)}. \quad (10.33)$$

- **Resolution Matrices**

Assuming $\mathbf{d}^{\text{obs}} = \mathbf{G} \mathbf{p}^{\text{true}}$ and rewriting the third algorithm, equation (10.33), yields

$$\begin{aligned} \mathbf{p} &= \mathbf{C}_p \mathbf{G}^T \mathbf{A}^{-1} \mathbf{d}^{\text{obs}} + \left(\mathbf{I} - \mathbf{C}_p \mathbf{G}^T \mathbf{A}^{-1} \mathbf{G} \right) \mathbf{p}^{\text{prior}} \\ &= \mathbf{R}_p \mathbf{p}^{\text{true}} + \left(\mathbf{I} - \mathbf{R}_p \right) \mathbf{p}^{\text{prior}}, \end{aligned} \quad (10.34)$$

where

$$\boxed{\mathbf{R}_p = \mathbf{C}_p \mathbf{G}^T \mathbf{A}^{-1} \mathbf{G} = \mathbf{C}_{p'} \mathbf{G}^T \mathbf{C}_d^{-1} \mathbf{G} = \mathbf{I} - \mathbf{C}_{p'} \mathbf{C}_p^{-1}} \quad (10.35)$$

is the *parameter resolution matrix*. If $\mathbf{R}_p = \mathbf{I}$, the model parameters are perfectly resolved, i.e., $\mathbf{p} = \mathbf{p}^{\text{true}}$. If $\mathbf{p} = \mathbf{p}^{\text{prior}}$, then $\mathbf{R}_p = 0$ and $\mathbf{C}_{p'} = \mathbf{C}_p$.

The prediction based on equation (10.33) is

$$\begin{aligned} \mathbf{d} &= \mathbf{G} \mathbf{p} \\ &= \mathbf{G} \mathbf{C}_p \mathbf{G}^T \mathbf{A}^{-1} \mathbf{d}^{\text{obs}} + \left(\mathbf{I} - \mathbf{G} \mathbf{C}_p \mathbf{G}^T \mathbf{A}^{-1} \right) \mathbf{G} \mathbf{p}^{\text{prior}} \\ &= \mathbf{R}_d \mathbf{d}^{\text{obs}} + \left(\mathbf{I} - \mathbf{R}_d \right) \mathbf{G} \mathbf{p}^{\text{prior}}, \end{aligned} \quad (10.36)$$

where

$$\boxed{\mathbf{R}_d = \mathbf{G}\mathbf{C}_p\mathbf{G}^T\mathbf{A}^{-1}}. \tag{10.37}$$

If $\mathbf{R}_d = \mathbf{I}$, then $\mathbf{d} = \mathbf{d}^{\text{obs}}$, and a perfect match is achieved. $\mathbf{R}_d$ is thus the *data resolution matrix*. $\mathbf{R}_d$ bears information on the relative data independence.

If the second algorithm for $\mathbf{p}$ in equation (10.30) is used, then

$$\begin{aligned}
\mathbf{d} &= \mathbf{G}\mathbf{p} \\
&= \mathbf{G}\mathbf{C}_{p'}\mathbf{G}^T\mathbf{C}_d^{-1}\mathbf{d}^{\text{obs}} + \left(\mathbf{I} - \mathbf{G}\mathbf{C}_{p'}\mathbf{G}^T\mathbf{C}_d^{-1}\right)\mathbf{G}\mathbf{p}^{\text{prior}}. \tag{10.38}
\end{aligned}$$

Comparing the last two relations for $\mathbf{d} = \mathbf{G}\mathbf{p}$, it is clear that the data resolution matrix is also related to

$$\begin{aligned}
\mathbf{R}_d &= \mathbf{G}\mathbf{C}_{p'}\mathbf{G}^T\mathbf{C}_d^{-1} = \mathbf{C}_{d'}\mathbf{C}_d^{-1}, \tag{10.39} \\
\mathbf{C}_{d'} &= \mathbf{G}\mathbf{C}_{p'}\mathbf{G}^T,
\end{aligned}$$

where $\mathbf{C}_{d'}$ is the postprocessed covariance matrix of data, providing a measure of the goodness of fitting. If $\mathbf{C}_{d'} = \mathbf{C}_d$, then $\mathbf{R}_d = \mathbf{I}$, and a perfect match is achieved ($\mathbf{d} = \mathbf{d}^{\text{obs}}$).

• **Partial Constraint**

If a preprocessing constraint is not imposed, we have the unconstrained least-squares solution

$$\mathbf{p} = \left(\mathbf{G}^T\mathbf{C}_d^{-1}\mathbf{G}\right)^{-1}\mathbf{G}^T\mathbf{C}_d^{-1}\mathbf{d}^{\text{obs}}, \tag{10.40}$$

which is equivalent to the unnormalized least-squares solution given by equation (10.17). It follows that constraints do not have to be imposed on all parameters. Partial constraints can be imposed by modifying equation (10.27) to obtain

$$\boxed{\mathbf{p} = \mathbf{C}_{p'}\left(\mathbf{G}^T\mathbf{C}_d^{-1}\mathbf{d}^{\text{obs}} + \mathbf{D}\mathbf{C}_p^{-1}\mathbf{p}^{\text{prior}}\right)}, \tag{10.41}$$

where

$$\mathbf{C}_{p'}^{-1} = \mathbf{G}^T\mathbf{C}_d^{-1}\mathbf{G} + \mathbf{D}\mathbf{C}_p^{-1} \tag{10.42}$$

and $\mathbf{D}$ is an $m \times m$ diagonal matrix of which the entries for the unconstrained parameters are zero and for the fully constrained parameters are one.

10.2.2 More on Biased Linear Inversions

If there is prior (biased) information $\mathbf{u}$ about parameters $\mathbf{p}$,

$$\mathbf{Bp} = \mathbf{u}, \tag{10.43}$$

we can define a new scalar misfit function

$$\delta^2 = \frac{1}{2} \left(\mathbf{d}^{\mathrm{obs}} - \mathbf{Gp}\right)^T \mathbf{C}_d^{-1} \left(\mathbf{d}^{\mathrm{obs}} - \mathbf{Gp}\right) + \frac{\beta}{2} \left(\mathbf{Bp} - \mathbf{u}\right)^T \mathbf{C}_u^{-1} \left(\mathbf{Bp} - \mathbf{u}\right), \tag{10.44}$$

where $\mathbf{B}$ is a matrix operator, β is a dimensionless *Lagrange multiplier,* and $\mathbf{C}_u$ is a diagonal covariance matrix that is needed to make the constraint dimensionless.

Minimization of δ^2 with respect to $\mathbf{p}$ yields

$$\frac{\partial \delta^2}{\partial \mathbf{p}} = -\mathbf{G}^T \mathbf{C}_d^{-1} \left(\mathbf{d}^{\mathrm{obs}} - \mathbf{Gp}\right) + \beta \mathbf{B}^T \mathbf{C}_u^{-1} \left(\mathbf{Bp} - \mathbf{u}\right) = 0. \tag{10.45}$$

Rearranging the terms yields the normal equation which, in turn, gives the biased inversion

$$\boxed{\mathbf{p} = \left(\mathbf{G}^T \mathbf{C}_d^{-1} \mathbf{G} + \beta \mathbf{B}^T \mathbf{C}_u^{-1} \mathbf{B}\right)^{-1} \left(\mathbf{G}^T \mathbf{C}_d^{-1} \mathbf{d}^{\mathrm{obs}} + \beta \mathbf{B}^T \mathbf{C}_u^{-1} \mathbf{u}\right)}. \tag{10.46}$$

Following are two examples for the usage of $\mathbf{B}$.

• **Case 1, $\mathbf{B} = \mathbf{I}$**

If $\mathbf{B}$ is an identity matrix, the biased inversion is simplified to

$$\mathbf{p} = \left(\mathbf{G}^T \mathbf{C}_d^{-1} \mathbf{G} + \beta \mathbf{C}_u^{-1}\right)^{-1} \left(\mathbf{G}^T \mathbf{C}_d^{-1} \mathbf{d}^{\mathrm{obs}} + \beta \mathbf{C}_u^{-1} \mathbf{u}\right). \tag{10.47}$$

This is identical with the maximum likelihood estimator in equation (10.27) if $\mathbf{u} = \mathbf{p}^{\mathrm{prior}}$ and $\mathbf{C}_u = \mathbf{C}_p$, except for the addition of a Lagrange multiplier β. This estimator is a damped least-squares solution. It adds positive constraints to the eigenvalues of $\mathbf{G}^T \mathbf{C}_d^{-1} \mathbf{G}$ for stabilization of matrix inversion. If $\mathbf{C}_d = \mathbf{I}$, $\mathbf{C}_p = \mathbf{I}$, and $\mathbf{u} = \mathbf{0}$, the estimator becomes

$$\mathbf{p} = \left(\mathbf{G}^T \mathbf{G} + \beta \mathbf{I}\right)^{-1} \mathbf{G}^T \mathbf{d}^{\mathrm{obs}}, \tag{10.48}$$

which constitutes the *Marquardt* (1970) solution.

• **Case 2, Smoothness**

If smoothness is a desirable constraint, it can be implemented by using **B** as a first difference operator. For example, the following constraint is imposed to smooth parameters p_3 through p_7,

$$
\begin{bmatrix}
1 & -1 & & & \\
& 1 & -1 & & \\
& & 1 & -1 & \\
& & & 1 & -1
\end{bmatrix}
\begin{Bmatrix}
p_3 \\
p_4 \\
p_5 \\
p_6 \\
p_7
\end{Bmatrix}
=
\begin{Bmatrix}
0 \\
0 \\
0 \\
0
\end{Bmatrix}.
\tag{10.49}
$$

This 4×5 coefficient matrix must be augmented with null row and column vectors to match the dimension of the parameter vector.

10.2.3   Goodness of Fit

There are several ways to evaluate the goodness of fit between predictions and observations. None can be used as the sole criterion for judging the acceptance of a model or a set of model parameters.

• **Chi-Square Density Distribution**

The χ^2 defined in equation (10.24) has a chi-square probability density distribution with ν degrees of freedom. The number of degrees of freedom is equal to the number n of data points (the dimension of the data vector) minus the number m_c of constraints ($m_c \leq m$, the dimension of the parameter vector), i.e., $\nu = n - m_c$. The probability density function is

$$
P_{\chi^2}[\chi^2] = \frac{(\chi^2)^{(\nu-2)/2}}{2^{\nu/2}\Gamma[\nu/2]} \exp\left[-\frac{\chi^2}{2}\right],
\tag{10.50}
$$

where the Gamma function,

$$
\Gamma[z] = \int_0^\infty t^{z-1} e^{-t} dt,
\tag{10.51}
$$

is related to factorial

$$
\Gamma[n+1] = n!
\tag{10.52}
$$

for an integer argument $n + 1$.

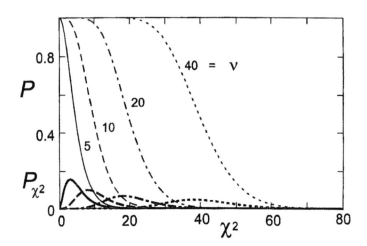

Figure 10.1: Chi-square probability density function $P_{\chi^2}[\chi^2]$ (dark curves) and probability $P[\chi^2]$ (light curves) for different degrees of freedom ν. Note that the probability density approaches the Gaussian distribution as ν increases and that the probability is approximately 0.5 when $\chi^2 = \nu$.

Figure 10.1 depicts the probability density P_{χ^2} for four different degrees of freedom ν. The distribution behaves like a normal distribution as ν increases. At large ν the mean of the distribution is ν and the distribution has a variance of 2ν.

- **Chi Square Probability**

The probability that any random set of n data points would yield an error as measured by chi square as large as or larger than χ^2 is

$$P[\chi^2] = \int_{\chi^2}^{\infty} P_{y^2}\, dy^2. \tag{10.53}$$

This function can be shown to be the incomplete gamma function Q of Press et al. (1992)

$$P = Q[\nu/2, \chi^2/2], \tag{10.54}$$

where

$$Q[a, y] = \frac{1}{\Gamma[a]} \int_y^{\infty} e^{-t} t^{a-1} dt. \tag{10.55}$$

Numerically, it is easier to evaluate $1 - Q$ or equivalently to integrate from 0 to y instead of from y to ∞. Figure 10.1 depicts the probability as a function of χ^2 for different values of ν.

• **Misfit**

After the model parameters are determined, the variance

$$\sigma^2 = \frac{\left(\mathbf{d}^{obs} - \mathbf{Gp}\right)^T \left(\mathbf{d}^{obs} - \mathbf{Gp}\right)}{n - m} \tag{10.56}$$

has been frequently computed to gauge the quality of fit. Since finding $\mathbf{p}$ is the target of inverse modeling, estimation of σ could be problematic especially when $\mathbf{C}_d$ needs to be constructed from the standard deviations for individual measurements. Using this postprocessed σ^2 as feedback, the recomputed χ^2 is expected to be $n - m$ (neglecting the misfit term for parameters). However, to redetermine $\mathbf{p}$ through feedback could violate the assumption that the parameters are independent of one another for the least-squares method.

If σ^2 is given and the data are independent, then the covariance matrix $\mathbf{C}_d$ is diagonal and its entries are composed of σ^2. Assuming each data point has the same σ, the expected mean value of χ^2 equals the number of degrees of freedom at large ν and its variance is 2ν. If $\chi^2 = \nu = n - m$ or when the reduced chi-square

$$\chi_\nu^2 = \chi^2 / \nu \tag{10.57}$$

is one, the chi-square probability P is about 0.5 and the model is said to have a good fit.

If $\chi_\nu^2 \gg 1$, the model underfits the data and the model may be defective. Large χ_ν^2 may also originate from underestimates of $\mathbf{C}_d$ which should include both experimental and model errors but are difficult to estimate for field conditions. A third possibility is that the misfit is not normally distributed; for example, a few erratic outliers can cause large χ_ν^2 and low P.

On the other hand, if $\chi_\nu^2 \ll 1$, the model overfits the data and the agreement between model and data is too good to be true, suggesting that the variances used in $\mathbf{C}_d$ or $\mathbf{C}_p$ may have been overestimated or perhaps the good fit is an artifact of computing.

The size of misfit, χ^2 or χ_ν^2, depends on the assigned values of $\mathbf{C}_d$ or $\mathbf{C}_p$. Frequently both values are not known in advance and guesstimates are used. If σ^2 is computed according to equation (10.56) following the determination of parameters, the independence of parameters could be compromised and the usage of chi-square probability as a measure of the goodness-of-fit for the model could be questionable. Therefore the chi-square probability should be carefully assessed; it cannot be used as the sole measure for the goodness of fit. Visual inspection of the misfit distribution is often desirable to spot potential bias in fitting.

• RMS

The root mean square error

$$\text{rms} = \left[\frac{\left(\mathbf{d}^{\text{obs}} - \mathbf{Gp} \right)^T \left(\mathbf{d}^{\text{obs}} - \mathbf{Gp} \right)}{n - m} \right]^{1/2} \tag{10.58}$$

is commonly used to represent the error of fitting. This rms is the σ for all parameter fitting, which is computed after parameters $\mathbf{p}$ have been determined. Note that the relation is applicable if all $\mathbf{d}^{\text{obs}}$ are dimensionally consistent; otherwise, $\mathbf{d}^{\text{obs}} - \mathbf{Gp}$ should be normalized by the respective errors (or standard deviations) associated with individual $\mathbf{d}^{\text{obs}}$.

If standard deviations σ_i are available, then we can define a dimensionless rms,

$$\text{rms} = \left[\frac{1}{\nu} \sum_{i=1}^{n} \left(d_i^{\text{obs}} - d_i \right)^2 / \sigma_i^2 \right]^{1/2}. \tag{10.59}$$

If the term $\left(\mathbf{p}^{\text{prior}} - \mathbf{p} \right)^T \mathbf{C}_p^{-1} \left(\mathbf{p}^{\text{prior}} - \mathbf{p} \right)$ is ignored in the computation of chi square, the reduced chi square is

$$\chi_\nu^2 = \frac{\chi^2}{\nu} = \frac{\sum_{i=1}^{n} \left(d_i^{\text{obs}} - d_i \right)^2 / \sigma_i^2}{\nu} = (\text{rms})^2. \tag{10.60}$$

Thus, if $\chi_\nu^2 = (\text{rms})^2 \approx 1$, the misfit is comparable in size to the standard deviation (e.g., average of σ_i) and the chi-square probability is 0.5 for large ν.

Another form

$$\text{rms} = \left[\frac{1}{n - m} \sum_i \sum_j \left(1 - \frac{G_{ij} p_j}{d_i^{\text{obs}}} \right)^2 \right]^{1/2} \tag{10.61}$$

has also been used. This computation is independent of units used for $\mathbf{d}^{\text{obs}}$.

If $\mathbf{d}^{\text{obs}} - \mathbf{d}$ is presented in a log-log or semilog plot, the choice of $\mathbf{p}$ at the minimum of rms for all trial sets of $\mathbf{p}$ may not give the best visible fit. Instead, minimization of

$$\text{rms} = \left[\frac{1}{n-m} \sum_i \left(\log d_i^{\text{pred}} - \log d_i^{\text{obs}} \right)^2 \right]^{1/2} \tag{10.62}$$

can usually provide a better visible fit but this choice of $\mathbf{p}$ does not necessarily give the minimum of $\sum (d_i^{\text{pred}} - d_i^{\text{obs}})^2$. However, programming for minimization of the logarithmic rms is more difficult than programming the linear curve-fitting methods. It is recommended only for fine-tuning $\mathbf{p}$ by scanning over small ranges of $\mathbf{p}$.

10.3   Nonlinear Inversion

A forward nonlinear problem can be represented by

$$\boxed{\mathbf{d} = \mathbf{g}[\mathbf{p}]}, \tag{10.63}$$

where $\mathbf{g}[\mathbf{p}]$ is a function of parameters $\mathbf{p}$. Because $\mathbf{p}$ cannot be factored out of function $\mathbf{g}$, iteration is needed to solve for $\mathbf{p}$. Our inversion will be restricted to problems of which $\mathbf{g}$ at iteration step $k+1$ can be represented by the first-order Taylor series around the set of parameters determined at step k

$$g_i^{k+1} \approx g_i^k + \sum_j G_{ij}^k \left(p_j^{k+1} - p_j^k \right), \quad G_{ij}^k = \left. \frac{\partial g_i}{\partial p_j} \right|_{\mathbf{p}^k}, \tag{10.64}$$

where $\mathbf{G}_k$ is an $n \times m$ matrix and n is the number of data $\mathbf{d}$ and m is the number of parameters $\mathbf{p}$ (see also Section 8.1 for definition of $\mathbf{G}$). In vector form

$$\underset{n \times 1}{\mathbf{g}_{k+1}} \approx \underset{n \times 1}{\mathbf{g}_k} + \underset{n \times m}{\mathbf{G}_k} \underset{m \times 1}{(\mathbf{p}_{k+1} - \mathbf{p}_k)}, \tag{10.65}$$

where an underlabel denotes the dimension of the underlabeled variable. Note that we have changed the notation by using subscript k to designate iteration step and reserved the usage of superscripts for other purposes.

Now, define the misfit or objective function at step k by the quadratic

$$S_k = \frac{1}{2}\left\{\left(\mathbf{d}^{\mathrm{obs}} - \mathbf{g}_k\right)^T \mathbf{C}_d^{-1}\left(\mathbf{d}^{\mathrm{obs}} - \mathbf{g}_k\right)\right.$$
$$\left. + \left(\mathbf{p}^{\mathrm{prior}} - \mathbf{p}_k\right)^T \mathbf{C}_p^{-1}\left(\mathbf{p}^{\mathrm{prior}} - \mathbf{p}_k\right)\right\} \tag{10.66}$$

which is similar in form to the χ^2 for linear problems, equation (10.24), except for the factor of $1/2$.

10.3.1   Gauss-Newton Method

If S_k is the misfit at iteration step k, then the misfit at the next iteration is

$$\underset{1\times 1}{S_{k+1}} = S_k + \underset{1\times m}{\left(\frac{\partial S}{\partial \mathbf{p}}\right)_k^T} \underset{m\times 1}{\underbrace{\left(\mathbf{p}_{k+1} - \mathbf{p}_k\right)}} . \tag{10.67}$$

We seek to minimize the misfit at step $k + 1$, i.e.,

$$\underset{m\times 1}{\frac{\partial S_{k+1}}{\partial \mathbf{p}}} = \underset{m\times 1}{\frac{\partial S_k}{\partial \mathbf{p}}} + \underset{m\times 1}{\frac{\partial}{\partial \mathbf{p}}}\left[\underset{1\times m}{\left(\frac{\partial S}{\partial \mathbf{p}}\right)_k^T}\underset{m\times 1}{\underbrace{\left(\mathbf{p}_{k+1} - \mathbf{p}_k\right)}}\right] = \mathbf{0}. \tag{10.68}$$

Rearrange the terms to give

$$\mathbf{p}_{k+1} = \mathbf{p}_k - \underset{m\times m}{\underbrace{\left[\frac{\partial}{\partial \mathbf{p}}\left(\frac{\partial S}{\partial \mathbf{p}}\right)^T\right]_k^{-1}}}\underset{m\times 1}{\left(\frac{\partial S_k}{\partial \mathbf{p}}\right)_k} . \tag{10.69}$$

According to the definition of S_k, the first derivative (the gradient $\boldsymbol{\gamma}_k$) is

$$\frac{\partial S_k}{\partial \mathbf{p}} = \underset{m\times 1}{\boldsymbol{\gamma}_k}$$
$$= -\mathbf{G}_k^T \mathbf{C}_d^{-1}\left(\mathbf{d}^{\mathrm{obs}} - \mathbf{g}_k\right) - \mathbf{C}_p^{-1}\left(\mathbf{p}^{\mathrm{prior}} - \mathbf{p}_k\right). \tag{10.70}$$

The second derivative of S_k (the Hessian $\mathbf{H}_k$) is

$$\frac{\partial}{\partial \mathbf{p}}\left(\frac{\partial S}{\partial \mathbf{p}}\right)^T \approx \underset{m\times m}{\mathbf{H}_k}$$
$$= \mathbf{C}_p^{-1} + \mathbf{G}_k^T \mathbf{C}_d^{-1}\mathbf{G}_k = \mathbf{C}_{p'k}^{-1} \tag{10.71}$$

(by neglecting the $\partial G/\partial \mathbf{p}$ term which is composed of second derivatives in $\mathbf{g}$). Note that this $\mathbf{C}_{p'k}$ is equivalent to covariance matrix $\mathbf{C}_{p'}$ of postprocessed parameters for linear problems.

Substituting the gradient and the Hessian of S_k into equation (10.69) yields an iteration algorithm by the Gauss-Newton method,

$$\boxed{\mathbf{p}_{k+1} = \mathbf{p}_k + \mathbf{C}_{p'k} \left\{ \mathbf{G}_k^T \mathbf{C}_d^{-1} \left(\mathbf{d}^{\text{obs}} - \mathbf{g}_k \right) + \mathbf{C}_p^{-1} \left(\mathbf{p}^{\text{prior}} - \mathbf{p}_k \right) \right\}} \qquad (10.72)$$

An alternative iteration scheme can be obtained by nullifying the first derivative at step k. Let $\partial S_k/\partial \mathbf{p} = 0$ in equation (10.70), then

$$\mathbf{C}_p^{-1} \left(\mathbf{p}^{\text{prior}} - \mathbf{p}_k \right) = -\mathbf{G}_k^T \mathbf{C}_d^{-1} \left(\mathbf{d}^{\text{obs}} - \mathbf{g}_k \right). \qquad (10.73)$$

Adding $\mathbf{G}_k^T \mathbf{C}_d^{-1} \mathbf{G}_k (\mathbf{p}^{\text{prior}} - \mathbf{p}_k)$ to both sides and regrouping the terms yields

$$\mathbf{p}_k = \mathbf{p}^{\text{prior}} + \mathbf{C}_{p'k} \mathbf{G}_k^T \mathbf{C}_d^{-1} \left\{ \mathbf{d}^{\text{obs}} - \mathbf{g}_k - \mathbf{G}_k \left(\mathbf{p}^{\text{prior}} - \mathbf{p}_k \right) \right\}. \qquad (10.74)$$

Since both sides contain the unknown $\mathbf{p}_k$, an iteration scheme is adopted to obtain the second formula (see Section 8.3.3)

$$\boxed{\mathbf{p}_{k+1} = \mathbf{p}^{\text{prior}} + \mathbf{C}_{p'k} \mathbf{G}_k^T \mathbf{C}_d^{-1} \left\{ \mathbf{d}^{\text{obs}} - \mathbf{g}_k - \mathbf{G}_k (\mathbf{p}^{\text{prior}} - \mathbf{p}_k) \right\}} \qquad (10.75)$$

Both of the above iteration schemes require inversion of an $m \times m$ matrix. A third one requiring inversion of an $n \times n$ matrix can be obtained by using the identity

$$\mathbf{C}_{p'k} \mathbf{G}_k^T \mathbf{C}_d^{-1} = \mathbf{C}_p \mathbf{G}_k^T \mathbf{A}_k^{-1}, \quad \mathbf{A}_k = \mathbf{G}_k \mathbf{C}_p \mathbf{G}_k^T + \mathbf{C}_d \qquad (10.76)$$

that was used to derive equation (10.33). Accordingly

$$\boxed{\mathbf{p}_{k+1} = \mathbf{p}^{\text{prior}} + \mathbf{C}_p \mathbf{G}_k^T \mathbf{A}^{-1} \left[\mathbf{d}^{\text{obs}} - \mathbf{g}_k - \mathbf{G}_k (\mathbf{p}^{\text{prior}} - \mathbf{p}_k) \right]} \qquad (10.77)$$

- **Remark**

All iterations require initializing $\mathbf{p}_1$, for example, let $\mathbf{p}_1 = \mathbf{p}^{\text{prior}}$.

10.3.2   Resolution

If the model is not defective and the data are free of error, then $\mathbf{d}^{\text{obs}} = \mathbf{g}^{\text{true}}$ and equation (10.77) at the final iteration becomes

$$
\begin{aligned}
\mathbf{p}_{k+1} &= \mathbf{p}^{\text{prior}} + \mathbf{C}_p \mathbf{G}_k^T \left(\mathbf{G}_k \mathbf{C}_p \mathbf{G}_k^T + \mathbf{C}_d \right)^{-1} \mathbf{G}_k (\mathbf{p}_k - \mathbf{p}^{\text{prior}}) \\
&= \mathbf{p}^{\text{prior}} + \mathbf{C}_{p'k} \mathbf{G}_k^T \mathbf{C}_d^{-1} \mathbf{G}_k (\mathbf{p}_k - \mathbf{p}^{\text{prior}}) \\
&= \mathbf{p}^{\text{prior}} + \mathbf{C}_{p'k} (\mathbf{C}_{p'k}^{-1} - \mathbf{C}_p^{-1})(\mathbf{p}_k - \mathbf{p}^{\text{prior}}) \\
&= \mathbf{R}_{pk} \mathbf{p}_k + (\mathbf{I} - \mathbf{R}_{pk}) \mathbf{p}^{\text{prior}},
\end{aligned}
\tag{10.78}
$$

where the parameter resolution matrix

$$
\mathbf{R}_{pk} = \mathbf{I} - \mathbf{C}_{p'k} \mathbf{C}_p^{-1}
\tag{10.79}
$$

is similar to that defined for the linear problems, equation (10.35).

10.3.3   Ridge Regression

Note that matrices $\mathbf{G}_k^T \mathbf{C}_d^{-1} \mathbf{G}_k$ or $\mathbf{G}_k \mathbf{C}_p \mathbf{G}_k^T$ in formulas 1 through 3 may be ill-conditioned for inversion. The addition of diagonal matrices $\mathbf{C}_p^{-1}$ or $\mathbf{C}_d$ is equivalent to adding positive weight to the diagonal of $\mathbf{G}_k^T \mathbf{C}_d^{-1} \mathbf{G}_k$ or $\mathbf{G}_k \mathbf{C}_p \mathbf{G}_k^T$. Therefore the addition can stabilize the matrix inversion. However, the algorithm listed above for the Gauss-Newton method may not converge at desirable rates. Various methods have been proposed and utilized. Here only the *ridge regression* or the *Marquardt-Levenberg method* is described because it is widely used and its program codes are readily available (Press et al., 1992).

Equation (10.72) can be rewritten as

$$
\mathbf{H}_k \Delta \mathbf{p}_k = -\boldsymbol{\gamma}_k,
\tag{10.80}
$$

where

$$
\begin{aligned}
\Delta \mathbf{p}_k &= \mathbf{p}_{k+1} - \mathbf{p}_k \\
&= -\mathbf{H}_k^{-1} \boldsymbol{\gamma}_k
\end{aligned}
\tag{10.81}
$$

is the increment of $\mathbf{p}$ from iterative steps k to $k + 1$. This formula ties the increment with the gradient $\boldsymbol{\gamma}_k$ through the Hessian $\mathbf{H}_k$. The Hessian represents an approximation and can be modified to increase the convergence rate. In addition to the Newton method, Tarantola (1987) has described the methods of steep descent, conjugate gradient, and variable metric and has tabulated the needed formulas for optimal performance.

The Marquardt-Levenberg method can be generalized by multiplying a modifier to the diagonal entries of the Hessian

$$(\lambda_k + 1)\mathbf{IH}_k\Delta\mathbf{p}_k = -\boldsymbol{\gamma}_k \tag{10.82}$$

which is given in a summation form by Press et al. Modifier λ_k is adjustable for increasing or decreasing increment $\Delta\mathbf{p}_k$ at iteration step k. For example,

$$\Delta\mathbf{p}_k = -\lambda_k^{-1}\mathbf{IH}_k^{-1}\boldsymbol{\gamma}_k \quad \text{as} \quad \lambda_k >> 1 \tag{10.83}$$

represents the method of steep descent, whereas

$$\Delta\mathbf{p}_k = -\mathbf{H}_k^{-1}\boldsymbol{\gamma}_k \quad \text{as} \quad \lambda_k << 1 \tag{10.84}$$

represents the Newton method. Equation (10.82) is therefore a hybrid of the two methods.

The Marquardt-Levenberg method begins with the steep descent. When the minimum is nearly approached (as judged by the change of S^2 or χ^2 between iteration steps), it switches to the Newton method. During the course of iterations, λ_k is adjusted to ensure that the misfit is decreasing. These features are implemented in subroutines provided by Press et al. At the final stage, λ_∞ is reset to zero for finding the postprocessed covariance matrix

$$\mathbf{C}_{p'} = \mathbf{H}_\infty^{-1}, \tag{10.85}$$

where subscript ∞ denotes the final iteration.

10.4 Example: A 1D Finite-Element Problem

• **General Statement**

A finite-element problem (see Chapter 9) is typically represented by

$$(\mathbf{C} + \omega\Delta t\mathbf{K})\,\mathbf{h}^{k+1} = [\mathbf{C} - (1-\omega)\Delta t\mathbf{K}]\,\mathbf{h}^k + \left[(1-\omega)\,\mathbf{f}^k + \omega\mathbf{f}^{k+1}\right]\Delta t \tag{10.86}$$

where **h** is the desired nodal values, **C** is the capacitance matrix, **K** is the conductance matrix (advective-dispersion matrix, depending on problems of interest), ω is the time-weighting factor between 0 and 1 (preferably greater than 0.5 for stability in time stepping), and Δt is the time interval between time steps $k + 1$ and k. The capacitance matrix **C** contains the storage information (e.g., specific yield, heat capacity) and is usually symmetric and banded. Matrix **K** contains transport properties (e.g., thermal conductivity, hydraulic conductivity, dispersion coefficient) and is symmetric for pure diffusion problems (conductive problems) but is asymmetric in the case of advective transport. Vector **f** bears information about specified flux at the model boundary (Neumann condition) and source/sink. The prescribed Dirichlet condition is implemented during equation solving.

Inversion for parameters usually involves the determination of transport and storage properties. Less often is the determination of source/sink strengths, boundary conditions, or locations of interfaces between different media. Let the desired parameters be **p**. The data kernel is thus represented by

$$G_{ij} = \frac{\partial h_i}{\partial p_j}, \quad i = 1, 2, ..., N; \quad j = 1, 2, ..., M, \tag{10.87}$$

where M is the number of desired parameters and N is the number of observations, $\mathbf{h}^{\text{obs}}[x, y, z, t]$. To numerically obtain G_{ij}, the finite element equation (10.86) is solved once for all nodal **h** for every incremental change in Δp_j. Altogether, equation (10.86) is solved $(M + 1)$ times for a suite of M trial $\Delta \mathbf{p}$ around a starting set of **p**. Only the nodal values at the nodes with observations are used and those values at all time steps are assembled to formulate **G**.

G_{ij} represents the sensitivity of a system to parameter changes. Sizing of $\Delta \mathbf{p}$ as a percentage of **p** for calculating G_{ij} is made by trial and error for a given set of **p**. Some of our guesstimates of a starting **p** could be far away from the true value. It is prudent to try several values of the guesstimate to minimize the chance of converging to a local minimum and to maximize the chance of finding **p** at the global minimum of $\chi^2[\mathbf{p}]$.

• Formulation of a Problem

Heat flow measurements at the ocean floor are usually made by inserting a probe of about 0.3 cm in diameter into the unconsolidated sediments. An

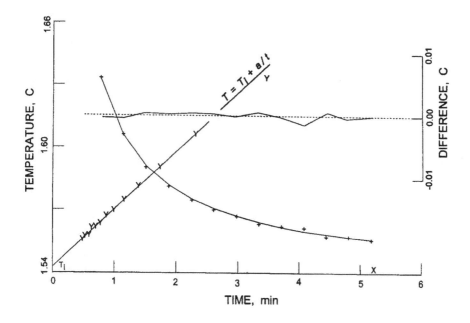

Figure 10.2: Results of inverse modeling for a cooling probe: observed temperatures (+) and predicted temperatures (connected by exponential-like line segments). Misfits (differences) are shown by solid line along with the dotted reference line. The computed equilibrium temperature is denoted by an ×, which is comparable with intercept time T_i of the temperature versus 1/time line (T versus $1/t$, marked by Y, time scale not shown).

equilibrium temperature is measured after the frictional heat generated during insertion has sufficiently dissipated. The geothermal gradient is calculated from two or more probe measurements. After the equilibrium temperature measurements are made, the probe is heated by an artificial power source. Thermal properties of the sediments are then measured from the changing probe temperature and the given heating rate (analogous to Jacob's semilog method, see Section 2.1.3). Alternatively, core samples can be retrieved and the conductivity is measured in the laboratory with the needle probe method. The product of thermal gradient and conductivity is the conductive heat flux.

Waiting for the frictional heat to dissipate and the time spent to measure conductivity or collect cores are costly and also increase the chance of data disturbance caused by ship movement. Lee and von Herzen (1994) proposed to determine the equilibrium temperature and conductivity from the cooling history immediately after the probe is inserted into the sediments.

The problem is assumed to be of one-dimensional axi-symmetric heat conduction. The equation is solved with a finite element method using Galerkin weighted residuals (see Section 9.1). The unknown parameters are thermal conductivity and heat capacity of the probe and the sediments, initial temperature, equilibrium temperature, and the total frictional heat generation. Also, the time zero could err by as much as the sampling interval. About 15 data points collected at intervals of 22 seconds are available for solving the inverse problem. Since the bounds for those properties can be reasonably bracketed within one order of magnitude, Newton's iterative scheme is used.

The results of using equation (10.75) are illustrated in Figure 10.2 for one probe insertion. The rms as calculated by equation (10.61) is 0.04%; the simulated equilibrium temperature differs $0.002°C$ from that obtained by the conventional method of extrapolation, which uses an asymptotic relation between temperature and inverse time; and the simulated thermal conductivity is off by 5.7% from the value measured in-situ with the conventional line-source method. All are within the limits of acceptance.

10.5   Suggested Readings

Inverse problems are of interest to scientists and engineers who engage in quantitative modeling. The literature on the subject is overwhelmingly abundant. A few formulas in Tarantola's (1987) *Inverse Problem Theory* are cited here. Sun's (1994) *Inverse Problems in Groundwater Modeling* is a good source for references in hydrogeological applications. *Numerical Recipes* by Press et al. (1992) provides subroutines that can be easily assembled for inversion of model parameters.

Appendix: Notes on Equation Solving

This book describes a few common methods for solving differential equations. The choice of methods varies with problems. With the exception of Chapter 3, the methods have not been presented in any systematic order. The following is a guide to equation-solving methods by category.

A.1   Integral Transforms

If differential equations are linear, applications of integral transforms always yield closed-form solutions. Some solutions are expressible in terms of well-established functions that can be easily evaluated with readily available software; most solutions require numerical integration by custom design.

• Laplace Transform

The method of Laplace transform is applicable to problems that are one-sided in time or space (e.g., $t \geq 0$, or $x \geq 0$). Most of the closed-form solutions in Chapter 7 on solute transport are derived by means of the Laplace transform. Also obtainable through the Laplace transform are all well functions (aquifer responses to pumping) presented in Chapters 4 and 5.

An application to semi-infinite medium (e.g., $x \geq 0$) is cited in Section 7.4.2 for three-dimensional problems of solute transport.

Many problems in heat transfer can be solved through the Laplace transform. For example, Section 6.4 on heating and fluid pressurization and Problem 5 of Chapter 6 with a radiative heat-transfer boundary (a Cauchy-type of boundary condition) are solved with the Laplace-transform technique.

There are numerous ways to obtain closed-form solutions for inverse Laplace transforms, each suitable for a particular class of problems. In several instances we have used two or three methods to invert the same transform although the results may be different in functional form. The convolution theorem together with parameter shifting method has been frequently applied (e.g., Section 7.2.1). Partial fraction is a technique employed in Section 7.2.6 to solve transport problems with advective boundary condition. Usage of integral or series representations can often simplify the inversion (Sections 3.2.9 and 3.2.12). Occasionally we can integrate or differentiate a given transform pair to obtain the desired solution (Sections 3.2.5 and 3.2.6, and Problem 10 in Chapter 3).

Inversion by means of the Bromwich integral requires application of residue theorem and contour integration. Numerous examples are given throughout the book, especially in Sections 3.1 and 3.2. Also, see *Laplace transform* in the Index for specific topics.

Root-finding is needed to determine residues for contour integration. If the roots cannot be determined analytically, inversion by numerical technique can always be performed. The Stehfest algorithm (Section 3.3) has been widely used for hydrogeological problems although it may not be applicable at extremely high or low values of drawdown.

• Fourier Transform

The method of Fourier transform is typically applied to problems with two-sided functions in time or space (e.g., $-\infty < t < \infty$, $-\infty < x < \infty$), or to data that are zero outside the recorded range or to records that are presumed to repeat periodically outside the recorded range. An example is cited in Section 7.4.2 for solute transport in steady groundwater flow.

A Fourier series is frequently used to represent a discontinuous function for matching boundary conditions. All analytical solutions for hydraulic responses to pumping in partial-penetration wells include a cosine series. In Section 4.4.5 a cosine series is used to represent drawdown Δh in Hantush's type of leaky aquifer while in Section 5.4.2 a cosine series is used to represent the radial flow, $r\partial\Delta h/\partial r$ (which is vertically discontinuous along the well face) in Neuman's type of water-table aquifer.

A solution, obtained by means of separation of variables, typically bears cosine functions that match upper- and lower-interface conditions even if the pumped well is full-penetrating. Such a solution can be regarded as a cosine series (Section 4.5.2) and the problem can indeed be solved by using a trial cosine series.

• Hankel Transform

The Hankel transform is valuable for axi-symmetric radial flow problems. Drawdown responses in confined aquifers with leaky boundary flux (Section 4.5) can be derived through the Hankel transform. By means of the Hankel transform, an attempt has been made to unify most of drawdown response functions in this book with one formula in Chapter 5, equation (5.120).

Section 3.4 emphasizes how to obtain inverse Hankel transforms analytically. If an analytical inversion is not attainable, numerical inversion techniques described in Section 3.5 are recommended.

Analytical inverse Hankel transforms result in solutions that bear the modified Bessel functions of the second kind. Applications of Fourier or Laplace transforms to a differential equation in cylindrical coordinates can also lead to modified Bessel equations. The choice depends on how easily boundary conditions can be implemented.

• Z-Transform

The z-transform has been widely used in geophysical data processing. It is introduced here to obtain transfer function from experimental data for which analytical relations may or may not be available (Section 2.2.5). In the z-transform domain, the data sequence can be operated like a polynomial. An example for finding system response to any input function is given in Section 7.6 and Problem 3 in Chapter 7.

A.2 Separation of Variables

The technique of separation of variables is probably the most common means to solve a partial differential equation (Sections 4.5.2, 6.3.2, and 7.3.1).

A.3 Series Solutions

In the unspoken hierarchy of solving differential equations, analytical solutions in terms of well-established functions rank at the top. Next come the closed-form solutions. The two categories are similar except that the latter contains no well-known integrals or functions. For complicated problems or more realistic models, numerical solutions by the techniques of finite element or finite difference are the last recourse in equation solving.

Before using numerical methods, try to find a series solution. Many special functions (e.g., Bessel functions, Legendre polynomials) are actually series solutions to differential equations. A series solution (the Airy function) to the Airy equation as derived for a radial dispersion problem is presented in Problem 5 of Chapter 7.

A.4   Linear Superposition

Chapter 2 is devoted to the application of linear superposition, including the convolution theorem and the method of images. Once a solution to a given differential equation is found, additional solutions can often be found to meet the boundary conditions without formally solving the equation again.

Examples for using the method of images are presented in Sections 2.4 and Problem 2 of Chapter 2. In the presence of groundwater flow, the method is demonstrated again in Section 6.2 for a moving heat source. It has also been applied to recovery-pumping tests (Section 2.3.6).

Examples for using the method of images and matching boundary conditions are given in Problem 3 of Chapter 2 and Section 7.3.3.

The perturbation method used in Section 6.3.3 for finding the disturbance of temperature distribution by groundwater flow is another example of linear superposition.

A.5   Numerical Methods

I have introduced the finite-element methods in Chapter 9 to readers who use existing software and want to know its formulation in general. The technique of finite difference was not covered except for a token introduction to solve a transient problem (Section 9.1.5).

A.6 Integration

Various analytical techniques for integration are indexed under Integration, which also indexes numerical integration techniques.

BIBLIOGRAPHY

Abramowitz, M., and Stegun, I. A., Ed., *Handbook of Mathematical Functions,* Dover Publications, Inc., New York, 1968.

Agnesi, A., Reali, G. C., Patrini, G., and Tomaselli, A., Numerical evaluation of the Hankel transform: remarks, *Journal of Optical Society of America, A.* 10, 1872, 1993.

Bathe, K-J, and Wilson, E. L., *Numerical Methods in Finite Element Analysis*, Prentice-Hall, Englewood Cliffs, 1976.

Bear, J., *Dynamics of Fluids in Porous Media*, American Elsevier, New York, 1972.

Beck, A. E., Garven, G., and Stegena, L., Ed., *Hydrogeological Regimes and Their Subsurface Thermal Effects*, American Geophysical Union Monograph 47, 1989.

Bownds, J. M., and Rizk, T. A., The inverse Laplace transform of an exact analytical solution for a BVP in groundwater theory, *Journal of Mathematical Analysis and Applications*, 165, 144, 1992.

Bracewell, R., *The Fourier Transform and its Application*, McGraw-Hill, Inc., New York, 1986.

Burnett, D. S., *Finite Element Analysis*, Addison-Wesley Publishing Co., Reading, 1987.

Butler, J. J., Jr., *The Design, Performance, and Analysis of Slug Tests*, Lewis Publishers, New York, 1988.

Carslaw, H. S., and Jaeger, J. C., *Conduction of Heat in Solids*, 2nd ed., Oxford Press, New York, 1959.

367

Chen, C. S., Analytical solutions for radial dispersion with Cauchy boundary at injection well, *Water Resources Research*, 23, 1217, 1987.

Cleary, R. W., and Ungs, M. J., Analytical Models for Groundwater Pollution and Hydrology, Princeton University, Water Resources Program, Report 78-WR-15, 1978.

Cooper, H. H., Jr., Bredehoeft, J. D., and Papadopulos, I. S., Response of a finite-diameter well to an instantaneous charge of water, *Water Resources Research*, 3, 263, 1967.

Davies, B., and Martin, B., Numerical inversion of the Laplace transform: a survey and comparison of methods, *Journal of Computational Physics*, 33, 1, 1979..

Dawson, J. K., and Istok, J. D., *Aquifer Test*, Lewis Publishers, New York, 1991.

de Marsily, G., *Quantitative Hydrogeology*, Academic Press, San Diego, 1986.

Domenico, P. A., and Palciauskas, V. V., Theoretical analysis of forced convective heat transfer in regional groundwater flow, *Geological Society of America Bulletin*, 84, 3803, 1973.

Domenico, P. A., and Robbins, G. A., A dispersion scale effect in model calibrations and field tracer experiments, *Journal of Hydrology*, 7, 121, 1984.

Domenico, P. A., and Schwartz, F. R., *Physical and Chemical Hydrogeology*, 2nd ed., John Wiley and Sons, New York, 1998.

Dougherty, D. E., and Babu, D. K., Flow to a partially penetrating well in a double-porosity reservoir, *Water Resources Research*, 20, 1116, 1984.

Driscoll, F. G., *Groundwater and Wells*, Johnson Division, St. Paul, 1986.

Endres, A. L., Clement, W. P., and Rudolph, D. L., Monitoring of a pumping test in an unconfined aquifer with ground penetrating radar, in *Proceedings of the Symposium on the Application of Geophysics to Engineering and Environmental Problems*, Environmental and Engineering Geophysical Society, 1, 483, 1997.

Evans, D. D., and Nicholson, T. J., Ed., *Flow and transport through unsaturated fractured rock*, American Geophysical Union Monograph 42, 1987.

Ferrari, J. A., Fast Hankel transform of order zero, *Journal of Optical Society of America*, 12, 1812, 1995.

Feshbach, P. M., and Morese, H., *Methods of Theoretical Physics*, McGraw-Hill, Inc., New York, 1953.

Fetter, C. W., *Applied Hydrogeology*, 3rd ed., Prentice-Hall, Englewood Cliffs, 1994.

Freeze, R. A., and Cherry, J. A., *Groundwater*, Prentice-Hall, Inc., Englewood Cliffs, 1979.

Furbish, D. J., *Fluid Physics in Geology*, Oxford University Press, New York, 1997.

George, A., and Liu, J. W., *Computer Solution of Large Sparse Positive Definite System*, Prentice Hall, Englewood Cliffs, 1981.

Gradshteyn, I. S., and Ryzhik, I. W., *Table of Integrals, Series, and Products*, 5th ed., Academic Press, San Diego, 1994.

Haines, G. V., and Jones, A. G., Logarithmic Fourier transformation, *Geophysical Journal*, 92, 171, 1988.

Hansen, E. W., Fast Hankel transform algorithm, *IEEE Transactions on Acoustics, Speeches, and Signal Processing*, 33, 666, 1985.

Hantush, M. S., Analysis of data from pumping tests in leaky aquifers, *Transactions, American Geophysical Union*, 37, 702, 1956.

Hantush, M. S., Hydraulics of wells, in *Advances in Hydrosciences*, 1, V. T. Chow, ed., Academic Press, San Diego, 1964.

Hantush, M. S., and Jacob, C. E., Non-steady Green's functions for an infinite strip of leaky aquifers, *Transactions, American Geophysical Union*, 36, 101, 1955.

Hsieh, P. A., A new formula for the analytical solution of the radial dispersion problems, *Water Resources Research*, 22, 1597, 1986.

Hunt, B., *Mathematical Analysis of Groundwater Resources*, Butterworths, Boston, 1983.

Huyakorn, P. S., and Pinder, G. F., *Computational Methods in Subsurface Flow*, Academic Press, San Diego, 1983.

Istok, J. D., *Groundwater Modeling by the Finite Element Method*, American Geophysical Union, Water Resources Monograph 13, 1989.

Javandel, I., Doughty, C., and Tsang, C. F., *Groundwater Transport: Handbook of Mathematical Models*, American Geophysical Union, Water Resources Monograph 10, 1984.

Jury, E. I., *Theory and Application of the Z-Transform Method*, John Wiley and Sons, Inc., New York, 1982.

Jury, W. A., and Roth, K., *Transfer Functions and Solute Movement through Soil*, Birkhauser Verlag, Boston, 1990.

Kabala, Z. J., Well response in a leaky aquifer and computational interpretation of pumping tests, in *Hydraulic Engineering, Proceedings of ASCE National Conference on Hydraulic Engineering and International Symposium on Engineering Hydrology*, Hsieh, W. S., Su, S. T., and Wen, F., Ed., 21, 1993

Lee, T. C., Thermal conductivity measured with a line source between two dissimilar media equals their mean conductivity. *Journal of Geophysical Research*, 94, 12,443, 1989.

Lee, T. C., Pore pressure rise, frictional strength, and fault slip: one dimensional interaction models, *Geophysical Journal International*, 125, 371, 1996.

Lee, T. C., and von Herzen, R. P., In situ determination of thermal properties of sediments using a friction-heated probe source, *Journal of Geophysical Research*, 99, 12121, 1994.

Leij, F. J., Skaggs, T. H., and van Genuchten, M. Th., Analytical solutions for solute transport in three-dimensional semi-infinite porous media, *Water Resources Research*, 27, 2719, 1991.

Lichtner, P. C., Steefel, C. I., and Oelkers, E. H., Ed., *Reactive Transport in Porous Media*, Mineralogical Society of America, 1996.

Mansure, A. J., and Reiter, M., A vertical groundwater movement correction for heat flow, *Journal of Geophysical Research*, 84, 3490, 1979.

Marquardt, D. W., Generalized inverse, ridge regression, biased linear estimation and nonlinear estimation, *Technometrics*, 12, 591, 1970.

Marschall, P., and Barczewski, B., The analysis of slug tests in the frequency domain, *Water Resources Research*, 25, 2388, 1989.

Mathews, J., and Walker R. L., *Mathematical Methods of Physics*, 2nd ed., W. A. Benjamin, Inc., New York, 1970.

Mercer, J. W., and Waddell, R. K., Contaminant transport in groundwater, in *Handbook of Hydrology*, Maidment, D. R., Ed., McGraw Hill, Inc., 1992, Chapter 16.

Moench, A. F., Computation of type curves for flow to partially penetrating wells in water-table aquifers, *Ground Water*, 31, 966, 1993.

Moench, A. F., Flow to a well in a water-table aquifer: an improved Laplace transform solution, *Ground Water*, 34, 593, 1996.

Moench, A. F., Flow to a well of finite diameter in a homogeneous, anisotropic water table aquifer, *Water Resources Research*, 33, 1397, 1997.

Moench, A. F., and Ogata, A., A numerical inversion of the Laplace transform solution to radial dispersion in a porous medium, *Water Resource Research*, 17, 250, 1981.

Neuman, S. H., Theory of flow in unconfined aquifers considering delayed response of the water table, *Water Resources Research*, 8, 1031, 1972.

Neuman, S. H., Effect of partial penetration on flow in unconfined aquifers considering delayed gravity response, *Water Resources Research*, 10, 303, 1974.

Neuman, S. H., Analysis of pumping test data from anisotropic unconfined aquifers considering delayed gravity response, *Water Resources Research*, 11, 329, 1975.

Nwankwor, G. I., Gillham, R. W., van der Kamp, G., and Akindunni F. F., Unsaturated and saturate flow in response to pumping of an unconfined aquifer: field evidence of delayed drainage, *Ground Water*, 30, 690, 1992.

Ogata, A., Theory of dispersion in a granular medium, *U. S. Geological Survey Professional Paper* 411-I, 1970.

Oppenheim, A. V., and Schafer, R. W., *Digital Signal Processing*, Prentice-Hall Inc., Englewood Cliffs, 1975.

Perrochet, P., and Berod, D., Stability of the standard Crank-Nicholson-Galerkin scheme applied to the diffusion-convection equation: some new insights, *Water Resources Research*, 29, 3291, 1993.

Press, W. H., Flannery, B. P., Teukolsky, S. A., and Vettering, W. T., *Numerical Recipes in FORTRAN, the Art of Scientific Computing*, Cambridge University Press, New York, 1992.

Robinson, E. A., and Treitel, S., *Geophysical Signal Analysis*, Prentice Hall, Inc., Englewood Cliffs, 1980.

Simunek, J., Huang, K., and van Genuchten, M. Th., The SWIMS_3D code for simulating water flow and solute transport in three-dimensional variably-saturated media, Salinity Laboratory, U. S. Department of Agriculture, Riverside, California, 1995.

Slattery, J. C., *Momentum, Energy, and Mass Transfer in Continua*, McGraw-Hill, Inc., New York, 1972.

Smythe, W. R., *Static and Dynamic Electricity*, 3rd ed., McGraw-Hill, Inc., New York, 1968.

Stehfest, H., Numerical inversion of Laplace transforms, *Communications of the Association of Computing Machinery*, 13, 47, 1970.

Strack, O. D. L., *Groundwater Mechanics*, Prentice-Hall, Inc., Englewood Cliffs, 1989.

Strout, A. H., and Secrest, D., *Gaussian Quadrature Formulas*, Prentice Hall, Inc., Englewood Cliffs, 1966.

Sun, N-Z., *Inverse Problems in Groundwater Modeling*, Kluwer Academic Publishers, Boston, 1994.

Tarantola, A., *Inverse Problem Theory*, Elsevier, New York, 1987.

Tseng, P. H., and Lee, T. C., Numerical evaluation of exponential integral: Theis well function approximation, *Journal of Hydrology*, 205, 38, 1998.

van Genuchten, M. Th., and Alves, W. J., Analytical solutions of the one-dimensional convective-dispersive solute transport equation, *U. S. Department of Agriculture Technical Bulletin* 1661, 1982.

van Nostrand, R. G., and Cook, K. L., Interpretation of resistivity data, *U. S. Geological Survey Professional Paper* 499, 1966.

Walton, W. C., *Groundwater Pumping Tests*, Lewis Publishers, New York, 1987.

Yeh, G. T., and Tripathi, V. S., HYDROGEOCHEM: a coupled model of HODROlogic transport and GEOCHEMical equilibria in reactive multicomponent systems. Environmental Science Division, Publication 3170, Oak Ridge National Laboratory, Oak Ridge, Tennessee, 1990.

Zaradny, H., *Groundwater Flow in Saturated and Unsaturated Soil*, A.A. Balkema, Brookfield, 1993.

Index

For Product Safety Concerns and Information please contact our EU
representative GPSR@taylorandfrancis.com
Taylor & Francis Verlag GmbH, Kaufingerstraße 24, 80331 München, Germany

www.ingramcontent.com/pod-product-compliance
Ingram Content Group UK Ltd.
Pitfield, Milton Keynes, MK11 3LW, UK
UKHW011457240425
457818UK00022B/879